GOD AND COSMOS IN STOICISM

God and Cosmos in Stoicism

Edited by
RICARDO SALLES

OXFORD
UNIVERSITY PRESS

This book has been printed digitally and produced in a standard specification in order to ensure its continuing availability

OXFORD
UNIVERSITY PRESS

Great Clarendon Street, Oxford OX2 6DP
United Kingdom

Oxford University Press is a department of the University of Oxford. It furthers the University's objective of excellence in research, scholarship, and education by publishing worldwide. Oxford is a registered trade mark of Oxford University Press in the UK and in certain other countries

Reprinted 2013

British Library Cataloguing in Publication Data
Data available

Library of Congress Cataloging in Publication Data
Data available

ISBN 978-0-19-955614-4

The manufacturer's authorised representative in the EU for product safety is Oxford University Press España S.A. of Parque Empresarial San Fernando de Henares, Avenida de Castilla, 2 - 28830 Madrid (www.oup.es/en or product.safety@oup.com). OUP España S.A. also acts as importer into Spain of products made by the manufacturer.

Acknowledgements

The project behind the present volume began at the conference 'Dios y el Cosmos en la Filosofía Estoica' held at the National Autonomous University of Mexico (UNAM) on 3–5 July 2006. Earlier versions of Chapters 1, 4, 5, 6, 8, and 9 were delivered at it. Given the large number of previously unexplored issues they addressed, it became clear that the interconnection between theology and cosmology in Stoic thought deserved intensive study through a volume devoted to it. Thus, the papers presented at the conference were thoroughly revised and Chapters 2, 3, and 7 were specially commissioned in order to provide a fuller treatment of special topics. Although each of the nine papers is self-contained, they all complement each other in that they were put together with the purpose of leaving no major gaps in the three fields covered by the volume. Often cross-references between them indicate how exactly they agree on particular issues, and also bring out major differences in the interpretation of the evidence. The editorial Introduction seeks to bring out the unity of the subject in Stoic philosophy and of the volume as a whole.

At the conference, most of the discussion centred on highly technical issues and I am grateful to the respondents of the papers—Juan Pablo Bermúdez, Laura Gómez, Enrique Hulsz, Luis Xavier López Farjeat, Andrea Lozano, Alejandro Tellkamp, and Héctor Zagal—for their help. I am also indebted to Thomas Bénatouïl, Marcelo Boeri, Brad Inwood, Andrea Lozano, Susan Meyer, and Teresa Rodríguez for their advice on the subsequent editorial work and their comments on the Introduction. The two anonymous readers appointed by Oxford University Press also provided encouraging and useful comments on the volume as a whole and on each of the individual papers, including the Introduction. The assistance and help received from Peter Momtchiloff, Victoria Patton, Tessa Eaton, and Catherine Berry at Oxford has also been extremely valuable.

The project received constant financial support from the UNAM and Mexico's National Science and Technology Council (CONACYT) through the projects IN401408 and 40891-H and through the Instituto de Investigaciones Filosóficas at UNAM. An earlier version of Chapter 2 was published in French as: Jean-Baptiste Gourinat, 'La Théorie stoïcienne de la matière: Entre le matérialisme et une relecture corporaliste du *Timée*', in Cristina Viano (ed.), *L'Alchimie et ses racines philosophiques* (Paris: Vrin, 2005). I wish to thank Cristina Viano and the Librairie Philosophique J. Vrin (http://www.vrin.fr/) for their permission to publish in this volume the present version of the chapter.

Contents

List of Contributors

KEIMPE ALGRA is Professor of Ancient and Medieval Philosophy at the Department of Philosophy of the University of Utrecht, the Netherlands

THOMAS BÉNATOUÏL is Maître de Conférences at the Department of Philosophy of the Université de Nancy, France and is a junior member of the Institut Universitaire de France

MARCELO BOERI is Professor at the Instituto de Filosofía of the Universidad de los Andes, Chile

JOHN COOPER is Stuart Professor of Philosophy at the Philosophy Department of Princeton University, United States

JEAN-BAPTISTE GOURINAT is Researcher at the Centre National de la Recherche Scientifique Paris, France

BRAD INWOOD is University Professor at the University of Toronto, Canada

INNA KUPREEVA is Lecturer in Philosophy at the Department of Philosophy of the University of Edinburgh, Scotland

SUSAN SAUVÉ MEYER is at the Department of Philosophy of the University of Pennsylvania, United States

RICARDO SALLES is Researcher at the Instituto de Investigaciones Filosóficas of the Universidad Nacional Autónoma de México

Introduction: God and Cosmos in Stoicism

Ricardo Salles

Theology and cosmology are closely interrelated in Stoic philosophy. The present volume brings together in nine chapters papers offering an in-depth study of this connection and its consequences in practical ethics and religious cult and myth. I begin with some brief general remarks on these subjects and on the overall structure of the volume. Subsequently, a detailed description is offered of the contents of each of the individual chapters. To avoid repetition, the bibliographical references given in this introduction are all (with few exceptions) to the chapters in this volume.

The Stoics express the interrelation between god and cosmos through the idea that the current state of the cosmos, as well as its creation, and even its destruction, are fully rational in the sense of being intelligently organized. Its rationality derives from god's all pervading reason, which physically penetrates the cosmos through and through. In so doing, the Stoic god actively controls the behaviour of every existing body and is, therefore, responsible for everything it does and undergoes. In this respect, god is best described as the single active physical principle that governs the whole cosmos. He does so in exactly the same way living entities are governed by their soul and, especially, by its ruling part (τὸ ἡγεμονικόν). In consequence, the Stoics assumed that the cosmos itself was a living entity and that god was its soul. In fact, some sources even identify god with the cosmos itself insofar as the identity of a living entity is fixed primarily by its soul.[1] Moreover, given the control exerted by god over the whole cosmos and on everything there is in it, everything behaves coordinately and according to a single plan. For this reason, the Stoics identified fate, and ultimately god himself, with the single, overarching set of causal relations through which every body is connected to the cosmos as a whole and to other individual bodies in it. In this respect, their god is not only the cohesive, unifying power that holds together the whole cosmos and each of its individual bodies, but also the power through which these bodies may act upon one another through other, non-cohesive, kinds

[1] See notably Cicero, *ND* 29–30, commented on in Ch. 5 at pp. 121–4.

of causation. One further implication of everything's being planned is, of course, that the Stoic cosmos, unlike for example the Epicurean cosmos, is teleologically aimed at an end, even though the Stoics famously denied that ends could have in themselves a causal power.

God, however, is not the only principle (ἀρχή) in the cosmology of the Stoics. Their cosmos is the result of the action of god upon a certain passive principle, identified in some sources as the matter (ὕλη) that underlies the change undergone by the cosmos in its various phases. These two principles are irreducibly *two*. Thus, even though the Stoic cosmos is material in that its being involves matter as one of its physical constituents, the Stoics do not espouse materialist monism, the view that everything, including god, proceeds from, and is ultimately reducible to, matter as the sole basic constituent of reality. But, on the other hand, the Stoics are not strict dualists either. For they affirm that matter and god have in common crucial physical properties. Most notably, they both are bodies, and, hence, they both are entities that occupy space and are resistant to touch, which are two landmarks of corporeality.[2] The complexity of this ontological view leads to the metaphysical question of how the two principles are individuated. If both are bodies and, for this reason, corporeality cannot serve to distinguish them from each other, what is the basis of the distinction active/passive? Is it primitive (the two principles are individuated by the very fact that one is active and the other passive) or is one principle active and the other passive in virtue of other notions that the Stoics bring forward to establish the distinction?

In orthodox Stoicism, the single most distinctive feature of the cosmos is perhaps its cyclical nature and its infinite repetition. The present cosmos will totally burn up, but the substance left by this mighty conflagration will give rise to a new cosmogony. In orthodox Stoic philosophy, the cosmos that will be created in this way will be indiscernible from the present one. But it will itself burn up through a new conflagration and a new, identical, cosmos will rise, and so on *ad infinitum*. How the substance left by the conflagration gradually transforms itself into a new cosmos, and the exact mechanism by which the cosmos ends up being completely burnt up, were subjects of controversy within early Stoic philosophy. With the possible exception of Cleanthes, it was agreed by all parties that elemental reciprocal change—together with the transformation of the four elements into complex bodies by mixture (which are eventually dissolved back into the four elements)—was the key to understanding these two processes. In particular, the initial stages of the cosmogony were thought of as a change of fire into the other elements, and the last stages of the conflagration as a change of these other elements back into fire. The main controversy centred on the nature of this original and final fire: what kind of substance is it, what is its relation to ordinary fire, and what is exactly the process by which this fire is transformed into

[2] See DL 7. 135. (See also the power of acting and being acted upon in Plutarch, *CN* 1073E.) This issue is discussed in Chapters 2 and 4.

the other elements? There also was some dispute over the question of whether the cosmos is destroyed when it is consumed by it. According to Chrysippus, for instance, the conflagration is a positive phenomenon that does not really involve any substantial change. The initial and final fire of any cosmic cycle is nothing but god himself in a completely undifferentiated state.

The study of Stoic cosmo-theology requires paying close attention to its origins in earlier thinkers, as well as to its impact on, and sensitivity to, rival contemporary cosmological theories. Some parallels may be drawn between Stoic cosmologists and Heraclitus on fire, Anaximenes on elemental change, and the Pythagoreans on cosmic recurrence.[3] But the clearest antecedent of Stoic cosmology is no doubt Plato's *Timaeus*. The Stoics give the most prominent place in their cosmology to central notions in the cosmology of the *Timaeus*, notably, the concept of a benevolent cosmic demiurge, the idea that the cosmos is a living entity with soul and body, and the notion (more conspicuous in some interpretations of the *Timaeus* in the Old Academy and the Lyceum than in the *Timaeus* itself[4]) that cosmology is governed by two principles: one active and the other passive. At the same time, however, the Stoics adopt in their cosmo-theology a position diametrically opposed to Plato's regarding other equally fundamental issues in metaphysics such as paradigmatism, divine transcendence, corporealism, and causation. This may lead us into thinking that Stoic cosmology is a conscious effort to adapt central notions of the cosmology of the *Timaeus* to a new, radically different, metaphysics. As for other influences on Stoicism, there is a notorious problem in the assumption that the early Stoics knew Aristotle's school treatises. Yet there are striking parallels between Stoic and Peripatetic cosmologies and one may wonder whether they reflect a strong mutual influence or are merely superficial similarities exaggerated by our sources. The final theme in the present volume is the practical consequences of Stoic cosmo-theology in the field of ethics and religion. Is our knowledge of cosmology required in order for us to lead a life that will give us happiness? If it is, how? And if not, why should we bother about cosmology? Similar questions may be raised in connection with religion. Given Stoic theology, should traditional religious myth and cult be adopted or rejected? Or is Stoic theology neutral with regard to religious belief and practice?

The volume is divided into three sections. Part I—'God, Providence, and Fate'—contains the first three chapters and covers three essential topics in Stoic theology: the active and demiurgical character of god, his corporeal nature and irreducibility to matter, and fate as the network of causes through which god acts upon the cosmos. Part II—'Elements, Cosmogony and Conflagration'—includes Chapters 4 to 6 and deals with these three special topics in Stoic

[3] See respectively DK 22 B 30–1, DK 13 A 5–7, and DK 58 B 34.

[4] For this interpretation of the *Timaeus*, see pp. 24, 49–51, and 52–3. However, the active–passive distinction is already present elsewhere in Plato: see *Theaetetus* 157A–B, *Sophist* 247D–E, and (in a cosmological context) *Philebus* 26B–27B.

cosmology and with how Stoic cosmology in general relates to contemporary Peripatetic cosmologies. Part III—'The Ethics and Religion of Stoic Cosmo-Theology'—is composed of the final three chapters and closes the volume by addressing the problem mentioned above about the ethical and religious consequences of these Stoic theories.

CHAPTER 1

The Stoic god is providential and benevolent in that he purposively acts on the cosmos and cares for it. This sharply differentiates Stoic theology from at least three other major ancient theologies: (*a*) that of Aristotle, according to which sole activity of the prime mover is self-intellection, (*b*) that of Epicurus and the Epicureans, who argue that god does not intervene in the cosmos—the cosmos cannot be accounted for by appeal to divine craftmanship and teleology, and (*c*) that of Plato in the *Timaeus*, according to whom god is benevolent but transcendent in that he acts on the cosmos from outside.[5]

A central question concerning Stoic divine providence is its nature and scope. What kind of action is it? And which entities are affected by it? The latter issue is the main subject of Thomas Bénatouïl's 'How Industrious can Zeus be?' (Chapter 1). Sources hostile to the Stoics often report the *industriousness* of their god, i.e. his ceaseless activity upon absolutely everything there is in the cosmos. But is the Stoic god really industrious in this sense? Is not this feature of his an exaggeration resulting from biased critics whose ultimate purpose is to ridicule Stoic theology? In fact, five possible qualifications could be envisaged: (1) during the ordered phase of the cosmos, god is not always active insofar as his intervention is limited to cosmogony and zoogony; (2) at the end of the ordered phase, god is not active either, inasmuch as he does not act upon anything during the conflagration; (3) god's agency is limited to the celestial region of the cosmos and only affects the earth in a secondary, indirect way, (4) god is not equally benevolent towards all individuals, some of whom are neglected by him, which implies that his action is not evenly distributed throughout the cosmos, and (5) there are petty affairs that are unworthy of god's attention, a fact that would in itself justify a limitation of his cosmic activity.

According to Bénatouïl, none of these qualifications is really envisaged by the Stoics themselves. As is proved by evidence independent from the sources hostile to Stoicism, god's activity is not limited in any of these five senses. In the course of his argument, Bénatouïl also provides a detailed analysis of what kind of action Stoic providence is. It is not merely (i) the action by which god creates the cosmos

[5] A fourth point of contrast is Gnosticism, with which, however, the Stoics were not in dialogue. See Mansfeld 1981 (for full bibliographical references, see the References of Ch. 5).

at the cosmogony, but also (ii) the continuous activity by which god brings about every single change that takes place in the cosmos once the cosmogony has been completed, and (iii) the sustaining or cohesive action by which god secures the endurance of every single entity that exists in the cosmos, but also of the cosmos itself as a whole before it is destroyed by the conflagration.[6] As Bénatouïl observes, there is one source—Cicero, *ND* 3.92—that actually identifies providence with at least (i) and (ii) when it describes providence as 'moulder and manipulator' (*fictricem et moderatricem*).[7] A further activity that god performs in addition to (i)–(iii) is the production of the conflagration. According to the orthodox version of the doctrine, the conflagration is not due to something distinct and independent from god, but to god himself. It is, paradoxically, an inevitable by-product of his sustaining activity upon the cosmos.[8] I return to the theme of the conflagration—the main subject of Chapter 5—later on.

The analysis of providence in terms of (i)–(iii) brings out the specificity of the Stoic position within ancient providentialism and, notably, with respect to Plato's providentialism in the *Timaeus*. The demiurge of the *Timaeus* is also a provident god, but his providence seems to be limited to the creation of the cosmos as a whole, i.e. of its general structure. In particular, Plato's demiurge seems to leave to the soul of the cosmos the task of keeping the cosmos alive and changing.[9] In other words, in contrast with the Stoic god, the Platonic demiurge is not himself the agent of activities (ii) and (iii) mentioned above. In these two respects, Stoic providentialism is wider in range and more complex in nature than its Platonic counterpart. The principal difference between the Platonic demiurge and the Stoic god is perhaps the immanence of the latter: he exerts his providential activity from within the matter it moulds and manipulates. This is also brought out in some detail in Chapter 1, and I shall return to it below since it is also one of the main themes of Chapter 2, Jean-Baptiste Gourinat's 'The Stoics on Matter and Prime Matter'.

CHAPTER 2

Given the immanence of the Stoic god in matter, the question arises of how god is to be distinguished from matter. The question is especially relevant if we look into the physical mechanism by which god is present in matter. According to our sources, he pervades matter by being mixed with it through and through in such a way as to be totally coextended with it. In consequence, god is present everywhere in this mixture.[10] But given the complete blending between god and

[6] See esp. pp. 27–8. [7] Cited at pp. 26–7. [8] See pp. 28–31.

[9] See Ch. 1 n. 4.

[10] See Alexander of Aphrodisias, *De Mixtione* 225. 1–2 Todd and more generally 224. 32–225. 3, cited by Gourinat at p. 57. See also Ch. 5 (pp. 121 n. 6).

matter, how different can they be? As Gourinat points out, some critics of the Stoics argued that no substantive distinction of god and matter was really possible in this system and, in particular, that the Stoic god was bound to be material and to proceed from matter.[11] This reductive materialist view does not coincide with what the Stoics actually claim or with anything they are logically committed to. On their view, which is anti-materialist, both god and matter are bodies, but they form nevertheless an irreducible pair. In this vein, Gourinat provides in the first section of the chapter (48–58), a detailed analysis of why the Stoics reached this somewhat paradoxical position, which is neither a materialist monism nor, however, a strict dualism that totally distinguishes god from matter by setting them in two separate realms. For, as Calcidius rightly points out, in Stoicism 'god is what matter is' (*deum hoc esse quod silva sit*),[12] namely a body. The argument of Gourinat proceeds through a discussion of how this mixed conception reveals a reflection upon, but also a reaction to, other conceptions of the principles of the cosmos in earlier Greek philosophy, mainly Plato, the Old Academy, and Aristotle. In what follows, I focus on the former two.[13]

The idea that there are two cosmological principles and that these are god and matter seems to go back (*a*) to an interpretation of the *Timaeus* advanced by Theophrastus, and (*b*) to the Old Academy under Polemo according to a testimony of Antiochus.[14] But the *Stoic* version of this dualism, as defended by Zeno, differs greatly from its Academic version. In a close debate with David Sedley and Michael Frede on this topic,[15] Gourinat argues that there are two significant innovations of Zeno with respect to Polemo and the Old Academy: in the Stoic version, god is corporeal and his action upon matter is not guided by a model or paradigm. These two tenets of Zeno's physics are closely linked to each other in that they are aspects of the Stoic rejection of Platonic paradigmatism. The corporeality of god must be understood as a reaction to the Platonist belief in immaterial principles (principles cannot be immaterial if they are to possess causal efficacy). But the abandonment of an action-guiding model, Gourinat maintains, is a consequence of 'Zeno's most distinctive innovation': the notion that god acts on matter from its interior and not from outside as an artisan in the production of artefacts.[16] The activity the Stoic god exerts upon the cosmos is in a way analogous to that performed by the semen upon the living entity that proceeds from it. Just as the growth and development of the living thing is predetermined by information contained in the semen, so too the cosmogony and development of the cosmos as a whole consists in a series of events whose sequence unfolds in an orderly fashion from 'seminal reasons'

[11] The evidence is also cited below. [12] Cited by Gourinat at pp. 47 and 68.

[13] For the relation to Aristotle see pp. 49–50.

[14] Both cited by Gourinat. See also Diogenes Laertius 3. 69 and 3. 75 cited by Gourinat at p. 51.

[15] Cf. Sedley 2002 and Frede 2005 in the References to Ch. 2. [16] See p. 50.

(σπερματικοὶ λόγοι) in which god is present.[17] This biological model departs sharply from the model of *Timaeus*, according to which the demiurge acts from without the matter that he uses to create the cosmos, in the same manner as an artisan relates to the artefact that he produces. By contrast, in the biological model, the natural objects created and sustained by god are acted upon inside out. To be sure, the idea of the cosmos as a living entity is already present in the *Timaeus* (notably at 30c6–9). But in the Stoic cosmos the changes that take place in it are themselves determined biologically rather than by an external demiurge. In sum, corporeality and immanence in matter are features of the Stoic god that have no clear trace in earlier Platonism. They are a reflective reaction to Platonism, and Stoic physics, rather than logic (or ontology), is where the main reasons lie for this reaction.

I leave for later the role of god in cosmogony, also studied by Gourinat, since it is a theme extensively studied in Chapter 4 (and so is the question of the nature of Stoic corporealism—in what sense both god and matter are *bodies*). Before I move on, let me simply mention how Gourinat explains why, contrary to what is alleged by one source hostile to Stoicism,[18] two central theses in Stoic physics—that (*a*) cosmic 'breath' (πνεῦμα) is the physical substrate through which god penetrates the cosmos (see Chapter 3 commented on below), and that (*b*) πνεῦμα in general is a composite of air and fire—do not jointly imply that the Stoic god is itself a composite of more basic material substances and, hence, a product of matter. In fact, contrary to the objection, these two theses entail at most that one of the forms adopted by god is composite. But the Stoic god is not reducible to any of the forms he adopts during the different phases of the cosmos. In particular, cosmic breath is an instrument or vehicle through which he acts upon the cosmos. As is explained later on in this volume (Chapter 4), it is 'god or reason's immediate vehicle for controlling the world's constitution and behavior'.[19] In consequence, we cannot conclude that god is a composite entity—which is false in Stoic theory—merely from the fact that πνεῦμα is itself composite.[20]

CHAPTER 3

As the chain of causes present throughout the cosmos, fate is the instrument by which the Stoic god exerts its providential activity. In fact, the Stoic god is often *identified* with fate understood as the chain of causes.[21] How is this chain of causes to be understood? Chapter 3, by Susan Sauvé Meyer, is devoted to

[17] See the evidence cited at pp. 54 and 60. See also pp. 64–5 and Ch. 4 n. 19.
[18] See Alexander of Aphrodisias, *De Mixtione* 225. 11–12 Todd cited by Gourinat.
[19] See p. 103. [20] See the evidence cited at pp. 62–6.
[21] See the discussion in Ch. 3, sect. 4, and mentioned below.

this question. As she observes, the metaphor of a chain of causes is often used nowadays by causal theorists. But she rightly argues that the *Stoic* usage of this metaphor differs greatly from the modern one. In modern theory, a chain of causes is a sequence of events ordered in temporal succession, each of whose members is the cause of its successor and the effect of its predecessor. This conception evinces two assumptions about causation: first, that it is a relation in which causes and effects are not simultaneous (and as such they cannot influence each other reciprocally), and, secondly, that it is a relation holding between events. Neither of these assumptions is present in the Stoic notion of cause. On the contrary: Stoic causes are typically simultaneous with their effects (even the so-called 'antecedent' causes are),[22] and causation is not a relation between events, but between bodies. In Stoic theory, causal relations are best construed as the production by a body A of a certain effect on a body B. To take a classic example: the scalpel is cause to the flesh of the effect *being cut*.[23] This conception of cause, so fundamentally different from ours, is reflected in what the Stoics mean by the chain of causes in terms of which they define fate. As Meyer observes, it is a complex system of reciprocal influence between all the bodies that exist in the cosmos.[24]

What is the nature of this influence? For the reasons mentioned above, the influence is certainly not causal in a modern sense. In other words, the Stoic metaphor of the chain of causes as pervading the cosmos does not mean that there is single temporally ordered sequence of events such that every body in the cosmos is involved in some of these events. The Stoic idea of fate as a chain of causes does not *preclude* the existence of this sequence. In fact, as is pointed out in Chapters 2 and 4, Stoic cosmogony is, under some description, a sequence of this sort: each body in the cosmos is caused to exist by the activity of another, more basic, body until we reach the four elements and ultimately god himself who creates the four elements out of himself by acting upon some absolutely basic matter. (In this view, the *activity* of god is an event that causes the *coming into bodies* of other bodies, which is also an event, caused by the former.) But this, Meyer contends, certainly does not capture the essence of the chain of causes envisaged by the Stoics when they define fate. Its essence lies rather in the idea of cohesion between all bodies into a unified network of reciprocal influence.

One key to understanding the meaning of this system is the concept of cosmic breath or πνεῦμα, also referred to in Chapter 2: the physical substrate that pervades the cosmos through and through and that holds it together.[25] According to Meyer, the link between the causes of the chain in the doctrine of fate as a chain of causes is precisely this cohesive cosmic breath. Given this physical connection between all bodies and between each body and the cosmos as a whole, the affections experienced by one body may be transmitted, either directly

[22] See the evidence cited by Meyer at pp. 85–9, which I comment on below.
[23] Cited at p. 74. [24] See pp. 78–80. [25] See pp. 74 and 79–81.

or indirectly, to all the other bodies and to the cosmos as a whole. This takes us to the Stoic doctrine of cosmic sympathy (συμπάθεια). As Meyer indicates, there is a strong connection between this doctrine and the Stoic idea of fate as a chain of causes: given the idea that all the bodies in the cosmos are connected to each other and to the whole by a physical substrate—this is an essential part of what it is for them to be part of a chain of causes—each body may in principle affect all the other bodies and the whole. And this being jointly affected is exactly what cosmic sympathy is. According to the doctrine (attested for Chrysippus and nearly all the major Stoics),[26] the cosmos as a whole possesses the same kind of unity as living organisms. It is one in which, given the interaction between the whole and its parts, the affections of the parts may be transmitted to other parts or to the whole. To give an example from Sextus Empiricus cited by Meyer at p. 82, and which I believe is central to the theory: 'in the case of unified things there is a kind of sympathy (συμπάθειά τις); for example, when the finger is cut, the whole body shares its condition. So the universe is a unified body.' It is not clear from our sources whether in Stoic theory any body is always directly influenced by every other body through sympathy. But, as Meyer rightly points out,[27] the organic conception of the cosmos does not require that it should be. Some parts of an organism may be related to other parts only indirectly. What is important is that the whole be unified by a substrate that allows interactions between its different parts. More importantly, the idea of an indirect influence is built in the very notion of the chain of causes as construed by the Stoics. Just as in a necklace every bead is connected to all the others through a string, but none of them acts directly upon all the others (for none of them touches all others), so too in the Stoic cosmic chain of causes every body is connected to all the others by breath without, however, acting upon all of them directly.

Meyer's argument involves an important reinterpretation of some central texts on Stoic cosmology. Before I turn to the next chapter, I should like briefly to refer to one especially striking claim in her account. If the chain of causes envisaged by the Stoics is a relation between bodies that act upon each other, ancient descriptions of this chain in terms of the idea that all things are brought about by 'prior' things, and bring about other things that 'follow' them, cannot mean a sequence of temporally ordered items in which each brings about its successor and is brought about by its predecessor. One case in point is the report given by Alexander of Aphrodisias in *De Fato* 192. 2–8 (cited by Meyer at p. 86). The notions of priority and posteriority used here correspond to the notion of antecedent causation attested elsewhere for the Stoics. But Stoic antecedent causation, she argues, is not essentially a relation of temporal succession. Something's being antecedently caused means rather that its cause is part of fate understood as the whole set of mutual influence between bodies. As Meyer puts it: '[t]o claim that something has an antecedent cause is to affirm

[26] See pp. 80–5. [27] Cited at pp. 87–8.

that its cause is part of the causal nexus'.[28] Evidence for this, she argues, is to be found in the account of antecedent causation in Chrysippus' cylinder analogy reported by Cicero at *De Fato* 41–4.

CHAPTER 4

Chapters 1–3 were concerned with the general relation between god and the cosmos. In contrast, Chapters 4–6 focus on particular aspects of the Stoic cosmos: the doctrine of the four physical elements (4–6), cosmogony (4) and conflagration (5). As will be seen, Chapter 6 also deals with important parallels between Stoic and Peripatetic cosmologies.

The main purpose of Chapter 4, John Cooper's 'Chrysippus on Physical Elements', is to advance a new interpretation of what is perhaps our main source for Chrysippus' theory of physical elements (fire, air, water, and earth): a passage from Stobaeus' fifth-century AD anthology, *Eclogae Physicae et Ethicae*, at 1. 129–30 Wachsmuth. This chapter also sheds light on (*a*) the nature of the corporealism of the Stoics in connection with their theory of principles—what it is for both god and matter to be *bodies*—and (*b*) Chrysippus' conception of cosmogony as distinct from that of Zeno and that of Cleanthes.

In contrast with earlier commentators, notably Long and Sedley, Cooper argues that three usages distinguished by Chrysippus of the term element (*στοιχεῖον*) are divided into (*a*) one according to which the four elements of the actual cosmos are all elements on a par with one another, (*b*) one in which the term applies to a certain fiery substance—a 'proto-fire' in Cooper's terminology—out of which the four elements, including ordinary fire, are generated during the cosmogony, and (*c*) one in which the term denotes a certain substance out of which the proto-fire alluded to in (*a*) is itself generated. This absolutely originary substance is composed of god and qualityless prime matter, which are bodies. But, unlike proto-fire (and any other material substance), it possesses no qualification other than those that are intrinsic to any body, namely, three-dimensional extension and resistance to touch (which are qualifications that god and matter themselves possess qua bodies), and also the property of being an interblending of god and matter. Notice that neither god not qualityless prime matter should be identified with this originary substance. They are just the two bodies that compose it by mixture.

Evidence for this originary substance comes in Diogenes Laertius 7. 136 and 137, where it is said that 'at the beginning' (*κατ' ἀρχάς*) god was 'by himself' (*καθ' αὑτόν*), 'having consumed all substance into himself' (*ἀναλίσκων εἰς ἑαυτὸν τὴν ἅπασαν οὐσίαν*). This refers to a pre-cosmic stage that comes immediately after the extinction of the fire of the conflagration

[28] Cited at p. 87.

of the previous cosmos. At this time, god has consumed all substance into himself, not in the sense that he has absorbed *prime matter* into himself and is now the only body in existence, but in the sense that, the conflagration having been completed, god has consumed all the *qualified* substances there existed in the previous cosmos. And he is 'by himself' because, even though he remains active in the sense that he keeps thinking to himself all there is to think about the design of the new cosmos, his activity is not yet directed at actually producing the new qualified substances required by the new cosmogony. As Cooper explains: 'although he or it in his active nature retains and keeps on thinking to himself all the thoughts that in the actual world get put into effect in introducing all the qualifications of matter that constitute all the different sorts of substance that there actually are, he is not then using those thoughts to *act* in any differential way upon particular expanses of matter so as to endow substances with their particular characters; he is therefore not then affecting matter with any of those qualifications'.[29] The absolutely originary substance is nothing but god pervading prime matter at this pre-cosmic stage. In Cooper's interpretation, the only qualification or character of this originary substance—in addition to three-dimensionality, resistance, and being a blend of god and matter—is one that god imposes on prime matter as a whole in virtue of pervading it at this non-productive stage. According to Chrysippus (*ap.* Philo Judaeus, *Aet.* 90), this character is a flash (αὐγή), which is the flash of light left by the flame of the conflagration once it has been extinguished and the conflagration is *over* (as is implied by the aorist—ἐκπυρωθέντα—used by Philo). In consequence, this originary pre-cosmic substance is not to be identified with the fire of the conflagration. This, as Cooper explains, is an innovation in Stoic cosmology. For the originary substance had been problematically identified by Zeno with some form of fire, namely 'designing fire' or πῦρ τεχνικόν, and in particular with the fire of the conflagration by Cleanthes, namely flame or φλόξ.[30]

The first stage of the cosmogony occurs when the originary substance transforms itself into a fiery substance, which, in turn, transforms itself into an airy substance (proto-air) and, then, into a watery substance (proto-water).[31] The second stage of the cosmogony takes place when the four actual elements are generated from proto-water and then combined with each other by mixture to produce the other natural substances. Cooper's reconstruction of the cosmogony differs in many respects from the one offered by Gourinat in Chapter 2.[32] But as Gourinat points out in the close discussion he offers of Cooper there is overall agreement. I believe that an issue that deserves further reflection is whether Stoic *zoogony* really occurs by a mixture of the elements, as is suggested in one source, of some other, radically different process, as some other source—not

[29] See pp. 102–3. [30] See the evidence cited at p. 102 n. 20.

[31] See the evidence discussed at pp. 101–5.

[32] See Ch. 2, pp. 57, 60 nn. 68 and 69, 61–2, and 66 n. 88.

mentioned by Cooper and Gourinat—seems to imply (Galen, *Caus. Cont.* 1. 1–2. 4 = LS 55F).

CHAPTER 5

Just as there is no uniform agreement within orthodox Stoicism about cosmogony, there is no uniform agreement either on the question of the conflagration. Some major Stoics suspended judgement about whether there will be a conflagration at all. But even among those who accepted that there will be one, there was no agreement as to its nature. In particular, Cleanthes and Chrysippus held different opinions on whether the conflagration entails a destruction of the cosmos. Both believed that it will be followed by a reconstitution of the cosmos. But is this reconstitution a reconstitution from a state of destruction caused by the conflagration? Cleanthes claimed that it is, and Chrysippus took issue with him on this question, as I try to establish in my own contribution to this volume, 'Chrysippus on Conflagration and the Indestructibility of the Cosmos'.

This disagreement is not purely verbal. It is rooted in two different elemental theories. According to Chrysippus, fire is the thinnest of the four elements, the other three being transformed into it by rarefaction. On this view, earth, water, and air are nothing but condensed or compressed fire. Thus, air, water, and earth are not destroyed by the flame of the conflagration. They are merely transformed into what each of them really is: fire. The only change is one in density and this is a simple qualitative change. No substance is destroyed. In Cleanthes, by contrast, the four elements are presented as different substances acting upon each other. He recognizes that fire is more basic than the other three, but only in the sense that it acts on them as matter in order to give them cohesion. In this view, air, water and earth are not conceived as substances that are *made of* fire as in Chrysippus.[33] Cleanthes' model is not necessarily incompatible with his. Think of the cohesive action of fire upon water. From a Chrysippean perspective, this is the action of a mass of uncompressed fire acting upon a mass of compressed fire. But none of this is suggested in the sources we have for Cleanthes or Chrysippus, who probably did not appreciate the compatibility between the two models.

As far as I can see, my interpretation of the conflagration is compatible with Cooper's interpretation of the cosmogony. In Cooper's reconstruction, the originary substance from which proto-fire is generated is a flash of light, an αὐγή, subsequent to the extinction of the flame of the conflagration. In my interpretation, the flame of the conflagration does not last forever. Once complex substances are broken down into the elements, and these are totally transformed into flame by rarefaction, this flame is extinguished because there is nothing

[33] See the evidence examined at p. 129.

left for it to consume. In Chapter 5, I do not deal with what happens next. But the process I describe is compatible with Cooper's hypothesis that once the conflagration process is over, all that is left is an αὐγή consisting in god's pervading prime matter and refraining from any substance-producing activity.

CHAPTER 6

There are substantive parallels between the cosmology of the Stoics and that of rival schools such as the Academy and the Lyceum. In Chapter 6, 'Stoic Themes in Peripatetic Physics?', Inna Kupreeva enquires into whether these parallels are only superficial or may reveal genuinely shared positions. They emerge most in certain authors. Kupreeva focuses primarily on four particular cases: (*a*) the Academic Antiochus of Ascalon (*c*.130–68 BC) reported by Cicero in *Academica* 1. 6. 24–7. 29, (*b*) the Peripatetic Critolaus (*c*.200–118 BC), head of the Lyceum and a contemporary of Diogenes of Babylon—pupil of Chrysippus, (*c*) the Stoicizing Peripatetic Xenarchus of Seleucia (late first century BC), and finally (*d*) Alexander of Aphrodisias (late second century AD), head of the Peripatetic School in Athens, in his discussion of the physical elements, god and soul. Given that Antiochus is also extensively discussed in Chapter 2, I focus here on the other cases.

Do Critolaus' cosmological fragments (especially frs. 12, 13, 15–18 Wehrli) reveal a clear Stoic influence? Although some ideas in them are indeed quite distant from Aristotle's own cosmology and bear some resemblance to Stoic theses, they may be accounted for as developments *within* the Peripatos. The ideas in question, as Kupreeva explains, are the postulation of two principles, the existence of providence and the ethereal constitution of the soul. Consider the latter example. The idea that the soul is made of this celestial substance is present in the *Academica* passage (as she clearly brings out on pp. 136–42, devoted to Cicero) and attested for Critolaus in two sources (frs. 17–18 Wehrli). Even though this has been interpreted as a Stoicizing attitude,[34] there is evidence in the Aristotelian corpus for a divine element in us, which, as Kupreeva explains, is an idea that may bear some close affinity with that of an ethereal soul.[35] The question of the influence of Stoic physics and cosmology in the Peripatos may also be raised in connection with Xenarchus' criticisms of the fifth substance and of Aristotle's argument against extracosmic void. If his criticism was indeed largely influenced by Stoicism, as is claimed by scholars such as Moraux,[36] Xenarchus is not on his own evidence that Stoics and *Aristotelians* agreed on certain cosmological question. But as Kupreeva demonstrates there is strong evidence that Xenarchus' criticisms are the result of an inner evolution of Peripatetic thought, begun

[34] See Mansfeld 1992: 139–40 cited by Kupreeva (Ch. 6 n. 44). [35] See pp. 149–50.
[36] See Moraux 1973: 203–4 (cf. Sharples 2002: 16–17) both cited by Kupreeva.

under Theophrastus and Strato, consisting in the critical analysis of Aristotle's treatises.[37] Analogous conclusions may be drawn in connection with the concept of prime matter as something formless and qualityless in Boethus of Sidon and Nicolas of Damascus.[38] In conclusion, some of the parallels between Stoic and Peripatetic cosmology reveal a genuine agreement between the two schools on central questions. More importantly, this agreement is ultimately explicable by a movement away from Aristotle begun in the earliest post-Aristotle Peripatos. But the critical attitude of these Peripatetics towards Aristotle's original system does not seem to amount to—and is not perceived by them as being—a Stoicizing rejection of the system. As Kupreeva herself explains: '[t]he critical tendencies within the school . . . do not necessarily amount to the rejection of the system, despite the fact that the system projected on the basis of criticisms, may be significantly (for some, perhaps irreconcilably) different from the criticized original'.[39]

Finally, a similar conclusion, Kupreeva contends, may be drawn in connection with Alexander of Aphrodisias. Her argument focuses upon his use of Stoic method and concepts in his discussion of the four elements in his treatise *De Anima* (and similar texts such as *Quaestio* 2. 3 and the last section of *De Intellectu* at *Mantissa* 112. 11–16 Bruns) and fr. 2 Vitelli reporting Alexander's criticism of a contemporary Stoic, Heraclides, in connection with the fifth element, and the nature of god and soul.[40] Both sets of texts seem to evince 'a common intellectual background with the Stoa', and one that goes much beyond the mere use of Stoic terminology and Stoic key concepts.[41] But if we look carefully at the place and function of this common background within each system, some deep differences emerge evincing that the two systems are in fact very different and stem from different origins. To conclude, Kupreeva's response to her original problem—'to what extent these parallels may amount to a genuinely shared position' (p. 135)—seems to be largely negative.

CHAPTERS 7 AND 8

The relation between cosmological and theological theory and practical ethics is the subject of the next two chapters (7 and 8): Marcelo Boeri's 'Does Cosmic Nature Matter?' and Brad Inwood's 'Why Physics?' The subject has been much debated in modern Stoic studies. As Boeri explains, some scholars and philosophers argue that in Stoicism cosmological knowledge is indeed required for leading our lives. This orthodox view has been called into question. The latest advocate of the alternative view is Julia Annas in her recent paper 'Ethics

[37] See the evidence on pp. 151–6. [38] See the argument on pp. 156–9.
[39] See p. 165. [40] See respectively pp. 165–6.
[41] For this conclusion see pp. 135 and 165.

in Stoic Philosophy' (Annas 2007). According to Boeri, this heterodox view cannot be right given the overwhelming amount of textual evidence against it. Accordingly, a thorough discussion of Annas's views in this paper is provided at pp. 184–6.[42] But some radical versions of the orthodox view cannot be right either. For example, the thesis that for the Stoics ethics is somehow subordinate to cosmology, a thesis defended by A. A. Long,[43] needs qualification. In fact, Boeri argues that the right method for defining the position of the Stoics on this question is through an analysis of their conception of philosophy as a whole and of ethics and cosmology (or physics) as two of its parts. None of its parts is preferred over the other (*προκεκρίσθαι*), and in consequence none of them is in any sense superior to the others. But precisely because of this, Boeri explains, our knowledge of any of them must have an impact on, and be required for, our knowledge of the others. Another contribution of Boeri is his argument to the effect that some of the views upheld by late Stoics such as Epictetus and Marcus Aurelius on the close relation between each of us as individual persons and the cosmos as a whole,[44] are already detectable in sources referring to early Stoicism and, notably, in the key passage from Diogenes Laertius 7. 86–9.[45]

It is easy to see how *theoretical* ethics may depend on such knowledge. To give an example, Stoic virtues have a physical basis: they necessarily involve a state of the soul consisting in its undergoing a certain appropriate degree of physical tension (*εὐτονία*).[46] In consequence, a theory of virtue will have to include at least this particular truth of physics. (In fact, even the knowledge of physics is in itself a virtue because it involves a certain tension in the person who possesses it, similar to the tension characteristic of moral virtues.[47]) But it is not immediately clear how *practical* ethics may depend on physical-cosmological knowledge. Considerable space is given by Boeri to this issue. In fact, in section 4 of his chapter he studies how certain particular tenets of Stoic practical ethics may relate to our knowledge of cosmic nature. One case in point is found at Cicero, *De Finibus* 3. 73.[48] As Boeri explains, 'Cato [Cicero's Stoic spokesman] underlines that no one can judge truly (*vere iudicare*) about good and evil unless one has known the whole plan or purpose (*ratio*) of nature, and also the life of the gods, as well as whether human nature is or is not in agreement with that of the universe'.[49] Thus, although cosmological knowledge might not be sufficient for making true value judgements, it is nevertheless a necessary condition for doing so. But why? The reason is that full cosmological knowledge involves knowledge of god's overall plan and, therefore, knowledge of whether some particular state and event that we come across in a given situation contributes to this plan (in which case, it is good), is against it (in which case, it is bad), or

[42] See also pp. 181–2 and 189–90.
[43] Cited by Boeri at p. 176.
[44] See the evidence cited in sect. 3.
[45] Quoted in full by Boeri at pp. 176–7.
[46] See the evidence cited at pp. 189–90.
[47] See the argument developed at pp. 189–90.
[48] Examined at pp. 203–5.
[49] See at pp. 190–1.

is indifferent to bringing it about (in which case, it is neither good nor bad but indifferent). This is one possible reading of Chrysippus' famous foot-example (*ap.* Epictetus, *Diss.* 2. 6. 9–10).[50] The metaphysical basis of this conception is that our rationality—our means to achieve this knowledge—is part of, and the same in nature as, the physical principle that unifies everything there is into a coherent whole and, thus, gives rational cohesion to the cosmos as a whole. This chapter ends with a passage from Marcus Aurelius 7. 9 that clearly illustrates this idea and which I quote in full: 'All things are reciprocally interwoven . . . , everything is coordinated and confers order to the same cosmos. For the cosmos is a unity made up of all things, and god is one, pervading all things. And there is only one reality and only one law, i.e. universal reason of all the living beings that are intelligent, and there is only one truth, since perfection of the living beings that are alike in kind and that participate in the same reason is one, too' (tr. Boeri).

In Chapter 8 Inwood focuses on Seneca and attributes to him a set of positions that is similar to the one attributed by Boeri to the Stoics in general. As Inwood explains, Seneca offers in different places different reasons for studying physics. One reason, conspicuous in *Ep.* 89–90 for example, is that all the disciplines constitutive of philosophy such as physics and ethics are indeed valuable subjects of study, even though they are most valuable not in isolation, but as parts of philosophy as an integrated body of knowledge, which is itself intrinsically valuable. Thus, a perfectly good reason for studying physics is that it is a necessary condition for studying philosophy as a whole (even if this latter study can never be completed in our lifetime). Another reason, provided in *De Providentia*, focuses on the relation between physics and ethics in particular. Although the answer to very specific ethical questions (e.g. why do bad things happen to good men if the world is governed by providence?) may not require demonstrative knowledge of certain truths of physics, the overall peace of mind and right attitude to life that ethics is supposed to provide does require extensive and comprehensive demonstrative knowledge of physics. It is important to notice that for Seneca knowledge of physics is a necessary, non-sufficient, condition for achieving the goal of Stoic ethics. In consequence, the objection that we could perhaps achieve this goal even without studying physics—an objection whose roots lie in Cynicism and in Aristonian Stoicism—is misguided. In fact, Inwood provides a careful analysis of Seneca's discussion of Demetrius the Cynic in *De Beneficiis* 7. 1. 3 and shows that, despite one's first impression, the extent to which Seneca departs from Demetrius' conception of the place of physics in philosophy is great. Thirdly, the study of physics is worthwhile in itself because it suits our nature, which is designed for contemplation and theoretical knowledge in general.[51] According to Inwood, the present reason underlines the intrinsic value of this activity much more clearly that the other two motivations, which

[50] Quoted by Boeri at pp. 191–2.

[51] See the evidence from *NQ* 6. 4. 2 and *De Otio* 4 and 5 quoted on p. 213.

suggest, however differently, an instrumental motivation. It expresses an idea that we find in earlier Stoics—'the rational animal was created by nature fit for action and contemplation' (DL 7. 130)—and that appears prominently in other Roman Stoics, e.g. Epictetus (*Diss.* 1. 6. 19–22). One question that may be asked, however, is whether from this perspective the study of physics is really valuable in itself or just as a necessary means to behaving fully in accordance with human nature. In Epictetus at least, I believe, and, according to Inwood, in Seneca too, the former view is suggested. Given that theoretical knowledge in general is constitutive of human nature (because it is part of what marks off humans from the lower animal species), and given that human nature is an end intrinsically valuable that ought to be pursued, this sort of knowledge is intrinsically, not instrumentally or derivatively, valuable. In the Epictetus passage cited by Inwood this train of thought is implicit: 'It is therefore shameful that man should begin and end where irrational creatures do. He ought rather to begin there, but to end where nature itself has fixed our end; and that is contemplation and understanding (*θεωρίαν καὶ παρακολούθησιν*) and a way of life in harmony with nature' (*Diss.* 1. 6. 20–1; tr. Gill and Hard).

In sum, as Inwood explains: 'Seneca recognizes two quite different reasons for studying physics and for attending to its doctrines, reasons which might well seem to be opposed to each other. Studying physics provides direct instrumental support to what we might call the enterprise of ethics, but it also fulfils something very important and fundamental in our natures, the built-in drive for contemplation of nature.'[52] This double motivation is most conspicuous in the two texts to which the final part of his chapter (pp. 215–22) are devoted: the *Natural Questions* as a whole and the later books of the *Letters to Lucillius.* As Inwood carefully brings out, these texts show how these apparently opposed motivations turn out to be complementary to each other. In particular, the study has an instrumental value for ethics insofar as it has an intrinsic value in the sense that if it were not valuable on its own, it would not be of any use for ethics. This complex account of the motivation for studying physics is indispensable for understanding Seneca's position on this question.

CHAPTER 9

In 'Stoic Philosophical Theology and Graeco-Roman Religion', Keimpe Algra carries out a thorough examination of early and late Stoic philosophical theology understood as the philosophical account of religious phenomena and its relation to traditional religious myth and cult. The project undertaken by Algra in connection is analogous to the one developed in Chapters 7 and 8 in connection

[52] See pp. 214–15.

with ethics. Just as our study of physics is essential to the conduct of our moral life, so too our philosophical knowledge of god is needed to clarify and articulate our natural preconception of him. Thereby, this knowledge will have an impact on our attitude towards traditional religious cult and myth (and hence on whether or not we accept traditional religious cult and myth in our *own* religious practice). Moreover, I should add, given the identity of god and cosmos, philosophical theology will ultimately have to help us to conduct our moral life, and the study of physics, in turn, will ultimately have to shape our natural preconception of god and to guide our religious life.

As Algra explains, the connection between Stoic philosophical theology and traditional religion is complex. Contrary to an interpretation recently advocated by Marie-Odile Goulet-Cazé, according to which, Algra observes, Chrysippus espoused an 'extremely conservative' position concerning traditional religion in that he adopted at least large parts of it without any criticism (thus departing sharply from Cynicism),[53] Algra points out that even when the Stoics accepted important parts of traditional religion, such as certain myths about the nature of god and its relation to the cosmos, they sought to provide them with a detailed rational basis. However, the Stoics also rejected certain aspects of religious and cosmological traditional myths, arguing that even though they originally proceeded from our natural preconceptions about god, these preconceptions have been contaminated to some extent by additional conceptions that distort our view of the true nature of god and the cosmos.[54] In consequence, the Stoic attitude to traditional religion is mixed. It is one of acceptance and adaptation of some of its elements, on the one hand, and one of objection and rejection of other elements, on the other. And the function of philosophical theology is to offer the adequate grounds for performing this double task. As Algra himself points out: 'mainstream Stoicism was committed to an interesting combination of primitivism (the "natural" world view of the people of old inevitably got corrupted), and progressivism (the subsequent development of philosophy can remedy this, and show us what can and cannot be salvaged)'.[55] Section 4 of this chapter is devoted to shedding light on various problems that arise in connection with this Stoic critical appropriation of traditional cosmo-theological myth. As for religious *cult* (sections 5 and 6), the role of philosophical theology in connection with it is analogous to its role in connection with myth. In myths, one central task of philosophical theology is eventually to lay out explicitly their implicit rational foundation. In cults, one of its tasks is to draw out and assess its conceptual content through a confrontation of this content with Stoic theological theory. As in the case of myth, there is no rejection of traditional cult *in toto*, but rather a complex combination of critique and adaptation of different parts and aspects of it. To take one of the five examples developed by Algra (Zeno on sanctuaries, Chrysippus, Diogenes of Babylon, Seneca, and Varro on the use of

[53] Cited by Algra at p. 213. [54] See p. 233. [55] See p. 234.

anthropomorphic images to represent god in religious rituals), Zeno was critical of the use of sanctuaries to worship the gods. But his point is probably not that, ideally, sanctuaries should be prohibited and removed from the cities. His concern is merely that they are superfluous and, in consequence, that one should not *promote* building them—the only proper way to honour the gods being 'by our own spiritual attitude, i.e. by imitating them through becoming virtuous'.[56]

To conclude this introduction, let me take up an issue raised by Algra in the penultimate section of Chapter 9. Why did Stoics not absolutely reject the use of anthropomorphic images to represent god in traditional cult? Absolute rejection seems to be required by their pantheism: the Stoic god, as a physical substance that permeates the whole cosmos, cannot have the human shape attributed to the gods in traditional representations of them. This is relevant for understanding how the Stoics conceived the relation between god and cosmos. Apparently, there is a tension between, on the one hand, their pantheism, which reflects how god physically interacts with the cosmos, but seems to rule out anthropomorphism, and, on the other hand, their conception of god as a person, which seems to rule out pantheism, but is apparently required by his demiurgical nature and the purposiveness of his action. Pantheism and this 'personalistic theism', to use Algra's expression, are two equally essential, but apparently incompatible, aspects of god's relation to the cosmos. How can this tension be resolved and, in consequence, why are the Stoics justified in not rejecting totally anthropomorphic representations of god despite their pantheism? The tension disappears if personhood does not necessarily require a human shape. Certain traits of a human body may help to express some of the attributes—moral or intellectual—of god. This is something, Algra points out, that Epictetus, *Diss.* 2. 8. 25–7 sharply brings out.[57] But even though the Stoic god has these attributes without having a human body or shape, it is difficult for us to imagine him as possessing them in this disembodied state. Hence the need to represent god through anthropomorphic shapes. In other words, it may be epistemically necessary (for us) that god's attributes be conveyed through human shape.[58] This fully justifies anthropomorphic representations of god in religious cult. But there is no metaphysical necessity in linking the concepts of person and of human shape, and therefore no contradiction between the two main theses of Stoic cosmo-theology: pantheism and personalistic theism.

[56] See p. 236. [57] Quoted at p. 245.

[58] See the evidence from Dio Chrysostomus and Varro discussed by Algra at pp. 245–7. Algra also mentions in connection with Dio, *Or.* 12. 60 as a further justificatory element that 'we want to conceive of [god] as an object of worship near to us (rather than as a remote astral or cosmic god) and as a father' (p. 247).

PART I

GOD, PROVIDENCE, AND FATE

1

How Industrious can Zeus be?

The Extent and Objects of Divine Activity in Stoicism

Thomas Bénatouïl

> [S]how me the sublime presence of the highest spiritual cause lurking, as always it does lurk, in these suburbs and extremities of nature; let me see every trifle bristling with the polarity that ranges it instantly on an eternal law; and the shop, the plough, and the leger, referred to the like cause by which light undulates and poets sing.
>
> (Emerson 1971: 68)

Epicurus has customarily been considered as the maverick of Greek philosophical theology, because his doctrine that gods have no influence on our world caused outrage in all the other philosophical schools engaged in theology, namely the Academy, the Lyceum, and the Stoa, the members or heirs of which all held that gods contribute to the ordering of the world. I would like to challenge one aspect of this widespread picture of the ancient theological arena. I grant that Epicurus was a maverick, and a daring one at that, but I think the Stoics were almost equally daring[1] when they claimed that god is directly present and active in absolutely everything that exists and happens in the world, from the most exalted beings or events to the lowliest ones.

The hostility aroused by this description of god has been, to my knowledge, underestimated and shows that, on some issues, Epicureans, Peripatetics,

I thank the participants in the Mexico conference 'Dios y el Cosmos en la Filosofía Estoica' (July 2006) for their helpful remarks on an earlier and longer version of this chapter, Ricardo Salles for his invitation to this conference and his comments on several versions of my paper, and Lionel Dahan for his thorough revision of my English.

[1] On the willingness of Stoics to break from tradition in theological matters, see below, Ch. 9, pp. 230–4.

Academics, and Platonists sided together against the Stoics.[2] Here is the kind of attack I have in mind:

Surely it demeans our preconception of the deity to say that god pervades the whole of the matter underlying everything and remains in it, whatever it may be like, and has as his principal task the perpetual generation and moulding of anything that can come to be from it; and for them to make god a craftsman of grubs and gnats, just like a modeler devoting himself to clay and making everything that can be created from it?[3]

We can distinguish four features of the Stoic god exposed here by Alexander of Aphrodisias as unworthy of any divine being: (*a*) his presence inside matter or in direct contact with it, (*b*) the fact that his activity is aimed at producing things or beings, (*c*) the fact that this immanent creation is Zeus' permanent and prevailing occupation, (*d*) the fact that features (*a*), (*b*), and (*c*) hold for *every* individual thing, and in particular *even* for base or evil things. These four features are closely connected to each other and define what I shall call the industriousness of the Stoic Zeus.[4] The various adversaries of the Stoa appear to object to at least one of these four features. Feature (*a*) is generally the focus of Platonist attacks,[5] feature (*b*) is quite specific to Alexander's refutation,[6] feature (*c*) is particularly targeted by Epicureans,[7] and feature (*d*) is mocked by all of these schools.[8]

All these texts could be used as testimonies about Stoic theology, but they are nonetheless polemical and may be suspected of being biased. Consequently, they cannot establish by themselves that the Stoic god was as industrious as they claim. This is the issue which I intend to explore here. There is no need to establish that Zeus is active, pervades matter and moulds it, or that his providence extends to the whole world, since these are well-known features of the Stoic god,

[2] This is obviously not to deny that there were also common grounds between Stoic and Peripatetic or Stoic and Platonist theology, especially against Epicureans. For a nuanced assessment of the affinity between Stoic and Peripatetic doctrines of principles (including providence) and its limits, see below, Ch. 6.

[3] Alexander, *De Mixtione* p. 226. 24–9 = *SVF* 2. 1048. R. Todd's translation (1976) was slightly modified.

[4] I cannot see any other god, in Greek philosophy, to which these four features (and especially the last one) apply. Empedocles' Cypris is 'busy' crafting living beings (*DK* 31 B 73), but she has a rival, Strife, which accounts for many events in the world (DK 31 B 26 or 35). The demiurge of the *Timaeus* is even less industrious, because he leaves to lower gods the task of fashioning mortal beings (*Tim.* 41a–d) and to the world-soul the task of keeping his creation alive; furthermore there is the likelihood that he himself is not the highest divine being. This is however not to deny that Stoic theology was influenced by the *Timaeus*: see Reydams-Schils (1999) and Ch. 2 below. On Empedocles' Cypris and Plato's demiurge as craftsmen, see Solmsen (1963: 476–84).

[5] See for instance Numenius, fr. 50 Des Places or Proclus, *In Tim.* 1. 413. 27–414. 7 = *SVF* 2. 1042.

[6] See Alexander, *De Providentia* 21. 5–20 Ruland for the idea that the Stoics turn god into a slave of men and other beings. Cf. Plutarch, *Defectu Oraculorum* 416–17.

[7] Cicero, *ND* 1. 52–3, but the objection is also Peripatetic: see Alexander, *Prov.* 19. 10–21. 1.

[8] See Philodemus, *Piet.* col. 356 for the Garden, Cicero, *Luc.* 120 for the New Academy, Porphyry, *Abst.* 3. 20. 2–3 for Platonism, and Alexander, *Prov.* 23. 6–25. 17 or *Fat.* 31. 203. 11–20 or *Mantissa* 2. 113. 12–14 for the Peripatos. On Alexander's criticism, see p. 163 below.

as is the fact that there is only one god according to the Stoics, all mentions of other gods being in fact references to specific parts or powers of Zeus.[9] What can still be disputed is how far or how seriously the Stoics upheld these positions: did they admit of exceptions or limitations to Zeus' industriousness, that their opponents would have overlooked? I am going to examine five of these, which seem to result from various aspects of Stoic cosmology. They will prove unable to qualify Zeus' industriousness as far as Stoic theology from Zeno to Chrysippus is concerned, but will, hopefully, shed some light on this odd notion of god.

1. IS ZEUS ALWAYS ACTIVE IN THE WORLD?

Testimonies about Zeus' activity inside matter usually present it as a process leading to the production of the world as it is. After having distinguished, named, and defined the two Stoic principles, Diogenes Laertius explains their relation and operation by saying: 'for this [= god], since it is everlasting, constructs (δημιουργεῖν) every single thing throughout all matter'.[10] This certainly is an industrious god in the sense I have just defined, namely a god actively taking care of all things in the world from inside their matter. The scope of its activity is however here limited to the *making* of things. If the industrious god is essentially a technical god or divine craftsman,[11] we have to wonder whether Zeus still has work to do after he has fashioned the world and all its inhabitants. We cannot simply assume that Zeus remains industrious once the world is produced: things might have been wrought in such a perfect way that they take care of themselves, or their life could be guided by another entity, and all Zeus would have to do would be to watch them from afar unfold the destiny he has ascribed to each of them.

This hypothesis is however based on too crude an analogy between god and craftsmen, and is hence not accepted by the Stoics, as shown by an argument preserved by Alexander of Aphrodisias:

Again one might enquire if it is possible to describe the god that has gone through matter and exists in it as a craftsman (οἷόν τε δημιουργὸν) of what comes to be from matter. For they offer in support of this the fact that artistic products do not come to be in the same way as what comes to be by nature; for the products of nature are not superficial but are completely formed and moulded and their interior parts are precisely crafted with care,

[9] DL 7. 147 and sect. 3 below. This doctrine prevents the Stoics from appealing to a Platonist hierarchy of divine beings, in which only the inferior ones perform the industrious tasks required by providence: see for instance Plutarch, *Defectu Oraculorum* 416e–417a or Plotinus, *Enneads* 4. 8. [6] 2. 24–53.

[10] DL 7. 134 = *SVF* 2. 300 = LS 44B (tr. Long and Sedley). Cf. Seneca, *Ep.* 65.2 = *SVF* 2. 303 and Alexander, *Mixt.* p. 225.1 = *SVF* 2. 310.

[11] See also DL 7. 147 and 156.

while those of the arts are formed as in the case of statues where the inner parts are not moulded (ἀδιάπλαστα). Therefore, they claimed that what creates artistic products was external and separate, while with things that come to be by nature, the power that forms and generates them was present in the matter.[12]

This argument emphasizes the limits of the analogy between god and craftsmen, in order to show that the divine craftsman of natural things has to work from the inside of these things, because they are crafted inside out or through and through, whereas artefacts are crafted only superficially and hence from the outside. This argument is thus crucial to our enquiry, since it shows (*a*) how the Stoics claimed to prove that an immanent natural universal craftsman, hence an industrious god, is required to account for the nature of things,[13] (*b*) that this divine craftsman and his products cannot be distinguished as simply and as sharply from each other as human craftsmen and their artefacts. This second point strongly suggests that Zeus' industriousness is not limited to the making of things.

This is borne out by another Stoic description of Zeus' relation to the world, which compares it to the relation our soul has to our body.[14] This analogy obviously implies that divine activity is responsible for the whole *life* of the world, and not only for its creation or birth. Since I shall later consider this analogy at length, I only mention its significance now and turn to one of its consequences concerning the making of the world. This activity should not be imagined as a difficult and tiresome task, during which Zeus, like a human craftsman, strives and struggles to force matter to take the shapes of the things he wants to produce:

for you yourselves are fond of saying that there is nothing that a god cannot accomplish, and that without any toil (*sine labore ullo*); as man's limb are effortlessly moved merely by his mind and will, so, as you say, the gods' power can mould and move and alter all things. Nor do you say this as some superstitious fable or old wives' tale, but you give a scientific and systematic account of it: you allege that matter, which constitutes and contains all things, is in its entirety flexible and subject to change (*flexibilem et commutabilem*), so that there is nothing that cannot be moulded and transmuted out of it however suddenly, but the moulder and manipulator (*fictricem et moderatricem*) of this universal substance is

[12] *SVF* 2. 1044 = Alexander, *Mixt.*, p. 225. 18–27 (tr. R. Todd). Note that Zeno's definition of nature as craftsmanlike (*artificiosum* = τεχνικὸν) or as a true craftsman in Cic. *ND* 2. 57 is similarly qualified by the mention of the technical superiority of nature *vis-à-vis* human crafts, and of the fact that the world contains all its products in itself. On the analogy between nature and craftsman in Greek theology or cosmology, from the Presocratics to the Stoics, see Solmsen (1963). On the priority of the biological over the technological model in Stoic theology, see J. B. Gourinat (pp. 50–1 below).

[13] This argument is obviously itself based on Stoic corporealism, which requires all actions and passions (such as moulding and being moulded) to occur between bodies in direct contact with each other (Cic. *Acad.* 1. 39). On this doctrine, which is crucial to the industriousness of the Stoic god, see Hahm (1977: 3–28) and Ch. 2 below.

[14] See e.g. Sextus, *M* 9.75 = *SVF* 2. 311 = LS 44C(3) and Plutarch, *De Stoic. Rep.* 1052c and 1053b = LS 46E and sect. 3 below.

divine providence, and therefore providence, whithersoever it moves, is able to perform whatever it will.[15]

Matter offers no resistance whatsoever to Zeus' activity and can even be said to welcome it, as Seneca nicely puts it: *Materia iacet iners, res ad omnia parata, cessatura, si nemo moueat.*[16] If the Stoics took their second principle, matter, from the *Timaeus*, they deprived it of any intrinsic movement or quality and made it *absolutely* passive.[17] This entails that Zeus' activity on matter is effortless,[18] and that matter cannot by itself have any quality or shape, because god is the cause of all the qualities or shapes matter takes.

Once we have acknowledged that divine activity is not defined (in nature or degree) by the effort or energy it requires, but by the simple fact of shaping matter from the inside, it becomes much easier to understand why Zeus remains industrious after the fashioning of the world. Note that 'providence' is here described as 'moulder and manipulator' of matter: this implies that taking care of the world after it is built is not something different from producing it. Both actions can be assigned to providence or to Zeus. If we need however to specify which activities are performed by Zeus once the world and its components have been produced, we might want to distinguish three types of activities, from the most evident one to the most invisible:

(1) God has to produce continuously all the changes happening inside the world as it is now: the regular movements of stars and planets,[19] the seasonal cycles of life and death on earth, the various movements of living beings, the complex developments of human history. This dimension of god's activity is only an extension of the initial making of the overall structure of the world and of the various beings in it, since the behaviour of each being is determined by its nature, and they are both inscribed in its 'seminal reason'.[20] This is why god is equated by the Stoics to the diachronic chain of fate, according to which all events happen.[21]

(2) Beneath or behind this first manifold but obvious activity lies another one. All objects depend on hidden activities constantly 'sustaining' them into existence,[22] namely preserving their unity, their structure, and their defining qualities through time. Nemesius describes them as produced by *τονική*

[15] Cic. *ND* 3. 92, where Cotta discusses Stoic theology (I quote H. Rackham's tr. publ. in the Loeb Classical Library).

[16] *Ep.* 65.2 = *SVF* 2. 303.

[17] Reydams-Schils 1999: 45; Frede 2005: 221.

[18] The Stoics hence claim against the Epicureans that Zeus can take care of the world without being *laboriosus* (*ND* 1. 52).

[19] These are offered by Stoics as evident proof of divine activity in the world, because they are both everlasting, regular, and beneficial: see *ND* 2. 43, 50–5, and 80.

[20] See DL 7. 148 and Hahm (1977: 75–6).

[21] See DL 7. 135 and Bobzien (1998: 45–53).

[22] On the Stoic notion of 'sustaining (*συνεκτικὸν*) cause', see LS 55 F, H, I. Diogenes Laertius says that 'the term Nature is used by them to mean sometimes that which sustains the world (*τὴν συνέχουσαν τὸν κόσμον*)' (DL 7. 148).

κίνησις,[23] namely a movement and not a static cause, and asks from which power this activity originates. I cannot go into the details of Stoic physics, but suffice it to say that it is fire or *pneuma*,[24] which are characterized by their perpetual movement[25] and equated to god.[26] Galen offers a nice comparison: a bird not moving in the air is only apparently motionless and would fall if its muscular tension did not counterbalance its weight.[27] We could similarly say that, if the world was to be frozen in a given state, the resulting absence of change in the world would not put an end to divine activity, because the latter would still be needed to preserve the various qualities and unity of each thing in the world as well as the order and unity of the world itself.

(3) Activities (1) and (2) last only as long as the world exists. According to the Stoics, the world as it is is not eternal and will be destroyed. But this destruction will be caused neither by something independent from god, nor by a punctual and sudden action of god: it will be a necessary or natural outcome of development which started when the world was born. The actual cosmic order is in fact a dynamic equilibrium between hot fire and humidity, which inevitably leads to the destruction of the world, because fire nourishes itself from humidity. It will thus end up burning everything in the world and bring about a state called 'conflagration' (ἐκπύρωσις), where there is only fire and from which a new world identical to the previous one will be produced.[28] Because of physical constraints imposed by the elements composing the world, the structuring and unifying of the world by Zeus are slowly evolving and, so to speak, deteriorating. As Chrysippus said in his *On Providence*, 'Zeus continues to grow until he has used up everything on himself'.[29] Although itself cyclical, this process is different from the natural cycles ordering the/each world mentioned in (1), and can be considered as a third dimension of Zeus' activity.

2. IS ZEUS ACTIVE DURING EACH CONFLAGRATION?

While this dynamic evolution of the cosmic order towards destruction is a sign of the divine activity at work in the world at all time, it leads also to a

[23] Nemesius, *De Natura Hominis* 70. 6–71. 4 = LS 47 J: 'there exists in bodies a kind of tensile movement which moves simultaneously inwards and outwards, the outward movement producing quantities and qualities and the inward one unity and substance'.

[24] On fire or heat as sustaining cause of the world and every being, see *ND* 2. 28 and Salles (2005: 63–6). On *pneuma* (air + fire) as sustaining cause in Chrysippus, see Long and Sedley (1987: s. 47) and Hahm (1977: 158–74).

[25] See e.g. *ND* 2. 23–4. [26] See Aetius 1.7.33 = *SVF* 2. 1027 = LS 46A.

[27] Galen, *Musc. Mot.* 4. 402. 12–403. 10 = *SVF* 2. 450 = LS 47K.

[28] See Mansfeld (1979) and Long (1985).

[29] Plutarch, *On Stoic Self-Contradictions* 1052c–d = *SVF* 2. 604 = LS 46E. For the specifics on the process leading to conflagration, which does not entail that the world is destroyed according to Chrysippus, see Ch. 5 below, by R. Salles.

second objection to the idea that Zeus is always active. During the phase of conflagration, after the world has been turned into pure fire and before another identical world-order is crafted, Zeus does not seem to be industrious, since he does not produce any beings or shapes with matter.[30] As Ricardo Salles clearly puts it: 'The conflagration thereby involves the elimination of any differentiation. Thus, in order for this elimination to occur, the sustaining cause of the world, or god, must stop its activity *qua* sustaining cause of the world' (Salles 2005: 67). This suspension of divine activity during each conflagration is accepted by Seneca: 'at the time when the world is dissolved and the gods have been blended together into one, when nature comes to a stop for a while, he reposes in himself (*adquiescit sibi*) given over to his thoughts'.[31]

However, if Seneca does not see any problem in the fact that god rests for a while, earlier Stoics did. Philo reports that Boethos and Panetius abandoned the Stoic doctrine of conflagration, and he then lists the various arguments against conflagration put forward by 'Boethos and his school'.[32] The third argument appeals to the constant activity of god:

> Moreover, if all things are as they say consumed in the conflagration, what will god be doing during that time? Will he do nothing at all? That surely is the natural inference. For at present he surveys each thing, guardian of all as though he were indeed their father, guiding in very truth the chariot and steering the bark of the universe, the defender of the sun and moon and stars whether fixed or wandering, and also the air and the other parts of the world, cooperating in all that is needful for the preservation of the whole and the faultless management of it according to right reason. But if all things are annihilated, inactivity and dire unemployment (ὑπ' ἀργίας καὶ ἀπραξίας δεινῆς) will render his life lifeless and what could be more monstrous than this?[33]

Conflagration appears to be incompatible with the idea that Zeus is always active: according to Boethos, one must choose between his own position, which discards conflagration, and Seneca's position, which upholds conflagration and is thus compelled to sacrifice Zeus' industriousness. Did the first Stoics, Zeno, Cleanthes, and Chrysippus, accept this dilemma and, since they are known to have upheld conflagration, grant that Zeus could be temporarily inactive? I agree with Ricardo Salles that they did not, because they conceived of god as 'essentially (and hence always) active'.[34] Boethos was clearly arguing from two Zenonian or Chrysippean tenets and tried to show their incompatibility, in order to convince his fellow Stoics to discard the less likely and the less crucial one (according to him), namely the destruction of the world. Early Stoics agreed about refusing divine idleness, but did not share Boethos' and

[30] In *ND* 1. 22, Velleius asks why the Stoic *Pronoia* remained inactive or idle *before* the fashioning of the world, meaning probably *during* each conflagration.

[31] Seneca, *Ep.* 9.16 = *SVF* 2. 1065 = LS 46O.

[32] Philo, *Aet.* 76–8.

[33] Philo, *Aet.* 83–4 (Colson's tr. for the Loeb Classical Library was slightly modified).

[34] Salles (2005: 67), referring to LS 44C, quoted below, p. 30.

Seneca's notion that god is not active during the conflagration. 'The idea is rather that at the conflagration god ceases to sustain the world, which is perfectly compatible with his doing something else, namely, causing the conflagration' (Salles 2005: 67).

Sextus argues in effect that the power moving and shaping the matter of the universe is everlasting:

So there exists a power which in itself is self-moving, and this must be divine and everlasting (ἀΐδιος). For either it will be in motion from eternity or from a definite time (ἀπό τινος χρόνου). But it will not be in motion from a definite time; for there will be no cause of its motion from a definite time. So, the power which moves matter and guides it in due order into generations and changes is everlasting.[35]

God, as self-moving cause, cannot stop moving and start again later. This refutes Seneca's position, unless the conflagration is not included in 'time'. But, since the Stoics held past and future to be infinite,[36] time cannot start after or stop before any conflagration. Moreover, Chrysippus defined time as 'the dimension of the world's motion'[37] and described the conflagration as something happening to the world.[38] It is therefore not out of time, and Sextus' argument applies to it:[39] god is thus moving and active during the fiery phase,[40] causing the whole of matter to take the qualities and movements of fire.

It could however be objected that, once everything in the world is burnt up and only pure fire exists, the 'world's motion' might come to a stop and time be, so to speak, suspended, since there is nothing left for god to act upon: he would just have to *be* (after all, god is fire) and stay with himself, as Seneca's text[41] and other testimonies seem to suggest:

[35] Sextus, *M* 9.75–6 = *SVF* 2. 311 = LS 44C.

[36] DL 7. 141= *SVF* 2. 520.

[37] Simplicius, *In Arist. Categ.* 350. 15–16 = LS 51A.

[38] See Plutarch, *De Stoic. Rep.* 1053b quoted below or Philo, *Aet.* 4 and 54, and R. Salles below (pp. 119–21). On this whole issue of the relationships between time, the conflagration, and the world, see already Algra (2004: 188–9).

[39] Long and Sedley (1987: i. 311) argue that 'time must be presumed to continue even when no world-order in any sense measurable by us exists' *because* 'god is continuously active during all states of the Stoic universe'. Although I agree with their conclusion, I deem preferable to establish it on other grounds, since Sextus' proof of god's permanent activity presupposes the notion of 'time'. They could however answer that the very *definition* of god as 'active cause' settles the matter and should spare us any further analysis.

[40] This leads however to another objection, raised by Velleius against the demiurge of the *Timaeus* and the Stoic god: 'why did these deities suddenly awake into activity as world-builders after countless ages of slumber?' (*ND* 1. 21). The Stoics could answer that the duration of each conflagration is physically determined by the maximum length of time during which fire can persist without any nourishment, as suggested in Mansfeld (1979: 161).

[41] See n. 31 above. Seneca adds however that god devotes his time alone to his own thoughts, as argued also by Epictetus, *Diss.* 3. 13. 2. Long (1985; repr. 2006: 270) quotes these texts as correct descriptions of Zeus during the conflagration. The notion of 'divine thoughts' is nonetheless absent from testimonies about early Stoic theology. On the meaning and role of 'divine thought' in Roman Stoicism, see Reydams-Schils (2005).

[god] originally being by himself (καθ' αὐτὸν ὄντα), turns the entire substance through air into water . . .[42]

In his *On Providence* book 1, [Chrysippus] says: 'When the world is fiery through and through, it is directly both its own soul and commanding-faculty . . .'[43]

However, the fact that god is here said to 'be by himself' does not mean that it is separated from matter. If matter was not present during the fiery phase, how could it be 'turned through air into water' in the following phase of cosmic evolution?[44] A distinction is to be made between god and fire.[45] During the conflagration, god causes the whole of matter to be fiery, and hence acts upon it.[46] That this is a true activity involving motion is further borne out by the fact, already mentioned in the preceding section (p. 28), that fire is described by the Stoics as an active and fast-moving element.

3. IS ZEUS EQUALLY ACTIVE FROM HEAVEN TO EARTH?

Once the world has been restored after a conflagration, how is divine activity distributed in the various parts of the world? Like Plato and Aristotle, the Stoics distinguished between the celestial region and the region around the earth. In Cicero's *De Natura Deorum*, after having proved the existence of the gods and the divinity and rationality of the world as a whole, the Stoic Balbus 'assigns the same divinity to the stars' (2. 39), and gives a long description of the complex but rational movements of celestial bodies. Here is the conclusion of this description (2. 56):

In the heavens therefore, there is nothing of chance or hazard, no error, no frustration, but absolute order, accuracy, calculation and regularity (*omnis ordo veritas ratio constantia*). Whatever lacks these qualities, whatever is false and spurious and full of error, belongs to the region between the earth and the moon (the last of all the heavenly bodies), and to the surface of the earth. Anyone therefore who thinks that the marvellous order and incredible regularity of the heavenly bodies, which is the sole source of preservation and safety for all things, is not rational (*mentis expers*), himself cannot be deemed a rational being.

[42] DL 7. 136 = *SVF* 1. 102 or 2. 580. I quote the translation in Hahm (1977: 34). On this text, see also Ch. 4 by J. Cooper (pp. 101–3).

[43] Plutarch, *De Stoic. Rep.* 1053b = *SVF* 2. 605 = LS 46F. Cf. LS 28O (4) = *Comm. Not.* 1077e.

[44] Besides, the very nature of matter as a principle requires that it be present during the fiery phase, since both principles, god and matter, are ungenerated and indestructible (DL 7. 134): see Hahm (1977: 33–4), and pp. 59 and 102 below.

[45] See Long (1985; repr. 2006: 268) and Gourinat (p. 63).

[46] There is a debate between Mansfeld (1979) and Long (1985) as to whether this fiery phase is a better state than the cosmic phase. I side with Long who answers in the negative. The fact that divine activity is less obvious during the conflagration might in fact be another argument against Mansfeld's position.

This sharp hierarchy between two regions of the world is quite surprising, because it does not seem to fit with Stoic pantheism, according to which the world in all its parts is pervaded by divine reason through and through and administered scrupulously according to it.[47]

Should we discard this passage as belonging to a late version of Stoic cosmology influenced by Plato or Aristotle, which are otherwise present in Balbus' speech? I prefer to offer a contextual interpretation of this dualism, as suggested by this other passage found a little later in Balbus' speech (2. 59):

> We have discussed the world as a whole, and we have also discussed the heavenly bodies; so that there now stands fairly well revealed to our view a vast company of gods who are neither idle nor yet perform their activities with irksome and laborious toil (*nec cessantium deorum nec ea quae agant molientium cum labore operoso ac molesto*). For they have no framework of veins and sinews and bones; nor do they consume such kinds of food and drink as to make them contract too sharp or too sluggish a condition of the humours; nor are their bodies such as to make them fear falls or blows or apprehend disease from exhaustion of their members—dangers which led Epicurus to invent his unsubstantial, do-nothing (*nihil agentes*) gods.

The contrast between divine activity and human toil is here explicitly aimed at refuting the Epicureans. Similarly, in the previous passage, the last sentence clearly targets the Epicureans who hold the world to be a product of chance. If these two texts emphasize the imperfection of life on earth to the point of sounding Platonist, it could be mainly because they are part of a defence of Stoic theology against philosophers who deny the existence of *any* physical heterogeneity or hierarchy between heavens and earth. Furthermore, Balbus' despising of life on earth and human frailty in these passages is largely made up for by his lavish praise of Nature and man in the rest of his speech.

By siding with Plato and Aristotle against Epicurus about the divinity of the heavens, the Stoic could nevertheless qualify the industriousness of their god. The text just quoted opposes only stars and planets to Epicurus' objection against divine industriousness, because these heavenly bodies are capable of moving constantly but effortlessly, contrary to terrestrial bodies. The continuous activity needed to preserve the cosmic order could thus be concentrated in the heavens and extend below the orbit of the moon either indirectly and from a distance, or to a very limited degree.

We need therefore to investigate the relationship or interaction between the celestial region and the rest of the Stoic world. It is captured by the notion of *ἡγεμονικον*, which implies an analogy between the world and an animal endowed with a soul:

> The world is directed by intelligence and providence, . . . since intelligence pervades every part of it, just like the soul in us. But it pervades some parts to a greater extent and others

[47] See e.g. Plutarch, *De Stoic. Rep.* 1056c = *SVF* 2. 997 = LS 55R quoting Chrysippus: 'no state or process is to the slightest degree other than in accordance with the rationale of Zeus'.

to a lesser degree. Through some parts it passes as tenor, as through bones and sinews. Through others as intelligence, as through the commanding faculty. So the whole world, which is an animal and animate and rational, has the aether as its commanding faculty, as Antipater of Tyre says in his *On the World* book 8. But Chrysippus in his *On Providence* book 1 and Posidonius in his *On Gods* say that the world's commanding faculty is the heaven, and Cleanthes the sun. Yet Chrysippus in the same book has a rather different account—the purest part of the aether; this they say, as primary god, passes perceptibly as it were through the things in the air and through all animals and plants, and through the earth itself by way of tenor.[48]

This testimony is very important because it claims to show how the analogy between god and our soul allows us to combine pantheism and cosmic hierarchy, by resorting to a gradation: god is present everywhere but more present in some parts of the world than in others. The exact nature of this gradation is however difficult to grasp.

To understand it, let us follow the various uses of the term *νοῦς* in Diogenes' testimony. (1) The world is ruled *κατὰ νοῦν*: *νοῦς* is the norm according to which the world is organized. (2) Νοῦς pervades every part of the world: *νοῦς* is the ubiquitous agent administering the world. (3)(*a*) God passes through the aether or the heaven *ὡς νοῦς*: there is a part of the world where *νοῦς* is present *as itself*, and (*b*) this part dominates and administers the rest of the world as its 'commanding' or 'ruling part' (*ἡγεμονικόν*). The reasoning seems fairly straightforward: the world is governed according to *νοῦς* (1), because one of its parts is made of *νοῦς* in its purest form (3*a*) *and νοῦς* is also present in some lower form in the other parts of the world (2), which are thus all under the control of the first part (3*b*).

Statement (2), however, is quite puzzling, because it implies that *νοῦς* is present *as something else than νοῦς* everywhere except in the ruling part of the world. This, and not the definition of the heaven as 'ruling part' of the world, is the main point of the analogy. But how can soul or intelligence[49] pass through

[48] DL 7. 138–9 = *SVF* 2. 634 = LS 47O: Τὸν δὴ *κόσμον διοικεῖσθαι κατὰ νοῦν καὶ πρόνοιαν, καθά φησι Χρύσιππός τ' ἐν τῷ πέμπτῳ* Περὶ *προνοίας καὶ* Ποσειδώνιος *ἐν τῷ τρισκαιδεκάτῳ* Περὶ *θεῶν, εἰς ἅπαν αὐτοῦ μέρος διήκοντος τοῦ νοῦ, καθάπερ ἐφ' ἡμῶν τῆς ψυχῆς· ἀλλ' ἤδη δι' ὧν μὲν μᾶλλον, δι' ὧν δὲ ἧττον. δι' ὧν μὲν γὰρ ὡς ἕξις κεχώρηκεν, ὡς διὰ τῶν ὀστῶν καὶ τῶν νεύρων· δι' ὧν δὲ ὡς' νοῦς, ὡς διὰ τοῦ ἡγεμονικοῦ. οὕτω δὴ καὶ τὸν ὅλον κόσμον ζῷον ὄντα καὶ ἔμψυχον καὶ λογικόν, ἔχειν ἡγεμονικὸν μὲν τὸν αἰθέρα, καθά φησιν* Ἀντιπατρος *ὁ* Τύριος *ἐν τῷ ὀγδόῳ* Περὶ *κόσμου.* Χρύσιππος *δ' ἐν τῷ πρώτῳ* Περὶ *προνοίας καὶ* Ποσειδώνιος *ἐν τῷ* Περὶ *θεῶν τὸν οὐρανόν φασι τὸ ἡγεμονικὸν τοῦ κόσμου,* Κλεάνθης *δὲ τὸν ἥλιον. ὁ μέντοι* Χρύσιππος *διαφορώτερον πάλιν τὸ καθαρώτατον τοῦ αἰθέρος ἐν ταὐτῷ, ὃ καὶ πρῶτον θεὸν ὃν λέγουσιν αἰσθητικῶς ὥσπερ κεχωρηκέναι διὰ τῶν ἐν ἀέρι καὶ διὰ τῶν ζῴων ἁπάντων καὶ φυτῶν· διὰ δὲ τῆς γῆς αὐτῆς καθ' ἕξιν.*

[49] Diogenes' second sentence ('Through some parts it passes as tenor, as through bones and sinews') is ambiguous: the subject of *κεχώρηκεν* seems at first to be cosmic intelligence (from the previous sentence), but 'bones and sinew' are parts of the body. Hence, the subject must be 'the soul', and the sentence describes an animal phenomenon, to be used as an example to understand what happens in the world, which is in turn described in the last sentence ('the primary god passes perceptibly through the things in the air'). Still, animals are included in the world and

'bones and sinews' *as something else than itself*, namely 'as tenor' or 'by way of tenor' (ὡς ἕξις or καθ' ἕξιν)? 'Tenor' is the unifying principle of non-living things like stones, which are placed at the lowest degree of the Stoic *scala naturae*, whereas 'soul' is the unifying principle of beings placed at the third degree of the *scala naturae*, namely animals.[50] How can the former be considered as a variety of the latter? The explanation may be that 'tenor' and 'soul' are both unifying principles of bodies, namely forms of breath (πνεῦμα), which is the active agent (made of fire and air) through which god produces and sustains all things.[51] Depending on its tension, breath can produce more or less complex qualities.

But saying that soul is a higher form of breath than tenor does not amount to saying that tenor is a lower form of soul. The solution can be drawn from the fact that 'bones and sinews', unlike stones, grow or change: they are parts of a complex living organism.[52] Hence, their tenor is not plain and simple tenor, which sustains stones, but soul taking the form of tenor[53] and giving them the stony features (hardness) needed by the animal as a whole. This analysis rests on the assumption that the same principle, the soul, is responsible for the growth of the animal and for the performing of its natural functions (perceiving, moving, procreating), since the former is aimed at the latter. This assumption holds even more firmly for the world, because it is produced by god, so to speak, *out of itself* (since everything is god during the fiery phase): god is the seed of the world and, once the world has been produced, part of this initial god in its purest form (νοῦς/aether) is concentrated in the heavens, and part of it pervades the rest of the world as the various sustaining powers of each thing, just as our soul has the

their 'bones and sinews' can also be considered as tiny (sub)parts of the world: cosmic intelligence can hence be said to 'pass through them as tenor'.

[50] See Origen, *Princ.* 3.1.2 = *SVF* 2. 988 = LS 53A and Philo, *Quod Deus Sit Immutabilis* 35–6 = *SVF* 2. 458 = LS 47Q.

[51] See Galen, *Intr. Med.* 14. 726. 7–11 = *SVF* 2.716 = *LS* 47N. Again, I leave aside the fact that Chrysippus substituted breath (air + fire) to fire or heat (used by Zeno and Cleanthes) as the all-pervading divine agent.

[52] I follow Long (1982: 40–1), who shows that there are two uses of the term 'soul' in Stoicism, soul as holding the body together and soul as ruling part. In his analysis, Anthony Long quotes our text (DL 7. 138–9) as implying this distinction. A neglected testimony confirming this interpretation is found in Augustine, *CD* 7. 23: Varro, here obviously influenced by Stoicism, 'states also that there are three degrees of soul in the whole universal nature: one which passes through all the living parts of the body which have no sensation, but only health in order to live, this force, he says, flows through our body in the bones, the nails and the hairs . . .; a second degree of soul, which has sensation, leads this force to the eyes, the ears, the nostrils, the mouth and touch; the third degree of soul is the highest [part] of the soul, which is called mind (*animus*), in which intelligence dominates'. This *scala animae* is then applied to the world-soul: god is its highest part, the stones and earth are 'like the bones, like the nails of god', the heavenly bodies are god's senses, and the aether is god's mind.

[53] Why is it not just 'tenor controlled by soul'? The argument must be the same as the one grounding Stoic pantheism (see sect. 1 above): bones and sinews are so well and entirely (on the surface and on the inside) crafted to perform their function in animal bodies that they have to be produced by an immanent and intelligent, moving agent.

double function of (1) ruling the body from and according to the mind, and (2) holding the body together, controlling its growth and its lower functions.[54]

There is however a discrepancy between the ruling part of the world and the ruling part of our soul. Animal souls *also* have secondary parts extending from the heart, where the ruling part is located, through the body to perform various functions (sensation, procreation, voice) by using specific parts of the body.[55] Unless we follow Varro in considering the heavenly bodies as the cosmic ruling part's sense organs,[56] there does not seem to be any direct and permanent extensions of the heavens to the sublunar region, as noted by R. Todd (1978: 146). But, in this case, how could the heavens *as such* effectively *rule* the rest of the world?

The last sentence of Diogenes' testimony could be alluding to the missing connection, since it uses the adverb αἰσθητικῶς to refer to the way in which god pervades the rest of the world, namely 'things in the air', animals and plants. This seems awkward at first, because plants and animals have different sustaining powers ('nature' and 'soul'), and only the latter are capable of sensation. Hence, the most simple interpretation of αἰσθητικῶς, as being on a par with καθ' ἕξιν and meaning κατὰ ψυχὴν, may not be the right one or, at least, not the only possible one. Αἰσθητικῶς might refer not only to god as sustaining power of specific beings capable of sensation, but rather to Nature as a whole, namely to the *sentient* power which binds together, produces, and nourishes all the things and beings in the world, especially through the interactions between its concentric regions (heavens made of pure fire, air, sea, earth).[57] Diogenes Laertius however refers to the beings living in these concentric regions ('things in the air'), rather than to these regions themselves, except in the case of the earth. This may be because he considers these various natural beings to be like god's senses, perceiving what happens in the parts of the world where they are.[58]

This finally allows us to come back to our initial question about the distribution of divine activity in the world. We can conclude (*a*) that everything in the world

[54] One could still ask about the exact meaning of the μᾶλλον/ἧττον gradation. It could be a simple matter of concentration: aether is all νοῦς, whereas plants or animals have only a small proportion of it in them. This interpretation is in fact implied by the Stoic identification of pure fire with λόγος or νοῦς. «Νοῦς passes through bones or earth as tenor (ὡς ἕξις or καθ' ἕξιν)» would thus mean that tenor is made of a (comparatively) limited amount of fire and of more air (this breath being in turn mixed with the water and the earth composing the body it sustains), whereas aether is pure fire. In nature and soul, the amount of fire would be greater than in tenor.

[55] Aetius, *Placita* 4. 21. 1–4 = *SVF* 2. 836 = LS 53G, which compares the stretching parts of the soul to the tentacles of an octopus.

[56] Augustine, *CD* 7. 23 (see n. 52 above): 'the sun, the moon, the stars, which we perceive and by which [god] himself perceives, are his senses'.

[57] See *ND* 2. 82–6, 91–2, and 115–18. In para. 91, air is considered a 'living (*animali*) and respirable substance'.

[58] See Diogenes Laertius' testimony that 'animal souls are parts (μέρη) of the soul of the universe' (DL 7. 157, cf. DL 7. 87), which could imply that animals are like god's senses. In *ND* 2. 86, the parts of the world are said to be 'like limbs' (*membra*), which is not incompatible with the idea that beings living in these parts are like sense-organs.

is ruled according to divine νοῦς and that the Aristotelian sharp distinction between the heaven and the sublunar world is blurred, if not erased, in the Stoic *cosmos* (Goldschmidt 1979: 84), (*b*) that divine νοῦς does not administer the earth from a distance, but is present and active everywhere in the world, albeit only *as sustaining power* of each thing, (*c*) that νοῦς is also present *in propria persona* in the heavens, where it is chiefly active as pure fire and as rotating around the centre of the world, (*d*) that this highest form of νοῦς also actively produces the structure and evolution of the rest of the world.

Activity (*d*) is obviously crucial to our problem, because it seems to link aether/νοῦς and the earth as an active and a passive part of the world. This impression might however be qualified, if we resort again to the animal model used by the Stoics to describe the world. Seneca reports the following Stoic definitions of action: 'Cleanthes said that [walking] was breath extending from the commanding-faculty to the feet, Chrysippus that it was the commanding-faculty itself.'[59] Applying these definitions to the world and its ruling part, we can infer that, according to Cleanthes, activity (*d*) was distributed evenly from the aether to the earth, whereas Chrysippus thought that it was concentrated in the aether, and must have granted only a passive function to the air and things on earth in cosmic administration. This could mean that Chrysippus—followed by later Stoics and Balbus—qualified statement (*a*) and was closer to Aristotelianism than Cleanthes, who resisted any cosmic dualism.[60]

4. DOES ZEUS NEGLECT INDIVIDUALS?

As a matter of fact, there may be other evidence pointing towards a qualification of divine industriousness on the part of Chrysippus, like this quotation by Plutarch:

> [Chrysippus] does concede certain blameworthy cases of negligence, in matters which are by no means minor or unimportant. For in *On Substance* book 3 he mentions that such things do befall good and honourable men, and asks: 'Is it because some things are neglected (ἀμελουμένων τινῶν), just as in larger houses the odd husk and a little wheat go astray even though the overall housekeeping is good? Or is it because the sort of matters in which real blameworthy cases of negligence occur have evil spirits in attendance? He says that there is also considerable involvement of necessity.[61]

59 *Ep.* 113.23 = *SVF* 2. 836 = LS 53L.

60 This squares with the fact that Cleanthes named the sun (which travels through the world, unlike the heavens) as the ruling part of the world (Cic. *Luc.* 126), and emphasized how heat or fire pervades the whole world from the sun to the earth (*ND* 2. 40–1). On the first tenet, see Bénatouïl (2005*b*: 213), on the second see Bénatouïl (2002: 309–14) and Salles (2005: 62–7). On Cleanthes' attachment to divine industriousness, see further sect. 5 below.

61 *De Stoic. Rep.* 1051b–c = *SVF* 2. 1178 = LS 54S (Long and Sedley).

The analogy with the housemaster neglecting small details, which is used by Alexander of Aphrodisias *against* the Stoic industrious god,[62] thus seems to have been already considered by Chrysippus himself. A similar doctrine is found at the end of Balbus' speech: 'the gods attend to great matter, they neglect small ones'.[63]

How are we to interpret these statements? As Robert Sharples has shown,[64] they should not be taken as undermining the Stoic doctrine that divine providence extends to and takes care of individuals. This is clearly established by the context of Balbus' statement, which comes *right after* three paragraphs arguing that the gods care for individuals.[65] Balbus' statement is in fact an answer to an objection using examples of individual misfortunes, like crops destroyed by a storm, as proofs of divine indifference towards men. Chrysippus was dealing with the same kind of examples,[66] since Plutarch mentions things 'befalling good men'. Stoic references to divine negligence were hence used to answer objections of the type: how can god, who controls everything and cares for individuals, let bad things happen to men?

Alexander of Aphrodisias on the other hand likens gods to high-minded housemasters or kings in the context of a more fundamental debate about the scope of providence: does it encompass all individuals and all beings, down to the lowest, or not? Chrysippus and Balbus never suggest that it is unworthy of Zeus or impossible for him to care about individuals or small details, but only that he is justified to take care of 'great matters' at the expense of small problems, for example those concerning only specific individuals. The two debates are on quite different levels, although obviously not independent from each other. As a matter of fact, Balbus' argument offers an apt answer to Alexander's objection

[62] Alex. *Prov.* 25. 2–18. Moreover, when he introduces the Stoic doctrine of providence (*Prov.* 5. 15–21), Alexander uses the opposite analogy: in houses where there are intelligent masters, nothing escapes their watch and foresight, and these houses are different from those administered by weak masters. Similarly, being the most intelligent beings, the gods have to take care of the world and all its aspects. On this household example in debates about providence, see Sharples (2003: 116–19), who traces it back to Plato's *Laws*.

[63] *ND* 2. 167. See also Cotta's comments on this statement in *ND* 3. 86 and 3. 90, quoted below, p. 38–9.

[64] (1983: 149–50) and (2003: 110–16). Contrast Frede (2003: 111).

[65] *ND* 2. 164–6. Cf. Cic. *Div.* 1. 117–18 = *SVF* 2. 1210, where Quintus says that the Stoics do not think that god takes interest in each and every divinatory sign, right after having stated that gods watch over human affairs *nec solum universis, uerum etiam singulis*. Quintus' qualification concerns in fact only the way in which god operates: the connection between signs and foreseen events has been arranged *a principio* during the fashioning of the world, and god does not need to watch over every particular sign when it happens.

[66] If he considered them seriously, Chrysippus' evil spirits should thus not be taken as equivalent to lower gods or demons (see n. 9 above). Chrysippus' problem is not with the administration of the earth or of human affairs as such, but with seemingly unfair rulings of this divine administration. Plutarch, *Quaest. Roman.* 276f–277a accordingly shows that the Stoics considered demons as having the ethical (rather than ontological) function of tormentor.

that taking care of individuals would turn god into the servant of men.[67] By stressing that universal providence is compatible with individual misfortunes, and by deeming them 'small matters' compared to Zeus' higher purposes, Balbus shows that watching over human affairs does not amount to obeying human wishes.[68]

It remains nevertheless that, in both testimonies, the idea of divine 'negligence' is contemplated (*ἀμελουμένων τινῶν*, *neglegunt*), whereas it seems excluded by Stoic physics and theology, as argued by Cotta (*ND* 3. 90): '"The gods do not take notice of everything, any more than do human rulers", says our friend [Balbus]. Where is the parallel? If human rulers knowingly overlook a fault, they are greatly to blame; but as for god, he cannot even offer the excuse of ignorance.'

This argument is not only based on a common notion of god as alien to mistakes or oversight; it targets also specifically the Stoic notion of an all-powerful god, which is described by Cotta two paragraphs below.[69] We must in effect grant Cotta that the definition of matter as absolutely passive prevents the Stoics from appealing to any external limitation on divine providence. If god did not produce a thing or event X which would have been better for men than what is actually existing in its place, Y, it can only be because X is logically or physically incompatible with another existing thing Z, which is itself better than X, and has thus been preferred to X and made to exist by god. In such cases, we may say (a) that god 'sacrificed' X, a would-be valuable thing, to Z, which is a more valuable feature of the world or (b) that god did not wish Y *in itself* or *as such*, since he only wished Z, which is beneficial but entailed Y as its necessary and harmful consequence (Z + Y being however better than X).[70] We should however not say that Zeus overlooked or was negligent about X, as if Y took place without him knowing about it or participating in its production.

The passage from Chrysippus might hence just have been a list of hypotheses quoted out of their context by Plutarch. I do not mean that we have to suppose that Chrysippus refuted these hypotheses later in his treatise, and that Plutarch 'neglected' to mention this fact.[71] It could also simply be that Chrysippus agreed

67 Alexander, *Prov.* 21. 14–25. This objection (*b*): see p. 24 above.

68 See further *ND* 2. 133 = LS 54N: 'the world was created . . . for those animate creatures which use reason: that is, for gods and men' (cf. Philo, *Prov.* 2. 64–5 for an application of this doctrine to the existence and nature of the sea). Divine industriousness does not aim at the well-being of humanity (or of some men) as such, but at the flourishing of reason.

69 See *ND* 3. 92, quoted on p. 26 above, which ends on this objection: 'Accordingly either providence does not know its own powers, or it does not regard human affairs, or it lacks power of judgement to discern what is the best.'

70 Aulus Gellius, *Noctes Atticae* 7. 1. 7–10 = *SVF* 2. 1170 = LS 54Q, reporting that Chrysippus explained in his *On Providence* that natural things harmful to humans, like illnesses, are *κατὰ παρακολούθησιν*, i.e. necessary consequences of good features of the world. See also Salles (2005) on the combination of good ends and bad 'concomitants' in Cleanthes.

71 As noted by Sharples (2003: 112), this interpretation was first put forward by D. Babut (1969: 291–2). In the commentary to his translation of *De Stoic. Rep.* (published in 2004 in the

to formulate these hypotheses in laymen's or in his opponents' terms,[72] as he is known to have done elsewhere,[73] and gave only later or elsewhere the accurate Stoic interpretation of 'negligence' along the lines we have just indicated.[74] Balbus' *di parva neglegunt* can be read in the same way, since Balbus himself condemned this phrase just before uttering it, by saying that we should not judge human misfortunes as cases of *neglectum a deo* (*ND* 2. 167). This blatant contradiction might also be explained away by the fact that the second sentence could be an implicit quotation[75] or a kind of proverb, the phrasing of which would not be fully upheld by Balbus.

5. ARE PETTY AFFAIRS UNWORTHY OF ZEUS' CARE?

We could however stress *parva* instead of *neglegunt* in Balbus' statement, and argue that, even if god does not—strictly speaking—involuntarily neglect anything in the world, a lot of petty things, events, or individuals are nevertheless unworthy of his attention. This seems to be Cotta's interpretation of Balbus' argument:

> But the gods disregard smaller matters, and do not pay attention to the petty farms and paltry vines of individuals, and any trifling damage done by blight or hail cannot have been a matter for the notice of Jupiter; even kings do not attend to all the petty affairs in their kingdoms: this is how you argue.[76]

When Balbus argues that providence extends to individuals, all his examples happen to be 'remarkable men': generals, statesmen, and Homeric heroes (*ND* 2. 165–7), not farmers. His point may hence be that important men worthy of Zeus' attention get it, whereas many others do not and justly so. As Callimachus put it in his *Hymn to Zeus*: 'So also of men you chose the greatest as yours, no seacaptains, no soldiers, no, not even poets—these you dismissed to lesser divinities, other wards for other gods—but reserved for yourself KINGS, rulers

Collection des Universités de France, Paris, Belles Lettres), D. Babut has however changed his position (see n. 448 *ad loc.*) and thinks the contradiction exposed by Plutarch is real—he offers Cicero's testimonies as proof—and gives evidence of a tension between two aspects of the Stoic notion of providence.

[72] Goldschmidt (1979: 85–7) interprets Balbus' statements as dialectical concessions.

[73] See Plutarch, *De Stoic. Rep.* 1040c–d and 1048a for Chrysippus' concessions to philosophical or ordinary statements, which, strictly speaking, contradict the Stoic doctrine that virtue is the only good. In fact, saying that god 'neglects' individual interests is, for a Stoic, tightly connected to saying that health or wealth are 'good': see sect. 5 below.

[74] Note Plutarch's allusion to Chrysippus saying 'that there is also considerable involvement of necessity' (πολὺ καὶ τὸ τῆς ἀνάγκης μεμῖχθαι), which might refer to παρακολούθησις (see n. 70 above).

[75] Cf. Plutarch, *Praecepta Gerendae Reipublicae* 811c, who quotes Euripides saying that 'god takes in hand great things and leaves small things to chance'.

[76] *ND* 3. 86. Cf. Alexander, *Prov.* 21, 21–3, 5, where a similar analogy with high-minded kings is used against the Stoic industrious god.

of cities...'[77] This is however probably a rebuttal of another famous hymn to Zeus, which surely gives a more accurate picture of the Stoic Zeus:

Let us begin with Zeus, whom we men never leave unspoken. Filled with Zeus are all highways and all meeting-places of people, filled are the sea and harbours; in all circumstances we are all dependent on Zeus. For we are his children, and he benignly gives helpful signs to men, and rouses people to work, reminding them of their livelihood, tells when the soil is best for oxen and mattocks, and tells when the seasons are right both for planting trees and for sowing every kind of seed.[78]

The omnipresence of god was a very common idea in Hellenistic times (Festugière 1949: 340), but Aratos gives democratic overtones to it: he mentions urban rather than natural places as divine dwellings, he shows Zeus taking care of simple men working daily on farms or boats, and giving them all the signs they need to anticipate sunrise or weather changes. Cleanthes spells out this egalitarian dimension of pantheism even more clearly in his own *Hymn to Zeus*: 'You love things unloved'.[79] As we have already noted in section 3, the existence of a hierarchy of natural beings does not imply that Zeus looks down on lower beings, avoids being in contact with them, or ignores them. Some ancient authors connect Stoic pantheism to Thales' famous saying, 'everything is full of gods',[80] but Heraclitus' similar but more provocative quote might be better suited: 'even in such places, there are gods'.[81]

I am however not claiming that the Stoics turned Zeus into a democratic god favouring working men over noble rulers, but only that the workings of his providence do not ratify, and may undermine, social or political hierarchies, because his criterion to distinguish between those who deserve his help and those who do not are wholly independent from any social distinctions. This criterion is spelt out in Balbus' conclusion (*ND* 2. 167): 'The gods attend to great matters; they neglect small ones. Now great men always prosper in all their affairs, assuming that the teachers of our school and Socrates, the prince of philosophy, have satisfactorily discoursed upon the bounteous abundance of wealth that virtue bestows.'

Great (*magna*) men favoured by Zeus are those who act virtuously, i.e. in agreement with nature or right reason, and not men with high social status or superior political or military skills, from which virtue and happiness are independent. Therefore, one should not assume that 'farms' or 'vines' are *parva*

[77] Callimachus, *Hymn I: To Zeus*, vv. 92–6 (tr. in Lombardo and Rayor 1987), which probably alludes to Hesiod, *Theogony*, vv. 81–4 and 96.

[78] Aratos, *Phaenomena* 1–9 (tr. in Kidd 1997). On Aratos' poetic rendering of the Stoic industrious god, see Bénatouïl (2005*a*: 139–41). On Callimachus' hymn as a reply to Aratos', see Cuypers (2004: 114).

[79] Stobaeus, *Ecl.* 1. 25. 3–27. 4 = *SVF* 1.537 = *LS* 54I: καὶ οὐ φίλα σοὶ φίλα ἐστίν.

[80] See Alexander, *Prov.* 5. 2–3 and Sophonias, *De Anima Paraphrasis, CG* 23.36.9 = *SVF* 2. 1046.

[81] Heraclitus *DK* A 9 = Aristotle, *Part. An.* 1. 5, 645a20.

and military victories are *magna* from the point of view of Zeus, whose value judgement is free from human prejudice. Zeus chooses in effect kings as his only protégés, but this is because only the wise are kings[82] and not because, as Callimachus would have it, kings in power are men of greater value.

This final appeal to ethics to defend Stoic theology could be seen as a weakness, but I surmise that such Stoic ethical doctrines are on the contrary an indispensable background to understand and vindicate divine industriousness, not only because they define the true aim or beneficiaries of Zeus' providence, but also because they help in proving that industrious activities are not unworthy of God himself.[83] Objections to divine industriousness target not only the fact that Zeus takes care of farmers and insects, but also the fact that, as active cause of the world, he performs base tasks analogous to those of craftsmen or slaves.[84] How could these tasks be worthy of a god? Stoic physics and theology can only show that Zeus has to be industrious in order to be able to account for the world as it is (see sections 1 and 3), but they cannot prove that this industriousness is compatible with the perfect life supposedly enjoyed by gods. This proof may on the other hand be found in the ethical doctrines mentioned by Balbus: the sufficiency of virtue for happiness entails that no specific occupation is incompatible with or required by the happy life.

Consider the examples offered by Philo in support of the famous Stoic doctrine that the wise man or woman is always free, even if he or she appears not to be free in the usual sense of the term, for example if he/she is in prison, or has to perform actions unworthy of a free man or is an actual slave:

> That services rendered (ὑπηρεσίαι) are not proof of enslavement (δουλείας) is very clearly shown in wartime. We see soldiers in the field all working on their own account, not only carrying all their weapons, but also laden like beasts with every necessary requirement, and then making expeditions to get water or firewood or fodder for the animals. As for labours required in defence against the enemy, such as cutting trenches or building walls or constructing triremes, and all other skilled or subsidiary operations in which the hands and the rest of the body are employed, there is no need to recount them at length. On the other hand, there is a peacetime war, no less grave than those fought with arms, a war set on foot by disrepute and poverty and dire lack of the necessaries of life, a war by which men are forced under duress to undertake the most servile tasks, digging and toiling on the land and practising menial crafts, labouring unceasingly to earn a meagre subsistence; often too, carrying burdens in the midst of the market place before the eyes of their fellows in age who were their associates in boyhood and in youth.[85]

[82] For this famous Stoic paradox, see e.g. DL 7. 122.

[83] Cf. Brunschwig (1994: 88–91), who claims that their ethical rigorism allowed the Stoics to define conjunction as rigorously (from a logical point of view) as they did. I similarly suggest that the cynical leanings of the first Stoics immunized them from prejudice against manual work, and thus allowed them to develop their (physically and theologically) radical conception of divine activity.

[84] See the texts referred to in nn. 3, 6, and 7, which expose features (*b*) and (*c*) of divine industriousness.

[85] Philo, *Quod Omnis Probus Liber Sit* 32–4 (Colson's tr. in the Loeb Classical Library).

The argument behind these descriptions can, I think, be reconstructed as follows. Some burdensome or subordinate activities are considered as unworthy of free men, but (*a*) they are praiseworthy when performed by specific types of (free) men in specific circumstances (soldiers during wartime), and (*b*) these conditions are not as exceptional as they seem, since similar activities appear necessary, legitimate, and praiseworthy for some men in normal circumstances (poor men[86]); therefore (*c*) these activities are not in themselves proof of enslavement, so much so that the wise man can undertake any of them and remain free.

There is an obvious parallel between these mundane and burdensome activities of soldiers or poor men and the involvement of Zeus in cosmic construction work mocked by the adversaries of the Stoa.[87] But did the Stoics really draw such a provocative parallel? We may have proof that Cleanthes did. In the chapter of his theological handbook devoted to Hercules, Cornutus reports: 'Next, even the twelve labours can be linked with god in a manner not improper [to god], as advanced by Cleanthes.'[88] Cornutus' phrasing implies that the attribution of the twelve labours to god is somewhat daring.[89] This is surely because they are toils: Hercules was forced to undertake them as a punishment. How could Cleanthes attribute to god such burdens, even allegorically? At the beginning of his chapter on Hercules, Cornutus offers the following general interpretation: 'Hercules is the reason in the universe, according to which nature is strong and powerful, being also unsurpassable,

[86] Note that Philo's final description could be an allusion to Cleanthes, who had to work hard at night to earn his living and be able to study philosophy (DL 7. 168–70).

[87] There are also differences between god and wise men, as noted by Philo (*Omnis Probus Liber Sit* 24): 'while what is god's has the honour of possessing eternal order and happiness, all mortal things are carried about in the tossing surge of circumstance'. The Stoic sage can be compared to soldiers or poor men bravely carrying their burdens, because he or she remains a part of the world and cannot choose what his or her life is made of. But how could Zeus be pictured as forced to do or undergo anything, since there is nothing outside him but void? True enough. However, Zeus could no more be pictured as free to do what he wants, like a leisured man devoting himself to the best possible occupation(s); god is identical to the laws of nature, and is therefore neither constrained by anything nor free to act otherwise: see Salles (2005: 68–9), Mansfeld (1979: 161), and Long (1985; repr. 2006: 268). Moreover, unlike other mortal things and contrary to what Philo suggests, the sage is not tossed about by circumstances, as shown by the fact that he is no less happy than Zeus (Plutarch, *Comm. Not.* 1075a–b). This allows analogies between the sage's intellectual and practical attitude towards circumstances and Zeus' activity in the world, as witnessed by Marcus Aurelius 8. 35. See further Bénatouïl (2006: 258–62) on these analogies.

[88] Cornutus, *Theolagiae Graecae Compendium* 31. 64. 15 Lang = *SVF* 1. 514: τοὺς δὲ δώδεκα ἄθλους ἐνδέχεται μὲν ἀναγαγεῖν οὐκ ἀλλοτρίως ἐπι τὸν θεόν, ὡς καὶ Κλεάνθης ἐποίησεν. Cornutus' next sentence is not explicitly linked with what our text deals with and has been read and interpreted in various ways. I leave it aside since it does not affect my argument.

[89] Καὶ Κλεάνθης ἐποίησεν could mean that other philosophers developed the same idea, or that Cleanthes made other suggestions concerning Hercules, but I think the best reading is that Cleanthes *even* proposed or *went as far as* proposing to link the twelve labours with god.

inasmuch as it gives strength and vigour also to singular parts.'[90] Not surprisingly, Hercules is connected with that most industrious part of Zeus which extends through the whole world and its every part. Cleanthes may therefore have interpreted the twelve labours as allegories of the various beneficial activities of this divine strength allowing the world and its parts to hold together in the infinite void.

If this is correct, the hypothesis offered at the end of section 3 about Cleanthes' conception of divine activity would be borne out. This radical position might be connected in some way to Cleanthes' social background and to his laborious way of life: he had to work at nights to earn his living, 'was called a second Heracles', and was praised by Zeno for that (DL 7. 170). He must therefore have been disinclined to despise industriousness. Chrysippus might have been less attached to the value of *πόνος*, but he is reported by Athenaeus to have praised Homer for showing the heroes of his *Iliad* busying themselves with cooking and serving dishes and being proud of it.[91] There are no testimonies about Chrysippus attributing allegorically these slavish activities to Zeus,[92] but the whole cosmic arrangement can be considered as an instance of divine cooking and serving food, since it is devised in order to allow the heavenly bodies to draw humidity from the Earth and *feed* on it.[93]

Just as good soldiers do not recoil from construction work, just as the wise do not fear for their freedom and happiness when they are sold as slaves, just as Hercules proved his greatness through his labours, just as Homeric heroes are not ashamed of cooking their meals, the Stoic god does not shy away from constantly performing analogous menial activities, because what makes an action good or bad, noble or base, free or servile is neither any of its intrinsic features nor its object or target, but its stemming from right reason or not, namely its being coherent or not with all the other actions performed by its agent. Taking care of stones, insects, or farmers would be base only if it led Zeus to contradict himself by neglecting more important cosmic matters. It would however be equally incoherent and base on the part of Zeus to overlook or look down on any aspect of the world he produced, so much so that his efficient micromanaging of

[90] Cornutus, *Theol.* 31.63.1 = *SVF* 1. 514: Ἡρακλῆς δ' ἐστὶν ὁ ἐν τοῖς ὅλοις λόγος καθ' ὃν ἡ φύσις ἰσχυρὰ καὶ κραταιά ἐστιν [καὶ ἀπεριγένητος οὖσα], μεταδοτικὸς ἰσχύος καὶ τοῖς κατὰ μέρος καὶ ἀλκῆς ὑπάρχων. Cf. Plutarch, *De Iside et Osiride* 367c, which suggests that Chrysippus maintained this interpretation of Hercules in the context of his doctrine of breath.

[91] *Deipnosoph.* 1. 18b = *SVF* 3. 708.

[92] Chrysippus was surely fond of attributing shameful activities to the wise man, in a Cynic fashion, and even to gods, but, contrary to Cleanthes, he seems to have favoured *aphrodisia* over *ponoi*, as shown by his (in)famous allegorical interpretation of a painting of Hera performing oral sex on Zeus (see DL 7. 187–8).

[93] See p. 28 below and e.g. Plutarch, *De Stoic. Rep.* 1052b–c for Chrysippus' analysis of divine nourishment (*τρέφεσθαι*).

the world proves in fact to be the most perfect instance of coherence, goodness, or freedom.[94]

REFERENCES

Algra, K. (2004) 'Eternity and the Concept of God in Early Stoicism', in G. Van Riel and C. Macé (eds.), *Platonic Ideas and Concept Formation in Ancient and Medieval Thought* (Leuven: Leuven UP), 173–90.

Babut, D. (1969) *Plutarque et le stoïcisme* (Paris: Presses Universitaires de France).

Bénatouïl, T. (2002) '*Logos* et *scala naturae* dans le stoïcisme de Zénon à Cléanthe', *Elenchos*, 23(2): 297–331.

—— (2005*a*) 'Les Signes de Zeus et leur observation dans les *Phénomènes* d'Aratos', in J. Kany-Turpin (ed.), *Signes et prédiction dans l'Antiquité* (Saint-Étienne: Publications de l'Université de Saint-Étienne), 129–44.

—— (2005*b*) 'Cléanthe contre Aristarque: Stoïcisme et astronomie à l'époque hellénistique', *Archives de Philosophie*, 68(2): 207–22.

—— (2006) *Faire usage: La Pratique du stoïcisme* (Paris: Vrin).

Bobzien, S. (1998) *Determinism and Freedom in Stoic Philosophy* (Oxford: Clarendon Press).

Brunschwig, J. (1994) 'The Conjunctive Model', in J. Brunschwig, *Papers in Hellenistic Philosophy* (Cambridge: CUP), 72–91.

Cuypers, M. P. (2004) 'Prince and Principle: The Philosophy of Callimachus' *Hymn to Zeus*', in M. A. Harder, R. F. Regtuit, and G. C. Wakker, *Callimachus II. Hellenistica Groningana: Proceedings of the Groningen Workshops on Hellenistic Poetry*, vii (Leuven: Peeters), 95–115.

Emerson, R. W. (1971) 'The American Scholar', in R. E. Spiller and A. R. Ferguson (eds.), *The Collected Works of Ralph Waldo Emerson*, i. *Nature, Addresses, and Lectures* (Cambridge, Mass.: Harvard UP), 52–70.

Festugière, A.-J. (1949) *La Révélation d'Hermès Trismégiste II: Le Dieu cosmique* (Paris: Gabalda).

Frede, D. (2002) 'Theodicy and Providential Care in Stoicism', in D. Frede and A. Laks, *Traditions of Theology: Studies in Hellenistic Theology, its Background and Aftermath* (Leiden: Brill), 85–117.

Frede, M. (2005) 'La Théologie stoïcienne', in G. Romeyer-Dherbey and J.-B. Gourinat (eds.), *Les Stoïciens* (Paris: Vrin), 213–32.

Goldschmidt, V. (1979) *Le Système stoïcien et l'idée de temps*, 4th edn. (Paris: Vrin).

Hahm, D. (1977) *The Origins of Stoic Cosmology* (Columbus, Ohio: Ohio State UP).

Kidd, D. (tr.) (1997) *Aratus: Phaenomena* (Cambridge: CUP).

Lombardo, S., and Rayor, D. (trs.) (1987) *Callimachus: Hymns, Epigrams, Select Fragments* (Baltimore, Md.: Johns Hopkins UP).

[94] Is divine industriousness a paradigm for ethics or is the definition of virtue presupposed by theology? I do not think we should try to assign logical primacy to any part of the Stoic system: they entail and reinforce each other, as argued by Goldschmidt (1979: 64–6). On this 'philosophical holism', see Ch. 7 by M. Boeri.

Long, A. A. (1982) 'Soul and Body in Stoicism', *Phronesis* 27(1): 34–57. Repr. in A. A. Long, *Stoic Studies* (Cambridge: CUP, 1996), 224–49.
—— (1985) 'The Stoics on World-Conflagration and Everlasting Recurrence', in R. E. Epp, *Spindel Conference 1984: Recovering the Stoics. Southern Journal of Philosophy,* 23 suppl.: 13–37. Repr. in A. A. Long, *From Epicurus to Epictetus* (Oxford: Clarendon Press, 2006), 256–82.
—— and Sedley, D. N. (1987) *The Hellenistic Philosophers* (Cambridge: CUP).
Mansfeld, J. (1979) 'Providence and the Destruction of the Universe in Early Stoic Thought', in M. J. Vermaseren (ed.), *Studies in Hellenistic Religion* (Leiden: Brill), 129–88.
Reydams-Schils, G. (1999) *Demiurge and Providence: Stoic and Platonist Readings of Plato's* Timaeus (Turnhout: Brepols).
—— (2005) 'Le Sage face à Zeus: Logique, éthique et physique dans le stoïcisme impérial', in T. Bénatouïl and P.-M. Morel, *Les Stoïciens et le monde. Revue de Métaphysique et de Morale,* 2005(4): 579–96.
Salles, R. (2005) '*ἐκπύρωσις* and the Goodness of God in Cleanthes', *Phronesis* 50(1): 56–78.
Sharples, R. W. (1983) 'Nemesius of Emesa and Some Theories of Divine Providence', *Vigiliae Christianae,* 37(2): 141–56.
—— (2003) 'Threefold Providence: The History and Background of a Doctrine', in R. W. Sharples and A. Sheppard, 'Ancient Approaches to Plato's *Timaeus*', *Bulletin of the Institute of Classical Studies,* 78: 107–27.
Solmsen, F. (1963) 'Nature as Craftsman in Greek Thought', *Journal of the History of Ideas,* 24(4): 473–96. Repr. in F. Solmsen, *Kleine Schriften* (Hildeshem: Olms, 1968), i. 332–55.
Todd, R. B. (1976) *Alexander of Aphrodisias on Stoic Physics* (Leiden: Brill).
—— (1978) 'Monism and Immanence: The Foundations of Stoic Physics', in J. M. Rist (ed.), *The Stoics* (Berkeley, Calif.: University of California Press), 137–60.

2

The Stoics on Matter and Prime Matter

'Corporealism' and the Imprint of Plato's *Timaeus*

Jean-Baptiste Gourinat

It is often claimed that the Stoics are materialists,[1] but for a long time scholars did not truly analyse the Stoic theory of matter: this was the case with Samuel Sambursky's classic work, *Physics of the Stoics.*[2] This book, which was the accepted reference work on Stoic physics, was one of the first ever to attempt to rehabilitate Stoic physics, by drawing systematic comparisons with contemporary physics, reminiscent of Łukasiewicz's attempts to rehabilitate Stoic logic by comparing it to modern logic. Nevertheless, Sambursky does not dedicate a chapter to the Stoic concept of matter, and the word 'matter' does not appear in his *index rerum.* For a long time, four notable exceptions were (in chronological order) Lapidge (1973), Hunt (1976), Hahm (1977), and Duhot (1989), each of whom devotes a chapter to the principles and hence to matter.[3] However if one wants to

An earlier version was published as 'La Théorie stoïcienne de la matière: Entre le matérialisme et une relecture "corporaliste" du *Timée*', in C. Viano (ed.), *L'Alchimie et ses racines philosophiques: La Tradition grecque et la tradition arabe* (Paris: Vrin, 2005), 37–62. I wish to express my gratitude to Cristina Viano and to Vrin for permission to publish this largely rewritten version. The previous article was extensively discussed with Michael Frede, when he was completing his 2005 'La Théologie stoïcienne', and also benefited from the meetings of the 'Projet Zénon', of which I am in charge at the Centre Léon Robin, Paris (editing the fragments of Zeno of Citium), esp. comments by Tiziano Dorandi, Serge Mouraviev, and Stéphane Toulouse. This chapter has benefited from comments on the French version by two anonymous readers, and by Ricardo Salles and Thomas Bénatouïl. To them all I express my gratitude. I also wish to thank John Thompson, for kindly checking my English. Finally, during the preparation of the volume, I presented some of the arguments here at a seminar held at Charles University, in Prague, and benefited from some last comments and suggestions by Filip Karfík, Vladimír Mikeš, and Gábor Betegh, whom I thank.

[1] See e.g. Bloch (1985: 48–52). Bloch (1985: 49–50) speaks of materialism, 'corporéisme', or corporealism, as if they were equivalent terms. According to him, since the Stoics claim that 'everything is corporeal', 'the principles of Stoic physics are essentially materialistic' ('les principes de la physique des stoïciens sont essentiellement de caractère matérialiste').

[2] Sambursky (1959).

[3] Hunt 1976: 17–25; Hahm 1977: 29–56; Duhot 1989: 73–86. See also Lapidge 1978*a*, 1978*b*; Long and Sedley 1987: i. 268–72. Stoic theory of matter has been recently dealt with by

know whether or not the Stoics were materialists, one must thoroughly study the Stoic theory of matter.[4] In his paper on the topic, the French philosopher Éric Weil prefers to use the word 'corporealism', since, in Stoicism, matter is only *one* of the *two corporeal* principles: it is true that, according to the Stoics, everything is corporeal (or it is *almost* true, since there are some incorporeal entities), but they never say that everything is material.[5] The problem is the following: ordinarily, 'materialist' historians of materialism tend to assume that a doctrine is materialist if this doctrine includes the tenet that 'everything is corporeal'.[6] Obviously, *c'est un peu court.* However, historians of philosophy, even if they do not hold materialism as an ideology, seem to share the same simplification. In Hahm's excellent book, for instance, the case is rather confused: he describes Stoic 'corporealism' in his first chapter, saying that 'no idea is more deeply ingrained in Stoic philosophy than the conviction that everything is corporeal';[7] but in the second paragraph of the same page, he explains this affirmation by saying that 'according to the Stoics the only things that really exist are material bodies'. However, as Robert Sharples rightly notes when he exposes the same Stoic tenet, 'the associations of that term [material] are likely to be misleading especially where the Stoics are concerned'.[8]

My working hypothesis will be the following: Stoics are materialists if, and only if, their theory fulfils the two following conditions: (1) their system must be materialistic monism and (2) the 'inferior reality' must account for the 'superior reality', according to the formula that encapsulates Auguste Comte's theory of materialism.[9] It is precisely the case that the Stoics have two principles, god and matter, and this seems to settle the question. However, the issue is more complicated than that, since the two principles are corporeal, and, as one of our sources says, *deum hoc esse quod silva sit*, 'god is what matter is'.[10] The Peripatetic scholarch Alexander of Aphrodisias, criticizing the Stoic theory of the two principles, even says that god, according to the Stoics, will be a 'body produced from matter' (*ἐκ τῆς ὕλης γεννώμενον σῶμα*) and will be 'secondary to matter' (*ὕστερον τῆς ὕλης*).[11] Therefore, the situation is the following: Stoic

Reydams-Schils 1999: esp. 42–60 and 89–100, and in two papers on Stoic theology: Sedley 2002; Frede 2005.

[4] Sambursky was simply not concerned with the problem: he was attempting to give an account of Stoic physics from the standpoint of modern physics.

[5] Weil 1964: 560: 'il n'existe, à notre connaissance, aucun texte de l'ancien stoïcisme qui permette de parler de matérialisme. Aucun stoïcien n'a affirmé que tout soit matière, tiré de la matière, réductible à la matière. Rien n'existe, enseignent-ils, qui ne contienne de la matière, qui ne contienne, en d'autres termes, à la fois le principe actif et le principe passif. S'il faut donner un nom à une telle doctrine, mieux vaudrait se servir du néologisme "corporalisme".'

[6] See Bloch 1985: 7. [7] Hahm 1977: 3. [8] Sharples 1996: 33.

[9] Comte's thesis is currently encapsulated in that formula, i.e. by Bloch 1985: 16 ('une explication du supérieur par l'inférieur'), even if Comte never expressed himself so clearly. I thank Michel Gourinat and Laurent Clauzade for clarifying this point for me.

[10] Calcidius, *In Timaeum* 294, Waszink (*SVF* 1. 87). See below p. 68.

[11] Alexander of Aphrodisias, *De Mixtione*, p. 225. 11–14 Bruns (tr. Todd).

theory of matter seems to be a disguised monism, i.e. a kind of materialism. Matter (as the 'inferior reality') accounts for god (as the 'superior reality'), since god appears to be engendered by matter, which is the only ultimate principle.

Basically, I shall argue, the Stoic doctrine of matter is a reinterpretation of the doctrine of matter or 'receptacle' in Plato's *Timaeus*[12] and of Aristotle's theory of matter.[13] If read like this, the Stoic theory of matter does not appear as materialistic, but nor is it dualist—it is not materialist in the sense that things are not explained by the movements and combination of passive matter, but it is not dualist in the sense that both principles are bodies.

1. THE TWO MEANINGS OF MATTER IN STOICISM

According to Diogenes Laertius,

> they say that primary matter (τὴν πρώτην ὕλην) is the substance (οὐσίαν) of all things: so Chrysippus in the first book of his *Physics*, and Zeno. Matter is that out of which anything whatsoever is produced. Both substance and matter are terms used in a twofold sense according as they signify (1) universal or (2) particular substance or matter. (DL 7. 150, tr. Hicks, modified)

The Stoics clearly distinguished between two meanings of matter (ὕλη). In one sense, matter is unqualified substance: it is the matter of the universe, called 'substance' (οὐσία) or 'prime matter'. In the other sense, 'matter' designates the qualified matter of particular realities. Prime matter is a principle, out of which qualified matter is engendered, through intermediate states. Many sources and some quotations refer this doctrine to Zeno himself (see *SVF* 1. 85–8).

A) Prime Matter as a Principle

Prime matter is one of the two principles, alongside god:

> [The Stoics] think that there are two principles of the universe, that which acts and that which is acted upon (τὸ ποιοῦν καὶ τὸ πάσχον). That which is acted upon is unqualified substance (τὴν ἄποιον οὐσίαν), i.e. matter; that which acts is the reason (λόγος) in it, i.e. god. For this, since it is everlasting, constructs (δημιουργεῖν) every single thing throughout all matter. . . . They say that there is a difference between principles and elements: the former are ungenerated and indestructible, whereas the elements pass

[12] There is no explicit reference to the *Timaeus* in our sources before Posidonius, who seems to have commented on Plato's dialogue—see Posidonius T 85 EK (Sextus, *M*. 7.93: φησὶν ὁ Ποσειδώνιος τὸν Πλάτωνος Τίμαιον ἐξηγούμενος) and Kidd's commentary in Kidd (1988): 339 (whether or not Posidonius wrote a separate and comprehensive commentary of the *Timaeus* is still an open question). This is not a sufficient argument to dismiss any Stoic comments on the *Timaeus* before Posidonius.

[13] See Reydams-Schils 1999: 44–8.

away at the conflagration. The principles are also bodies and without form, but the elements are endowed with form (ἀλλὰ καὶ σώματα εἶναι τὰς ἀρχὰς καὶ ἀμόρφους, τὰ δὲ μεμορφῶσθαι). (DL 7. 134, LS 44 B, tr. LS)

We do not know of any Stoic definition of 'principle' (ἀρχή): 'principle' is characterized only by being distinguished from the 'elements' (στοιχεῖα), as ungenerated, indestructible, and shapeless. To distinguish between principles and elements may seem innovative, when compared to Plato or Aristotle, since both of them qualify here and there the elements as principles (Plato, *Tim.* 48c; Aristotle, *Metaph.* Δ1, 1013a20). However, even if this is literally true, it is also true that Plato in the *Timaeus* draws a clear-cut distinction between the elements on the one hand, and the intelligible paradigm, the receptacle, and the demiurge as the first causes on the other. And it is also true that Aristotle criticizes his predecessors for having identified as principles only one or several of the elements (*Metaph.* A3, 983b7–8). It may be noted immediately that, as the Stoics do not identify the principles with the elements, they are not 'materialists' in the sense in which Aristotle's predecessors may be said to be.

The word and the concept of 'matter' are both borrowed from Aristotle.[14] So is the distinction between matter and reason, and the characterization of matter as substance (see *Metaph.* H1, 1042a26–30). However, it is clear that Aristotle himself built his theory inside a conceptual framework belonging to the Academy, especially when he interprets the doctrine of Plato's *Timaeus*. As is well known, in the *Timaeus*, it is the 'receptacle' (ὑποδοχή, 49a) that plays the role of matter: it is a 'space' (χώρα, 52a) that receives being. Plato never used the word 'matter' to describe his receptacle but, according to Calcidius, 'Plato's auditors' used the word.[15] Aristotle was evidently one of them, and it is a fact that he says that 'in the *Timaeus* Plato identifies "matter" and "space"'.[16] It is true that, in *Timaeus* 69a, Plato compares the elements and sensible qualities used by the divine craftsman to 'wood' (ὕλη) in the sense of the material used by carpenters, and this may be what Aristotle has in mind, and, furthermore, this may have influenced the use of the word in the Academy.[17] Therefore, it is plausible that this identification was widespread in the Academy thanks to Aristotle. However, the reverse explanation cannot be excluded completely: in this case, Aristotle may have used the word because it was widespread in the Academy due to Plato's use of the word—but this is less plausible, since Plato used the word only in passing.

In contrast to this influence of Plato and Aristotle, Zeno's position has its own distinctive features. For Aristotle says that substance is in one sense matter, in

[14] See Hahm 1977: 34–5; Long 1986: 154.

[15] *Nomen vero ei dederunt auditores Platoni*: Calcidius, *In Tim.* 308, p. 309. 4 Waszink.

[16] Aristotle, *Phys.* 4. 2, 209b11–17. See also *Gen. et corrupt.* 2. 1, 329a13–24.

[17] Frede 2005: 221, who also mentions the use of the word in *Philebus* 54c, on which see also Sedley 2002: 56.

another sense 'reason and shape' (ὁ λόγος καὶ ἡ μορφή), and, in a third sense, the compound of the two (*Metaph.* H1, 1042a26–30), but he is more inclined to identify substance with form and with the compound of matter and form (Z3, 1029a29–30), while Zeno resolutely identifies substance with prime matter, and never with *logos* or the compound.[18] Moreover, the kind of matter that Aristotle tends to consider as a possible candidate for substantiality is qualified matter, bronze for instance (1029a4), while bronze for Zeno is rather qualified matter as distinct from substance.[19] Nevertheless, the very notion of prime matter is borrowed from Aristotle: it is a rare expression in Aristotle, and most of the time it has a different meaning, but Aristotle used it in the same sense as Zeno would in *GC* 2. 1, 329a23. According to Aristotle, 329a24–6, this prime matter is the 'matter of perceptible bodies', and it is 'not separable, always accompanied by some contrariety', and different from the elements, since 'they are generated from it' (ἐξ ἧς γίνεται). Stoic prime matter also 'engenders the four elements' (DL 7. 136), and Stoic prime matter is always associated with a quality (Calcidius, *In Tim.* 292), so that it never exists in a 'pure' 'qualityless' state. Similarities between Aristotle and Zeno are quite striking, and it is very unlikely to be a pure coincidence.

The main difference between the two principles is the following: matter is the passive principle (τὸ πάσχον), god is the active principle—the producer (τὸ ποιοῦν). Once again, this is, at least partly, Aristotelian vocabulary, since τὸ ποιοῦν is the expression ordinarily translated by 'efficient cause' in the context of Aristotle's theory of the four causes[20] and, more significantly, the opposition of active and passive principles, the former being identified with matter, may also be found in Aristotle.[21] However, as I shall argue below, the πάσχον/ποιοῦν pair is central to Plato's conception of being in the *Sophist*, and this is decisive for our topic. Moreover, the doctrine of the active principle reflects Plato's craftsman in the *Timaeus*. God within matter is described in DL 7. 134 as 'producing like a craftsman' (δημιουργεῖν) 'every single thing throughout all matter' (διὰ πάσης αὐτῆς ἕκαστα). Nevertheless, the divine craftsman in Zeno is immanent: he pervades matter, he is not transcendent, and he does not shape the world by taking inspiration from a transcendent paradigm. He works matter from the inside, biologically, like semen in animal reproduction.[22] In the *Timaeus*, there are two models closely imbricated: the technological model of craftsmanship and the biological model of animality, since the world is an ensouled living being (30b), and the receptacle a female principle, a 'mother'

[18] See Hahm 1977: 40–1. [19] See Calcidius, *In Tim.* 290 (*SVF* 1. 86).

[20] See Seneca, *Ep.* 65. 4 (who acknowledges only three causes in Aristotle, since he does not mention the final cause): 'the Stoics believe in one cause only, the maker, but Aristotle thinks that the word can be used in three ways (*Stoicis placet unam causam esse, id quod facit. Aristoteles putat causam tribus modis dici*)' (tr. Gummere).

[21] See texts and their discussion in Hahm 1977: 43–6.

[22] Calcidius, *In Tim.* 294 (*SVF* 1. 87).

(51a), as opposed to the craftsman, who is the 'father' (41a) of the universe.[23] The Stoics seem to take the technological model as a pure metaphor, and to take the biological model literally. From that point of view, their doctrine works as a coherent reinterpretation of the myth of the *Timaeus*: they were probably no more inclined to take the myth of the *Timaeus* literally than they were any other myth.[24]

Moreover, the doctrine of the two principles, matter and god, seems to have been attributed to Plato himself. For instance, Aristocles (first century AD) says that Zeno has two principles, 'god and matter, like Plato' and, according to him, the only difference is that 'Zeno says that they are both bodies, . . . whereas Plato says that the first active cause is incorporeal'.[25] According to Simplicius, this doctrine was attributed to Plato as early as Theophrastus:[26]

> Here [= in the inquiry concerning nature] he [Plato] wished to make the principles two in number, one underlying [things] as matter—and this he calls 'receptive of all things'; the other being cause and source of movement (αἴτιον καὶ κινοῦν), and this he attaches to the power of god and the good.[27]

There are several traces of this doctrine, especially in Diogenes Laertius,[28] and mainly in a long passage of Cicero's *Academica Posteriora* reporting the physical doctrine of the Old Academy as depicted by Antiochus of Ascalon and put by Cicero in the mouth of Varro.[29] Both Diogenes' account of Plato's physics (3. 67–77) and Cicero's account are probably more or less dependent on the same tradition, a late syncretic tradition influenced by Stoicism.[30] Cicero's Antiochean account of Platonic physics has many points in common with Zeno's theory of the principles: according to Antiochus, the Old Academy acknowledged two 'principles', the one active, 'force', and the other passive, 'matter', and those two principles combine in 'body'. The active principle is also said to be god and the world-soul. From this Sedley has recently inferred that the Stoic theory of the two principles has its origin in the Old Academy, since, in fact, according to him, Antiochus' account is a genuine and trustworthy account of the doctrine of the early Academy: it could have been Polemo's doctrine, and Zeno, having been his pupil,[31] could have learnt this doctrine from his Platonic master.[32] Furthermore, Sedley provides an explanation of the reason why early Platonists

[23] See also *Tim.* 50d.

[24] On Stoic interpretation of mythology, see Gourinat 2005.

[25] Eusebius, *EP* 15. 14. 1 (*SVF* 1. 98, LS 45 G).

[26] See Pearson 1891: 86; Reydams-Schils 1999: 44; Sedley 2002: 42.

[27] Simplicius, *In Ar. Phys.* 26. 11–13 Diels, tr. Fortenbaugh (Theophrastus, fr. 230 FHS and G; *Platonismus in der Antike*, test. 119. 1 Dörrie Baltes).

[28] DL 3. 69 (*Plat. Ant.* 119. 2 DB); DL 3. 75 (*Plat. Ant.* 119. 3 DB).

[29] Cic. *Ac. Post.* 1. 24–9.

[30] See the references given by Sedley 2002: 48 n. 17.

[31] DL 7. 2 maintains that Zeno was Xenocrates' and Polemo's pupil, but given the fact that Zeno did not come to Athens before 312 BC, it is highly improbable that he was Xenocrates's pupil, since Xenocrates presumably died in 314/313 BC. See (among others) Hahm 1977: 223.

[32] Sedley 2002: 47. On the influence of the Academy, see also Reydams-Schils 1999: 52–6.

extracted a two-principle theory from the *Timaeus*. As he notes, in antiquity, Plato's *Timaeus* was often depicted as presenting a three-principle theory:[33] for instance, according to Simplicius, Alexander of Aphrodisias, presumably refuting Theophrastus' counting of two principles, argued that Plato in the *Timaeus* had three principles, i.e. matter, the efficient cause (τὸ ποιοῦν) and the intelligible paradigm.[34] In this interpretation, the paradigm is added as a third principle. Sedley thinks that this a rather natural interpretation of the *Timaeus*, and, according to him, the reduction to two principles is an attempt towards harmonization with one of the famous 'unwritten doctrines' of Plato, stating that there are two principles, the One and the Indefinite Dyad, the latter being usually identified with matter,[35] and it is clear that Aristotle was presumably not the only auditor of Plato who believed Plato's Dyad to be identical with the receptacle of the *Timaeus*.[36] Therefore, it is plausible that the Old Academy may have wished to find two principles in the *Timaeus* as in the 'unwritten doctrines'.[37]

Sedley's hypothesis is quite attractive, but I wish to make two remarks to limit the extent of my agreement with his hypothesis. First, as Frede rightly points out, Polemo seems to have shown little interest in physics and one may have serious doubts that he developed a physical theory comparable to the theory sketched out in Cicero. Presumably, Cicero's account is just what it is intended to be, i.e. a composite account by Antiochus of various trends inside the Old Academy:[38] in the course of his account, Antiochus includes Aristotle's point of view on the fifth element (*Ac. Post.* 1. 26) and the whole account is described later by Atticus as the doctrine of the Old Academy and Aristotle.[39] It is clear, all in all, and is so stated in Cicero that Antiochus wants to *demonstrate* that 'the Stoic theory should be deemed a correction of the Old Academy rather than actually a new system' (*Ac. Post.* 1. 43). Given Theophrastus' account of the two-principle doctrine as reported by Simplicius, and given Cicero's ascription of a two-principle theory to the Old Academy, it seems clear that Plato's *Timaeus* was interpreted very early on as endorsing a two-principle theory and that such an interpretation influenced Zeno's theory. However, it cannot be considered as certain that all the details of Cicero's account accurately represent the doctrine of the Old Academy, and it is doubtful that it is Polemo's doctrine.

Secondly, the origins of the Platonic account of the two-principle theory may have a different origin than the one alleged by Sedley. This is suggested by

33 Sedley 2002: 60–1. 34 Simplicius, *In Ar. Phys.* 26. 13–16.

35 Sedley 2002: 61. Testimonies in Richard 1986: 248–69 and in *Platonismus in der Antike* (Dörrie and Baltes 1996: 154–62). See in particular Aristotle, *Metaph.* A6, 988a7–15 (test. 120.0 DB, 34 Richard); Alexander of Aphrodisias, *In Ar. Metaph.* 59. 28–60. 2 (test. 120. 1 DB, 17 R); Simplicius, *In Ar. Phys.* 151. 6–19 (test. 120. 2 DB, 13 R.); Plutarch, *De Def. Orac.* 35. 428 E–F (test. 120. 3 DB); id., *De An. Procr.* 24. 1024 D–F. (test. 120. 4 DB).

36 Aristotle, *Phys.* 4. 2, 209b11–17; see Simplicius, *In Ar. Phys.*, p. 542. 9–12 (quoting Aristotle).

37 Sedley 2002: 61. 38 Frede 2005: 217–18.

39 'He is giving a brilliant exposition of the doctrine of the Peripatetics and the Old Academy' (Cic. *Ac. Post.* 1. 33).

Diogenes Laertius' account in 3. 75–6 and by Calcidius' commentary on the *Timaeus,*[40] since they both link the two-principle doctrine with passage 47e–48a of the *Timaeus*, which acknowledges two causes of the generation of the universe, intellect and necessity.[41] This is a different tradition from the theory of the One and the Dyad, which are not described only as the causes of the sensible world, but also as the causes of the Forms.[42] Moreover, if one looks more closely at this tradition as it appears both in Diogenes Laertius and Calcidius, it is clear that, in this interpretation of *Timaeus*, the two-principle theory does not eradicate the paradigm and the Forms. According to Diogenes Laertius 3. 76, god and matter are 'the principles and the causes of the things of which there is a paradigm',[43] It is also the case in Calcidius who does not consider the paradigm as being a principle or cause, since it does not have any efficiency. Antiochus himself, in the exposition of the physical doctrine of the Academy, only mentions Aristotle as rejecting Forms (Cic. *Ac. Post.* 1. 33), and, in Antiochus' account, Forms are dealt with in the exposition of Logic (1. 30–1). Accordingly, the fact that the doctrine of the two-principles excludes the Forms or the paradigm does not mean that this doctrine rejected the existence of the Forms, but only that it did not grant them the status of principles. The interpretation according to which the paradigm is a principle or a cause may have appeared later: it may, for instance, have been influenced by Aristotle's system of the four causes, which includes formal cause.

This naturally leads me to a third remark, concerning the extent of Zeno's innovations within the framework of the two-principles theory. According to Varro's Antiochean account of Zeno's innovations, one of his major innovations was his denial of the efficiency of such incorporeal entities as the mind (Cic. *Ac. Post.* 1. 40). From this, one could infer that Zeno's major innovation was to consider the principles as bodies. However, according to Sedley, one may think that 'the final generation of the old Academy under Polemo' no longer did believe in the efficiency of the incorporeals, since Cicero's remark that the defence of the efficiency of incorporeal entities was due to 'Xenocrates and his predecessors', could virtually imply that Xenocrates' successor, Polemo, no longer accepted such a view.[44] Hence, according to Sedley, even treating the principles as bodies[45] may not be an innovation of Zeno, but may be due to Polemo.

Here again, I wish to make two remarks. First of all, one may wonder whether Plato himself did assume the position ascribed to Xenocrates in Varro's account, since Aristotle criticizes Plato for having composed the world-soul from material

[40] Calcidius, *In Tim.* 278–9. [41] See also Plato, *Tim.* 46d–e, 68e, 69a.

[42] Aristotle, *Phys.* 1. 6, 988a8 ff. (test. 34 R.). In this tradition, Dyad as matter becomes a principle of Forms, and this clearly generates a problem, since in *Timaeus* 'matter' is supposed to belong only to becoming: see Simplicius, *In Ar. Phys.* 151. 6–19 (test. 13 R.). This is precisely what led Plotinus to assume the existence of intelligible matter: see *Enn.* 2. 4 (12).

[43] Contrary to Baltes's correction, which deletes *μὲν ὧν παράδειγμα* (test. 119. 4).

[44] Sedley 2002: 81. [45] More on this below.

elements and for having attributed extension to it,[46] and it is clear that the *Timaeus* maintains that the soul is composed from body and the 'indivisible [i.e. intelligible?] substance'.[47] And one might also wonder in a more general way whether Plato's god in the *Timaeus* is incorporeal or corporeal. Therefore, Polemo may not be needed in order to account for Zeno's rejection of incorporeal entities within the framework of the *Timaeus*.[48] On the other hand, it is clearly introduced in Varro's account as Zeno's innovation in physics, and if one looks closely at this account, the Academic doctrine of the two principles has a rather different ring from Zeno's doctrine, since, in Varro's account, body is composed of both matter and the active force (*Ac. Pr.* 1. 24): instead of being bodies, they are the constituents of body. As Sedley acknowledges, 'to that extent, the Academic theory does differ from Stoic physics'.[49]

Secondly, I think that one has to stress that Zeno's theory also departed from Academic physics in eradicating the paradigm and the Forms of the *Timaeus*, even if none of them was treated by the Old Academy as a principle or a cause. Rejection of the Forms is attributed by Varro to Aristotle (*Ac. Pr.* 1. 34), but of course Zeno also rejected the Forms and the paradigm, since he replaced them by an immanent *logos*. And this is clearly Zeno's most distinctive innovation: the conflation of the transcendent paradigm with the immanent craftsman as a biological principle, as Calcidius maintains:

The Stoics also criticize Plato for saying that, since exemplars of all things exist of old in another sublime and most excellent substrate, the sensible world was made by god after an immortal exemplar. They say that no exemplar is needed, since seminal reason, pervading a certain nature, which holds and contains it, has created the whole world and everything existing in it. (Calcidius, *In Tim.* 294, p. 296. 11–16 W.)

Rejection of the Forms is clearly attested as one of Zeno's innovations. In Von Arnim's collection of fragments, it appears in the logic section of Zeno's fragments (*SVF* 1. 65), not in the physics. This choice is inspired by Diogenes Laertius' exposition of Stoic philosophy and, in like manner, the doctrine of the Forms appears in Cicero's exposition of the doctrine of the Old Academy, since in both cases it pertains to logic, not to physics.[50] This has some sense, even for the Stoics, since ideas are precisely seen from an epistemological point of view as pure figments of the mind (DL 7. 61). However, in Stobaeus and in the *Placita*, the testimonies pertaining to Zeno's rejection of the Form belong to the exposition of physics, in a chapter that immediately follows the chapter on matter.[51] So, these testimonies seem to imply that, in Zeno, rejection of the Forms belonged

[46] Aristotle, *De An.* 1. 2, 404b16–27; 1. 3, 407a3–22. [47] See Plato, *Tim.* 35a–b.
[48] See Reydams-Schils 1999: 59. [49] See Sedley 2002: 81.
[50] DL 7. 61; Cic. *Ac. Post.* 1. 30.
[51] Ps.-Plutarch, *Plac.* 1. 10. 882 E (*SVF* 1. 65, 2. 360; LS 30 B); Stobaeus, *Ecl.* 1. 12. 3 (*SVF* 1. 65, LS 30 A, *FDS* 316). Zeno was followed by Cleanthes, Chrysippus, and Archedemos (*SVF* 1. 494, 2. 364; *FDS* 318 A). Once again, Academic discussions presumably influenced Zeno, since he used the word ἐννόημα to describe figments of the soul, a word which is very close to the word

to physics. In that case, it is highly plausible that Zeno's theory of the principles proceeded from a reading of the *Timaeus* from the viewpoint of the deletion of the intelligible paradigm. The importance of references to Stoicism in Calcidius' commentary to the *Timaeus* seems to indicate the Platonic origin of Zeno's theory and the close relationship between the two theories. The main point is clearly Zeno's renunciation of incorporeal realities and transcendent paradigm.

This is why it is important to read in the text of Diogenes Laertius that the two principles are 'bodies', *σώματα*, as in the latest edition of the text (Marcovich 1999: i. 523), and not that they are 'incorporeal', *ἀσωμάτους*.[52] The first reading is the one attested in the best manuscripts of Diogenes Laertius (BPF). This reading has often been corrected into *ἀσωμάτους*, which is the reading of the *Suda*: this is the case in Traversari's and Aldobrandini's Latin translations, in Cobet's and H. S. Long's editions, and in Von Arnim's *SVF*. This correction has been criticized as early as Lipsius,[53] contrary to what is maintained in the apparatus of both Long and Marcovich. A similar reading, *ἀσώματα* is attested in manuscript D (*versio vulgata*, later than BPF), and in the *Magnum excerptum* (Φ), but there is no particular reason to privilege those manuscripts. The only real argument in favour of *ἀσώματα* could be, as Frede pointed out, that incorporeality may appear as an element of the opposition between principles and elements, where it is clear that the elements are bodies.[54] However, it is more plausible that the contrast is between two kinds of bodies, the principles being devoid of any form, while the elements are bodies with a form. And many other sources maintain that the Stoic principles are bodies.[55] Basically, the principles are 'that which acts and that which is acted upon', and Zeno argued that 'that which is devoid of body . . . cannot act or be acted upon' (Cicero, *Ac. Post.* 1. 39).

However, there is a problem with the definition of body that immediately follows in Diogenes Laertius, and this may have influenced readings with 'incorporeal'. Diogenes 7. 135 gives the following definition: 'Body is defined by Apollodorus in his *Physics* as that which is extended in three dimensions, length, breadth, and depth. This is also called solid body' (tr. Hicks). Clearly, this definition is not a proper definition of bodies, according to Stoic standards, for (1) it appears in a list of mathematical definitions in 7.135 (surface, line, point), and (2) some of the Stoic incorporeals, namely void and place also have three dimensions.[56] Accordingly, three-dimensionality may be considered as a

used in Plato's *Parmenides* 132b, when Plato refutes the possibility that the Form may be a pure concept (*νόημα*). See Montoneri 1993: 245–6. In this case, it may have appeared as in Cicero in the context of the discussion of the criterion of truth.

[52] See the opposite view of Lapidge 1978*a*: 139–40.

[53] See Lipsius 1623: 78–9 (2. 5).

[54] Frede 2005: 215.

[55] For Zeno: Aristocles in Eusebius, *EP* 15. 14 (*SVF* 1. 98, LS 45 G); for the Stoics in general: Origen, *Adv. Cels.* 6. 7 (*SVF* 2. 1051); see *SVF* 1. 153; 2. 318, 323, 380. See Hunt 1976: 23–4.

[56] See Galen, *On Incorporeal Qualities* 464. 10–14 (*SVF* 2. 502, LS. 49 E).

property of bodies only when it applies to solid bodies as contrasted to other geometrical entities, which only have one or two dimensions.[57] This is why one perhaps has to assume that this definition of body is misplaced in the context of the principles. When mentioning Zeno's renunciation to the efficiency of incorporeal entities in *Ac. Post.* 1. 39, Cicero seems to point to a more general definition of body as 'that which acts or is acted upon'. This definition would apply disjunctively in the case of the principles, respectively to one efficient body (reason or god) and to a passive one (matter).

However, it is not clear at all whether 'that which acts or is acted upon' is a more general definition of body or a property of bodies. For instance, the Epicurean Lucretius 1. 443 argues that 'nothing can act or be acted upon without body' in such a way that 'being able to act or being acted upon' appears as a property of body. On the other hand, Sextus Empiricus, *P.* 3. 38–9, seems to consider 'that which is capable of acting or being acting upon' as a definition of body, different from a definition by three-dimensionality and resistance. Therefore, it is clear that, within the Stoic system including incorporeals, it is necessary either to add resistance to three-dimensionality,[58] or to give a more general definition of body as 'that which is capable of acting or being acting upon'. A striking feature of incorporeals is precisely that they are not defined by non-dimensionality, since they all have dimensions (with the exception of the sayables[59]), but by being unable to act or be acted upon.[60] From that point of view, it is clear that three-dimensionality plus resistance cannot be a general definition of body, since the reverse is not a general definition of incorporeals. This does not exclude that bodies may all be three-dimensional and resistant, but three-dimensionality and resistance may be considered as properties shared by all bodies, rather than the elements of their definition. On the other hand, as it is probable that incorporeals like the sayables were introduced later than Zeno, it is also probable that Zeno had a more specific definition of bodies by three-dimensionality and resistance. Nevertheless, the only conception of body clearly attributed by Zeno in our sources is not three-dimensionality and resistance but the capacity of acting and being acted upon.

This is not surprising since, once again, Plato discussed this pair of properties in connection with principles of being and corporeality. From this point of view, the question whether an incorporeal may act and be acted upon seems to belong to Zeno's rejection of Platonic incorporeal entities. In Plato's *Sophist*, the Stranger explains that the capacity of acting or being acted upon is a capacity of being in general, not of bodies:

They will have to tell what that is which is inherent in the incorporeal and the corporeal alike, and which they have in mind when they say that both exist. . . . I suggest that

[57] See Hahm 1977: 10–11.

[58] This is explicitly acknowledged by Galen in the text mentioned in n. 56 above.

[59] There are four incorporeals: sayables, void, space, and time according to Sextus, *M.* 10. 218 (LS 27 D). Void and space have three dimensions, time is supposedly two-dimensional.

[60] Sextus, *M.* 8. 263 (LS 45 B).

everything which has a power of any kind, either to produce a change (ποιεῖν) in anything of any nature or to be affected (παθεῖν), even in the least degree by the slightest cause, though it be only on one occasion, has real existence (ὄντως εἶναι).[61]

The Stranger criticizes the 'Sons of the Earth', who maintain that 'nothing which they cannot squeeze with their hands has any existence at all' (*Sophist* 247c). They identify being and body and they define them by 'touch and contact' (προσβολὴν καὶ ἐπαφήν) (246a[62]). In opposition to Plato's position in the *Sophist*, Zeno and the Stoics seem to have restricted the capacity of acting or being acted upon to bodies, and to have identified being with body, as the 'Sons of the Earth'. It leaves open the question whether or not they considered all bodies, including the principles, to be three-dimensional and resistant or not. Might one ask if it makes any sense to say that matter and god are three-dimensional, since they are shapeless (ἄμορφα)? Is it not the case that if something has dimensions, it has a shape? We have to assume here that 'shape' refers to qualities and not to dimensional quantity, and that something may have an extension without this property constituting a 'shape', as Cooper argues in Chapter 4. In fact, this is what is implied by the notion of 'qualityless matter'. On the other hand, is it not precisely the case that god is the principle that gives shape to matter? Is there any sense in saying that god is 'resistant'? God, as the active principle, does not seem to offer any resistance: it is rather matter that seems to offer resistance to him. On the other hand, if body is reduced to the capacity of acting and being acted upon, it seems to be a rather abstract entity; then it is not surprising that Stoic principles may have been considered 'incorporeal' in a certain sense[63]—except that this is not in any Stoic sense.

In his *De Mixtione* 224. 32–225. 3 Alexander of Aphrodisias clearly says that god gives shape to matter:

Entering the argument at this point, one might reasonably challenge them with also claiming the existence of two universal principles, matter and god, of which the latter is active, the former passive; and with saying that god is mixed with matter (μεμῖχθαι τῇ ὕλῃ), and pervades the whole of it (διὰ πάσης αὐτῆς διήκοντα), in this way shaping and forming it (καὶ σχηματίζοντα αὐτήν, καὶ μορφοῦντα) and creating the universe (κοσμοποιοῦντα). (Tr. Todd)

From that point of view, matter seems shapeless only in the sense that it has no other shape than the one given to it by god, not in the sense that it is strictly shapeless. It is shapeless also in the sense that it has no shape of itself, but can take any shape whatsoever, contrary to individual matter, which has its own

[61] Plato, *Sophist* 247d–e, tr. Fowler; see among others Pearson 1891: 85; Long 1976: 153.

[62] Note that the two words may be found in *SVF* 2. 359, so that one may think that the Stoics borrowed them from Plato, but if one checks Von Arnim's source for the fragment (Clement of Alexandria, *Strom.* 2. 4. 15), one realizes that Clement is just quoting Plato's *Sophist*, not the Stoics.

[63] See Lapidge 1978*a*: 139–40.

shape, i.e. the sum of its own qualities. Above all, the main difference between the principles and individual matter and individual bodies lies in the fact that the principles never perish nor diminish:

From Zeno. Substance is the prime matter of all beings, and this in its entirety is everlasting, without increase or decrease, but its parts do not always remain the same, but divide themselves and mingle together. (Stobaeus, *Eclogae* 1. 11. 5, p. 132. 27–133. 3 = *SVF* 1. 87)

B) Particular Matter

Even if his wording is not perfectly clear, Diogenes Laertius 7. 150 seems to maintain that both 'matter' and 'substance' have a twofold sense, according to which they signify universal or particular substance or matter. This twofold sense of substance is also acknowledged by Calcidius, *In Tim.* 289. But, according to him and to Stobaeus, most of the Stoics, including Zeno and Chrysippus, were more specific, and applied the word 'substance' only to prime matter:

However, most of them [the Stoics] distinguish matter from substance, as did Zeno and Chrysippus. They say that matter is that which underlies everything having qualities, whereas substance is the primary matter or oldest foundation of everything, which by itself is shapeless and devoid of form. Thus, bronze, gold, iron, and such like are the matter of all things fabricated from them, but not their substance, whereas that which is the cause of these things as well as of anything that exists is the substance. (Calcidius, *In Tim.* 290, p. 294. 6–18 W. = *SVF* 1. 86)

According to Zeno and Chrysippus then, bronze, gold, and iron are not substances, but informed particular matter. Stobaeus (at 1. 11. 5, p. 133. 1–3, quoted above) and Diogenes Laertius 7. 150 agree that, contrary to prime matter, individual matter may increase or decrease, be divided, and mingle with another part of another matter. This clearly distinguishes Stoicism from Epicureanism, since atomic matter is indivisible. As far as the four basic elements are concerned, they may be considered as made of informed matter, but the reverse is also the case, since each particular matter is said to be made out of the elements. The text of Calcidius, among others, may lead a reader to believe that each individual matter is only a part of prime matter, as if prime matter were dividing itself into bits of informed matter. However, this is not the way the Stoics conceived the relation between prime matter and individual matter, since what they think is that, in each cycle of the universe, the elements are generated first, and then individual bodies and their matter are composed from the combination of the elements. Therefore, there is an intermediate step. It is clear that, to have a more precise view of matter in its two senses, it is necessary to have a look into the process of world-formation.

2. MATTER AND GOD IN THE GENERATION AND CORRUPTION OF THE WORLD

The main difficulties associated with Stoic theory of matter result from the Stoic account of the generation and corruption of the world, and from the relationship between prime matter and god as the two principles on the one hand and individual matter and god as posterior to the generation of the elements on the other. A first difficulty consists in the fact that elements seem to be generated from pre-existing elements. A second difficulty is that god may appear as proceeding from matter and as identical with one or several elements.

A) World-Generation in Stoicism

The process of world-generation according to the Stoics is sketched in two passages of Diogenes Laertius 7. 136 and 142. In these two texts of Diogenes, substance turns into water through a preliminary state of air:[64]

[God] in the beginning by himself turned all substance through air into water. (7. 136)

The world is generated when substance is turned from fire through air into moisture. (7. 142)

From the preceding definitions of matter in 7. 134 and from context, substance in 7. 136 could be interpreted as prime matter. The first engendered elements could seem to emanate directly from prime matter through the action of god. However, the second passage, though not mentioning god, makes rather clear that the turning of the substance into water and air proceeds from an initial state of fire. In that case, it means that prime matter never exists by itself in a 'pure' state of qualityless matter, but is always already informed. This is clearly stated in Calcidius, *In Tim.* 292, p. 295. 4–6 W. (*SVF* 1. 88, LS 44 D):

[Zeno] thinks that matter, the basis of everything, has no shape or any form or any quality whatsoever, yet it is always joined to a certain quality and inseparable from it (*cunjunctam tamen esse semper et inseparabiliter cohaerere alicui qualitati*).

Of course, this is coherent with the famous Stoic doctrine of conflagration and everlasting recurrence, which does not appear explicitly in Diogenes Laertius but is well-attested in many other sources.[65] According to this doctrine, there is not only one generation of the world from chaos, as in the *Timaeus*, but the process repeats itself an infinite number of times, so that at the end of each process the

[64] I have been convinced by a discussion with S. Mouraviev and S. Toulouse that this standard interpretation of DL 7. 136 and 142 (δι' ἀέρος εἰς ὕδωρ) is the right one.

[65] See Salles in Ch. 5 and, among others, Gourinat (2002), and Salles (2005: 19–29), both with references to previous secondary literature.

world turns into fire and disintegrates, while at the beginning of each process the world is to be reformed again from this state of fire.[66] This initial state of fire is compared to semen, from which everything else is generated.[67] If we turn back to the first of the two texts by Diogenes Laertius mentioned above, we see that the transformation of fire into water is precisely explained by the fact that humidity is naturally contained in a liquid:

Just as the sperm is enveloped in the seminal fluid, so god, who is the seminal *reason* of the world, stays behind as such in the moisture, making matter serviceable to himself for the successive stages of creation. (DL 7. 136, tr. Long and Sedley, slightly modified)

So god, here, being *logos*, seems to be one of the two principles. Therefore the process seems to be the following: god, being the seminal *reason* of the world, informs matter, and transforms it from fire into water through air. Then, the four elements are progressively generated from this primitive state of substance:

He then creates first of all the four elements, fire, water, air, earth. (DL 7. 136)

Then the thicker parts of the moisture condense and end up as earth, but the finer parts are thoroughly rarefied, and when they have been thinned still further, they produce fire. (DL 7. 142, tr. Long and Sedley)

However, this is not without problems. For in the first stage of the world, it is fire as the substance of the universe that is described as semen (*τὴν οὐσίαν μεταβάλλειν οἷον εἰς σπέρμα τὸ πῦρ*). And in the second stage, it is not the substance or matter which is described as semen, but god itself as the reason of the universe. Therefore, the semen seems to be alternatively each of the two principles, and this does not appear coherent.

Moreover, if the primitive fire is semen, it should perhaps be different from fire as one of the four elements. And, if this primitive fire is different from fire as an element engendered from primitive fire, the three other primitive states of substance may also be different from the elements generated afterwards.[68] In fact, either the four elements are generated from a pre-existing element, and this pre-existing element is not an element in the same sense as the other elements (in this case, the primitive fire is not identical with the elemental fire), or the primary fire is the ultimate element, out of which the three others are generated. Given the fact that there are different meanings of the word 'element' in Stoicism, it seems to be the case that the first hypothesis is the right one.[69] This is not a subject I will

[66] See e.g. Aristocles in Eusebius, *EP* 15. 14 (*SVF* 1. 98, LS 46 G).

[67] Ibid: 'The primary fire is as it were a sperm which possesses the principles of all things and the causes of past, present and future events'. Ibid. 15. 8. 3 (*SVF* 1. 107): 'the whole of substance turns into fire as if it were into sperm'; Stob. *Ecl.* 1. 20, p. 171. 2, W. (*SVF* 1. 107; *SVF* 2. 596). Of course it is true that Zeno, in claiming that fire is the element *par excellence*, as he does according to Stob. *Ecl.* 1, pp. 129–30 (*SVF* 2. 413, LS 47A, see below), was deeply influenced by Heraclitus, but, though this is a major question in the understanding of Zeno's cosmogony, I shall not treat it here. On Heraclitus' influence on Stoicism, see Long (1975/6).

[68] On this, see Ch. 4 by Cooper.

[69] This is basically Cooper's argument in this volume.

discuss at length here, since it is discussed in Chapter 4. According to Cooper, especially when it comes to Chrysippus, who refined Zeno's original theory and attempted to eradicate its incoherencies, one has to distinguish between (1) the primitive state of the universe, which Chrysippus preferred to describe as a 'flash' or 'light' (αὐγή), (2) the proto-fire of the universe, which turns out to proto-moisture through proto-air, and (3) the four elements generated by condensation and rarefaction of the proto-moisture.

However, I am not certain that things are in the end as clear-cut as Cooper assumes them to be, even in Chrysippus. It is clear that fire, when the whole of substance has turned into it fire, cannot be exactly the same as it is when the other elements exist in the same proportion, or at least exist as well. But fire, as an element, seems to be nothing else than matter plus the quality of heat, or even nothing other than heat itself (DL 7. 137). And in Stobaeus, *Eclogae* 1, pp. 129–30 (*SVF* 2. 413, LS 47 A), when fire is described as the element *par excellence* (κατ' ἐξοχὴν), it is not distinguished from fire as one of the elements. There are different chronological states, and different meanings of 'element' in the Stobaean passage, but Stobaeus speaks as if the primitive fire was not different from the elemental fire emanating from it in the subsequent stages. Basically, according to Philo, in a text that is inspired by Stoic distinctions,[70] 'light' or 'flash' (αὐγή), which is the primitive state of the universe according to Chrysippus,[71] is nothing other than one of the three forms of fire, alongside coal and flame: coal is 'fire embodied in the earthy substance', flame is 'what rises up from that which feeds it', and light is an emanation from flame, which 'co-operates with the eyes to allow them to see'.[72] Then, it may be the case that Chrysippus defined the final and primitive state of substance as 'light' or 'flash' not to distinguish it from fire, but to explain in what sense of fire it was a fire. Evidently, as Philo points out, it cannot be coal or flame, since that would imply that it is mixed with earth in the first case, or consuming something in the second case[73]—more precisely, when the world turns into fire, it means that everything else is consumed,[74] so that it cannot be anything else than fire in its purest and final state. According to Philo, *On the Indestructibility of the World* 88 and 92, nor can it be light, since it cannot be coal or flame, and light cannot exist without flame. However, it seems that Chrysippus thought that it should be light, since it cannot be coal or flame, so that 'light' is something like embers, where the seminal principle

[70] See the parallel texts in *SVF* 2. 427 and 432, the latter (from Alexander of Aphrodisias) asserting that 'light', according to the Stoics, is the 'emanation of fire' and its third form.

[71] Philo, *On the Indestructibility of the World* 90 (*SVF* 1. 511, *SVF* 2. 611; LS 46 M).

[72] Philo, *On the Indestructibility of the World,* 86 (*SVF* 2. 612). Cooper, p. 104 n. 24, prefers to assume that the classification of light as a form of fire is 'polemically motivated'.

[73] Philo, *On the Indestructibility of the World* 87–8, 91.

[74] See for Zeno Alexander of Lycopolis, *Contr. Manich.* 12 (LS 46 I) and for Chrysippus, Plut. *De Stoic. Rep.* 39. 1052 C (*SVF* 2. 604, LS 46 E): the soul of the universe 'completely absorbs matter'.

is still smouldering.[75] Moerover, he may take his vocabulary from a passage by Heraclitus, who says that 'light' or 'flash' is a 'dry soul, the wisest and the best' (Heraclitus, fr. B 118). Therefore, the identification of this state of the universe with light corresponds to the most divine form of fire, and this is presumably why Chrysippus chose that word,[76] since, as mentioned above, it is clear that the importance of fire in the Stoic doctrine of the elements was greatly influenced by Heraclitus. Then, if the substance of the world has turned into light, and if light is a form of fire, it has turned into fire. A form of something is not identical with what it is a form of, since what it is a form of has other forms—but it does not mean that a form of something is something *other* than what it is a form of.[77] Therefore, it may be the case that 'light' is a form of fire, and that fire as an engendered element has more forms than primitive fire—but 'light' still may be a fire, in a more final or primitive stage than the forms under which it develops afterwards. I think Cooper is perfectly right in distinguishing primitive fire from elemental fire, but I think both of them are nothing other than matter with the quality of heat, or heat itself.

In any case, even if the four elements are generated from a pre-existent element, the primitive fire, this does not preclude the priority of the principles over the elements. The principles are principles, they are not elements. Things can be truly divided into elements, but they cannot be divided into principles. As a consequence, the problem is rather the question of the substance of god. For god is supposed to be the active principle, as opposed to matter, but god is also identified with one of the elements (fire, air) or a compound of these elements (*pneuma* or breath).[78] This problem finds an echo in the above-mentioned ambiguity of semen, which, in the first stage of cosmogony, is identified with fire as the substance of the universe, while, at a later stage, it is identified with god as the reason of the universe. Now, if god may be identified with an element, or a compound of elements, the Stoics have a major problem, which has been pointed out from antiquity onwards: if god is a compound, he is chronologically and ontologically posterior to the elements and to matter. In other words, the Stoic theory of matter and god is typically materialistic, though it claims not to be.

B) The Problem of the Substance of God

This criticism of the Stoic theory goes back at least to Alexander of Aphrodisias:

For if god is on their view body—an intelligent and eternal breath—and matter is body, first there will be again body through body; then this breath will certainly be one of the

[75] See the denial by Philo, *On the Indestructibility of the World* 93, who says, 'no embers of the seminal principle are mouldering' (μηδενὸς ἐντυφομένου σπερματικοῦ λόγου): I think we may assume that Philo is denying precisely what Chrysippus was asserting.

[76] This reference and its interpretation were suggested to me by Gábor Betegh.

[77] See the classic definition of form as what is contained in a genre in DL 7. 61.

[78] Compare *SVF* 1. 88, 154; 2. 310, 423, 1009, 1027, 1075, 1100. There is an excellent tabular presentation in Duhot 1989: 85. See also Duhot 1989: 73–86.

four simple bodies which they say are also elements, or a compound of them (as of course they themselves say; for certainly suppose that breath has the substance of air and fire) or, if something else, the divine body will certainly be a fifth substance. . . . But if it were one of the four bodies or a compound of them, then the body that is produced from matter will have pervaded it before it comes to be and will generate itself too from it just like other things. Again, god would be secondary to matter if all enmattered body is secondary to matter; for what is derived from a principle is secondary to it, and god is such a body since he is not identical with matter. (*De Mixtione* 225. 3–15, tr. Todd)

God will be either an element, or a body derived from the elements and made out of them. In this case, god is derived from the elements, and hence posterior to them. The doctrine of the principles collapses. Alexander accuses the Stoics of materialism and circularity. They are monists because everything is derived from matter: in his mind, it is clearly mistaken. One is not obliged to follow him on this point. On the other hand, his second criticism aims at what is certainly a vicious circle: elements derive from principles, but god, if it is a compound, derives from the elements, as the most simple bodies, and hence god is not a principle. God is simply the soul of the animal-world, borne out of matter. Therefore, concludes Alexander, the Stoic god 'would be eternal in name alone (μέχρι φωνῆς μόνης)' (*De Mixtione*, 225.16). It cannot be an everlasting principle.

However, Alexander's criticism seems rather unfair, since god seems in this respect similar to matter with its two meanings. In one sense, 'matter' is a principle and, in this sense, it is the unqualified substance of the universe, and in the other sense, 'matter' is the qualified matter of particular realities. Similarly, 'god' is a principle, which has neither form nor shape but informs matter, and he is also fire, air or breath, world-soul, and the nature of the world. He is the 'peculiarly qualified individual made out of the whole substance' (DL 7. 137). In the first sense, god *may be compared to* the soul of an individual body (see Alexander, *Mixt.* 226. 11–12, who applies this comparison to matter and god as principles). In the second sense, *he is* the soul of the world. God is shapeless but may incorporate himself under various forms, and, similarly, matter is shapeless as prime matter, but may take various forms and, as individual matter, matter is that out of which concrete bodies are made. The Stoic position is very subtle and, in a certain way, rather intricate, since the Stoics clearly distinguish principles from concrete bodies, but they also tend to assimilate both, since principles never exist separately. It is said in the *Placita* 1. 6. 897C that god, as a breath, 'has no form of his own, but may take any form whatsoever'. This alludes to the fact that god may incorporate himself primarily in breath, fire, or air, but may disseminate himself even in water or earth (DL 7. 147). God as a principle is to be distinguished from the forms he may take as the active world-soul or nature of the world.

In the *Placita*, god is described in different ways, as fire, breath, and the world itself and its supreme intellect:

The Stoics made god out to be intelligent (νοερὸν), a designing fire which proceeds towards creation of the world according to a certain order (πῦρ τεχνικὸν ὁδῷ βαδίζον ἐπὶ γένεσιν

> κόσμου), and encompasses all the seminal reasons (πάντας τοὺς σπερματικοὺς λόγους) according to which everything comes about according to fate, and a breath (πνεῦμα) pervading the whole world, which takes on different names owing to the alterations of the matter through which it passes (κατὰ τὰς τῆς ὕλης δ' ἧς κεχώρηκε παραλλάξεις). God is also the world and the stars and the earth, and the supreme intellect in aether. (Ps.-Plutarch, *Placita* 1. 7. 881F–882A, tr. based on LS, 46 A)

It is clear that these descriptions no longer apply to god as a principle, but as a certain kind of animating (or animated) body. However, god under these forms has the same properties as god as a principle. The identification of god with a 'designing fire' (πῦρ τεχνικόν) reminds not only of Zeno's definition of nature as a 'designing fire' engendering the world according to a certain order[79] but also of the description of god as a principle producing every single reality (δημιουργεῖν ἕκαστα) in the Stoic doctrine of the principles in DL 7. 134. The analogy is all the more striking in the description of god as a rational animal in 7. 147, where god is also said to be 'the craftsman of the universe and a kind of father of everything' (δημιουργὸν τῶν ὅλων καὶ ὥσπερ πατέρα πάντων). There again, the influence of the *Timaeus* is evident even in the wording, but what is puzzling is that the same description applies to god as a principle and to god as a world-animal or as the world-soul. It is clear that when god is described as fire, the emphasis lies in his craftsmanlike capacity, while when he is described as breath, he is considered to be a soul. It is again as a fire, i.e. as it is at the origin of the universe, that his function is closer to the function of god as a principle, i.e. the active capacity to produce and give form to matter. On the one hand, god as a breath is not nature but soul, as Calcidius says: 'This mobile breath is, in his opinion, not nature, but a soul, and a rational soul' (Calcidius, *In Tim.* 292 = *SVF* 1. 88). Therefore, god as a creating nature under the form of fire seems to be different from god as a world-soul under the form of breath. On the other hand, god and nature are often equated in Stoicism, even perhaps by Zeno himself.[80] In DL 7. 156, the designing fire of nature is identified with a 'fiery and designing breath'—but note that this is the definition of god by Posidonius,[81] not by Zeno, so that there may be a conflation of sources.

In any case, since generation is a continuing process in the world, designing nature is still operating in the world once it is formed, so that god must be simultaneously psychic breath and designing fire. It is also mentioned in the above-quoted passage of the *Placita* 1. 7. 881F–882A that god is 'the supreme

[79] See Cic. *ND* 2. 57 (*SVF* 1. 171).

[80] Themistius, *De An.* 35. 32–43 Heinze (*SVF* 1. 158), Plut. *De Stoic. Rep.* 34. 1050B (*SVF* 2. 937), Alexander of Aphrodisias, *De Fato* 22, p. 192. 25–6 Bruns (*SVF* 2. 945, LS 55 N 4), Seneca, *Benef.* 4. 7 (*SVF* 2. 1024), Philodemus, *De Pietate* 11 (*SVF* 2. 1076).

[81] In the *Plac.* 1. 6. 879 C (*SVF* 2. 1009), this definition is attributed to the Stoics in general, but according to Stobaeus, 1. 1. 29b, p. 34. 26 W. (source for Aetius, *Plac.* 1. 7. 19, Diels), it is the definition of Posidonius (Posidonius F 101 EK). The attribution to Posidonius may also be found in a Lucan scholium (Posidonius F 100 EK).

intellect in aether'. According to other testimonies, the identification of the ruling part of the soul (ἡγεμονικόν) with aether seems to have been one of Chrysippus' theses.[82] Aether is the highest part of the sky (DL 7. 137). This seems to imply that god, as the ruling part of the world-soul, is concentrated in the highest part of the sky while the other parts of this same soul diffuse into the world under the form of breath and fire.[83]

It is a Stoic tenet that the soul of individual animals only appears at birth: at birth, the quality of the breath of the embryo changes from 'nature' into soul.[84] God, under the form of fire, seems to entertain a similar change of quality when the world is born, and to turn into world-soul, from a kind of original pure fiery intellect, which is close to what nature is as 'designing fire'. However, even before the rebirth of the world, this fire contains the divine *logos* as semen: the 'designing fire . . . encompasses all the seminal reasons' (Ps.-Plutarch, *Placita* 1. 7. 882A). The seminal principle is contained in fire during the fiery stage of the universe, as it will be in liquid when the world is proceeding through generation (DL 7. 136). Accordingly, god is an active principle always present in matter during all the phases of the history of universe, but he may incorporate himself in particular elements (fire or air) or compounds (breath), through which he exerts his action during the phases of the universe. What is engendered after the elements and the particular matter is not god as a principle, but god as the incorporation of this principle in various phases of world-history and in various parts of the substance of the world.

What throws some confusion into the Stoic doctrine is that god as a principle has the same active properties as god as fire, air, or breath. Similarly, prime matter is passive like particular matter. Similarly again, among the four elements, two are active, fire and air, and two are passive, water and earth.[85] So it may appear that in fact the principles are identical to the elements: the active principle, god or reason, being identical with fire and air and their immediate compound, breath, while the passive principle, matter or substance, is identical with earth and water. In this case, there seems to be no difference between the principles and the elements, and the doctrine seems very confused. However, what the Stoics have in mind is rather that each element is a mixture of the two principles, the active and the passive, some elements (fire and air) have a higher concentration of the active principle, while there is a higher concentration of the passive principle in others (earth and water).[86] Each of the elements contains both the active quality and the passive matter and, when the elements mix with each other to produce specific bodies, they follow the same pattern as the principles by mixing one or

[82] See Arius Didymus, fr. 29 Diels (*SVF* 2. 642) and DL 7. 156 (*SVF* 2. 774).

[83] See Cic. *Ac. Pr.* 2. 126; DL 7. 140.

[84] See Gourinat 2008.

[85] Nemesius, *Nat. Hom.* 5 (*SVF* 2. 418, LS 47 D); see Galen, *On Natural Faculties* 106. 13–17 (*SVF* 2. 406, LS 47 E); Galen, *On Bodily Mass* 7. 529. 9–14 (*SVF* 2. 439, LS 47 F); Plut. *De Stoic. Rep.* 49, 1085C–D (*SVF* 2. 444, LS 47 G).

[86] This was suggested to me by Thomas Bénatouïl.

two passive elements with one or two active elements. According to Alexander,[87] there is no such thing in the Stoic universe as simple bodies (i.e. elements), since in fact each body is sustained and informed by breath and, therefore, the elements themselves are compounded. However, once again, this is not the way things are described by the Stoics. What they describe is rather a transformation of the original fire during the process of world-formation thanks to the active power of the semen it contains, and this obviously alludes to the fact that primitive fire is a blend of the active semen (god, reason) and of the passive matter. It is rather in the developed stages of the universe that god acts through the active elements and the active compound (breath) that diffuses itself through the whole of matter. Presumably, this is one of the reasons why Zeno needed a theory of total blending, against Aristotle's doctrine in the *De Generatione*. Obviously, this is partly another question—and I shall not explore it in this chapter. Here I shall only state the following: in every body, there is an active principle and a passive principle, and these principles are two bodies blended with each other;[88] the substance of the universe is so constituted, and, at the origin, it is pure fire—but still with an active principle and passive matter in it—and, from this, the semen or active principle contained in fire turns this originary substance into the other elements; and then again, these elements, some active, some passive, mix with each other to produce qualified bodies, and in each of these bodies, breath, i.e. a compound of the two active elements (fire and air), mix with water and earth to give them qualities. However there is no confusion between principles and elements—it is only the case that, since these two principles are the ultimate reality of everything, the more complex bodies are made of passive and active elements, just as simple bodies are solely made of these two principles in various proportions—and of nothing else.

C) The Contemporaneity of Matter and God

According to Zeno and the Stoics then, god is not engendered from and by matter, and he is not 'eternal in name alone'. Epiphanius says that Zeno 'speaks of matter calling it contemporaneous with god, just like the other sects' (*Haeres.* 1. 5 = *SVF* 1. 87). Epiphanius is not a trustworthy source is general, but, in this case, he may reproduce a genuine Stoic answer to criticisms similar to Alexander's criticism. The two principles never exist in separated or pure state, so despite the fact that matter is qualityless by itself, it never exists without any quality.[89] This

[87] See Alexander, *De Mixtione* 224. 14–17 (*SVF* 2. 442, LS 47 I).

[88] According to Lapidge (1978*a*: 139–40), the two principles are only two 'aspects of a single body'. See Lapidge (1973: 243–4 and *passim*). Against this interpretation, see Alex. *Mixt.*, p. 225. 1–2 (*SVF* 2. 310, LS 45 H) and the comments of Long and Sedley (1987: i. 273). See also Reydams-Schils (1999: 55–8) and Cooper (Ch. 4, p. 99 n. 16 below).

[89] See Calcidius, *In Tim.* 292 (*SVF* 1. 88, LS 44 D) quoted above.

is presumably the reason why Posidonius maintains that the difference between matter and substance exists in name alone:

From Posidonius. Posidonius said that the substance of the whole, i.e. matter was without quality and without shape (ἄποιον καὶ ἄμορφον), in so far as in no way has it a form detached of its own, nor quality by itself either (οὐδὲν ἀποτεταγμένον ἴδιον ἔχει σχῆμα οὐδὲ ποιότητα καθ' αὑτήν), but always is in some form and quality (ἀεὶ δ' ἔν τινι σχήματι καὶ ποιότητι εἶναι). He said that substance differs from matter, being the same in reality, in thought only.[90]

The same point of view seems to appear in a parallel text by Calcidius, which may reflect Posidonius' point of view, especially since it seems to be contrasted with the point of view of Zeno and Chrysippus stated in the previous paragraph:

A great many also distinguish between matter and substance in the following way: they say that substance is the foundation of a work, so that we may rightly speak and think of a substance of the world, but that it is called matter with respect to the craftsman, who moulds and forms it.[91]

As van Winden (1959: 96) maintains, commenting on Calcidius, according to these unidentified Stoics (who may be Posidonius), 'the two terms denote the same reality but seen from different viewpoints'. This is clearly different from the doctrines of Zeno and Chrysippus, but this may have been a way to reformulate their point of view. Substance and matter differ only in thought, since qualityless matter never exists in a separate state.[92] Diogenes Laertius 7. 137 states that 'the four elements taken together are qualityless substance, i.e. matter', and this is also acknowledged by Plutarch as a Stoic tenet, when he says that 'substance is given by some Stoics the epithet "without quality" not because it is devoid of every quality but because it has all qualities'.[93] The idea seems to be that the pairs of contradictory qualities (hot + cold, liquid + dry) add up to a null sum, so that prime matter is nothing else than the sum of the elements taken together and does not differ in reality from the sum of particular matters.

The four elements, since they are the last constituents into which everything else dissolves at the end of a cycle (εἰς ὃ ἔσχατον ἀναλύεται) cannot be melted so as to constitute qualityless matter as existing independently from them. If this is the case, this is because the principles are immanent. They cannot be seen, they

[90] Stob. *Ecl.* 1. 11, p. 133.18 W. (Posidonius F 92 EK, Kidd translation).

[91] Calcidius, *In Tim.* 291, p. 294. 13–16 (my tr., partly based on van Winden). According to Reydams-Schils (1999: 93–4), it is Waszink's opinion that the 'in thought only' (ἐπινοίᾳ) of Posidonius has been transformed by Calcidius into 'in the thought of a craftsman' (*contemplatione opificis*), which would allude to the thought of the craftsman. However, I think this is not the correct interpretation of the passage: like van Winden (1959: 96–7), I understand *contemplatione opificis* as meaning 'with respect to the Maker', in the usual sense of *contemplatione alicuius rei*, 'with regard to something', so that there is no allusion to the thoughts of the craftsman, but only to a standpoint.

[92] See Hunt 1976: 21.

[93] Plut. *De Stoic. Rep.* 50. 1086A (*SVF* 2. 380).

cannot be touched, they are not three-dimensional visible and tangible bodies but, meanwhile, they only exist in the form of such bodies, including the four elements taken together, in the case of matter, and a particular body like air, fire, or breath, in the case of god.

CONCLUSIONS

If the Stoic theory may appear materialistic, i.e. monist, it is because everything is corporeal and the two principles are always mixed together, so that they do not constitute two *separate and independent* principles. They are two bodies, but never exist separately from each other. From that point of view, god may appear identical with matter, as Calcidius maintains:

They arrived at the impious opinion, *viz.* that god is identical with or even an inseparable quality of matter (*hoc esse quod silva sit vel etiam qualitatem inseparabilem deum silvae*), that he passes through it as seed through the genitals.[94]

God is what matter is. Like substance, which exists only under the form of particular matter, god himself only exists in matter, and the two principles never exist separately from each other. However, these two principles are two different bodies, even if they are always mixed together, and constitute by mixture a unified body. Moreover, everything is not explained mechanically by figures, movements, 'touch and contact' and rebound of an inert matter. In Epicureanism for instance, the two principles may be considered to be matter and void. By introducing an active principle, which is identical with a seed, the Stoics depart from materialism, and tend towards vitalism, even if this seminal principle is always blended with matter and inherent in it. Therefore, their doctrine is vitalistic, not materialistic. On this, the influence of Plato's biological model of the universe in the *Timaeus* was decisive. They departed from the *Timaeus* in that they took literally the biological model and, some way or another, renounced the technological model of the craftsman and, above all, the intelligible paradigm—the blueprint of the divine architect. They kept something of it, since they carried on defining nature as a 'designing fire'—but this was also under the influence of Heraclitus, who, according to Aristotle's classification, was a materialist. However, their fire did its work by itself and from the inside—without any divine blacksmith or potter, contrary to the case of Plato's craftsman. God is nature, and nature, as in Aristotle, is an inner principle.[95] The craftsman is an inner principle, working like fire, engendering like seed, and ruling like soul. This is clearly not 'materialism' in the strict sense.

[94] Calcidius, *In Tim.* 294, p. 296. 19–297. 3 (*SVF* 1. 87), tr. van Winden.
[95] ἡ δὲ φύσις ἀρχὴ ἐν αὐτῷ: Aristotle, *Metaph.* Λ3, 1070a7–8.

REFERENCES

Bloch, O. (1985) *Le Matérialisme* (Paris: PUF).
Brisson, L., Congourdeau, M.-H., and Solère, J.-L. (eds.) (2008) *L'Embryon: Formation et animation* (Paris: Vrin).
Dahan, G., and Goulet, R. (eds.) (2005) *Allégorie des poètes, allégorie des philosophes: Études sur l'herméneutique de l'allégorie de l'Antiquité à la Réforme* (Paris: Vrin).
Döring, K., and Ebert, T. (eds.) (1993) *Dialektiker und Stoiker* (Stuttgart: Frank Steiner Verlag).
Dörrie, H., and Baltes, M. (1996) *Der Platonismus in der Antike* (Stuttgart: Frommann-Holzboog).
Duhot, J.-J. (1989) *La Conception stoïcienne de la causalité* (Paris: Vrin).
Frede, D., and Laks, A. (eds.) (2002) *Traditions of Theology* (Leiden: Brill).
Frede, M. (2005) 'La Théologie stoïcienne', in J.-B. Gourinat and G. Romeyer Dherbey (eds.), *Les Stoïciens* (Paris: Vrin), 213–32.
Gourinat, J.-B. (2002) 'Eternel retour et temps périodique dans la philosophie stoïcienne', *Revue philosophique*, 127: 213–27.
—— (2005) '*Explicatio fabularum*: La Place de l'allégorie dans l'interprétation stoïcienne de la mythologie', in G. Dahan and R. Goulet (eds.), *Allégorie des poètes, allégorie des philosophes* (Paris: Vrin), 9–34.
—— (2008) 'L'Embryon végétatif et la formation de l'âme selon les stoïciens', in L. Brisson *et al.* (eds.), *L'Embryon: Formation et animation* (Paris: Vrin), 59–77.
—— and Romeyer Dherbey, G. (eds.) (2005) *Les Stoïciens* (Paris: Vrin).
Hahm, D. (1977) *The Origins of Stoic Cosmology* (Columbus, Ohio: Columbus UP).
Hunt, H. A. K. (1976) *A Physical Interpretation of the Universe: The Doctrines of Zeno the Stoic* (Melbourne: Melbourne UP).
Kidd, I. G. (1988) *Posidonius: The Commentary* (Cambridge: CUP).
Lapidge, M. (1973) 'Ἀρχαί and στοιχεῖα: A Problem in Stoic Cosmology', *Phronesis*, 18: 240–78.
—— (1978*a*) 'Monism and Immanence', in J. Rist (ed.), *The Stoics* (Berkeley, Calif.: University of California Press), 137–60.
—— (1978*b*) 'Stoic Cosmology', in J. Rist (ed.), *The Stoics* (Berkeley, Calif.: University of California Press), 161–85.
Lipsius, J. (1623) *Physiologia Stoicorum Libri Tres,* in *Opera Omnia* (Antwerp: J. Moret; original edn. 1604).
Long, A. (1975/6) 'Heraclitus and Stoicism', *Philosophia* (Yearbook of the Research Center for Greek Philosophy at the Academy of Athens), 5/6: 132–53. Repr. in A. Long, *Stoic Studies* (Cambridge: CUP, 1996), 39–44.
—— (1986) *Hellenistic Philosophy,* 2nd edn. (London: Duckworth).
—— (1996) *Stoic Studies* (Cambridge: CUP).
—— and Sedley, D. (1987) *The Hellenistic Philosophers* (Cambridge: CUP).
Marcovich, M. (ed.) (1999) Diogenis Laertii *Vitae Philosophorum* (Stuttgart and Leipzig: Teubner).
Montoneri, L. (1993) 'Platon, die Ältere Akademie und die stoische Dialektik', in K. Döring and T. Ebert (eds.), *Dialektiker und Stoiker* (Stuttgart: Frank Steiner Verlag), 239–51.

Pearson, A. C. (1891) *The Fragments of Zeno and Cleanthes* (London: C. J. Clay & Sons).
Reydams-Schils, G. (1999) *Demiurge and Providence: Stoic and Platonic Readings of Plato's* Timaeus (Turnhout: Brepols).
Richard, M.-D. (1986) *L'Enseignement oral de Platon* (Paris: Le Cerf).
Rist, J. (ed.) (1978) *The Stoics* (Berkeley, Calif.: University of California Press).
Salles, R. (2005) *The Stoics on Determinism and Compatibilism* (Ashgate: Aldershot).
Sambursky, S. (1959) *Physics of the Stoics* (London: Routledge & Kegan Paul).
Sedley, D. (2002) 'The Origins of Stoic God', in D. Frede and A. Laks (eds.), *Traditions of Theology* (Leiden: Brill), 41–83.
Sharples, R. (1996) *Stoics, Epicureaus and Sceptics. An Introduction to Hellenistic Philosophy* (London and New York: Routledge).
Weil, É. (1964) 'Remarques sur le "matérialisme" des stoïciens', in *Mélanges Alexandre Koyré*, ii. *L'Aventure de l'esprit* (Paris: Hermann), 556–72.
Winden, J. C. M. van (1959) *Calcidius on Matter: His Doctrines and Sources* (Leiden: Brill).

3

Chain of Causes

What is Stoic Fate?

Susan Sauvé Meyer

One of the main theses of Stoic natural philosophy is that everything happens by fate (εἱμαρμένη; Latin, *fatum*):[1]

That everything comes to be by fate is stated by Chrysippus in his 'On Fate', by Posidonius in the second book of his 'On Fate', by Zeno, and by Boethus in the first book of his 'On Fate'. (Diogenes Laertius 7. 149 = *SVF* 2. 915)

But what is fate? The canonical Stoic answer is that it is a chain or string (εἱρμός; Latin: *series*) of causes:[2]

The Stoics say fate is a chain of causes (εἱρμὸν αἰτιῶν), that is an inviolable order and binding together. (Aetius, *Plac.* 1. 28. 4 = *SVF* 2. 917 = LS 55J; cf. Alex. *Mantissa* 185. 1–5 = *SVF* 2. 920)

By fate (*fatum*) I mean what the Greeks call εἱμαρμένη, that is, an order and string of causes (*ordinem seriemque causarum*), since the connection of cause to cause generates things from itself. (Cic. *Div.* 1. 125 = LS 55L; cf. *Fat.* 20)

Versions of this chapter have benefited from discussion by audiences at McGill University, Rutgers University, The Ohio State University, the Chicago Area Colloquium in Ancient Philosophy, and the University of Toronto. I am especially grateful to Brad Berman, Robert Bolton, Jim Hankinson, Brad Inwood, Richard Kraut, and Pierre Pellegrin for helpful criticisms and suggestions, and to Laura Gómez, Ricardo Salles, and two anonymous referees for detailed written comments. A preliminary version of my main claims appears in Meyer 1999.

[1] For other attributions of the thesis, see Aulus Gellius, *Noctes Atticae* 7. 2. 15 = *SVF* 2. 977; Cic. *Fat.* 21 = LS 38G; *Fat.* 41 = LS 62C5; *Div.* 1. 127 = LS 55O; Alexander, *Fat.* 164. 17–20, 171. 26–7, 181. 8–9; cf. 210. 15; Plutarch, *De Stoic. Rep.* 1050a–b.

[2] Other versions of the definition include: Nemesius, *De Nat. Hom.* 37 (*SVF* 2. 918): 'fate is an inviolable chain of causes (εἱρμός τις αἰτιῶν)—for this is how the Stoics define it, that is, as an order and binding together that is ineluctable'; Critolaus in Philo, *Aet. Mund.* (*SVF* 2. 913): 'Fate is without beginning or end, stringing together (εἴρουσα) unfailingly and seamlessly the causes of each thing . . .'; DL 7. 149 (*SVF* 2. 915): 'Fate, they define as a strung-together cause (αἰτία εἰρομένη) of things'; Cic. *Fat.* 27: 'One cause strung together with another (*causa causam serens*) from eternity.' See also Calcidius, *In Tim.* 144 (*SVF* 2. 933).

This picture of fate goes back at least as far as Chrysippus, who proposed an etymological connection between εἱμαρμένη ('fate') and εἰρομένη ('strung together') (Eusebius, *Praep. Evang.* 6. 8. 8–10 = *SVF* 2. 914).[3] Alexander of Aphrodisias, four centuries later, reports that it states the essence (οὐσία) of fate for the Stoics (Alex. *Fat.* 193. 4–8). Thus we have here an enduring piece of orthodox Stoic doctrine. To understand the thesis of fate we must grasp what they mean by the chain of causes.

Now, strictly speaking, an εἱρμός is not a chain, but a string, as of beads in a necklace. Later in antiquity, however, the term 'chain' (ἅλυσις; Latin *catena*) was used to refer to the Stoic εἱρμός,[4] and it is typical of readers today to refer to it as a 'chain of causes'. I shall defer to this usage, even though it invites a misconception which it is my project, in this chapter, to dispel. The potential for error arises because we too employ the metaphor of a causal chain, and thinking of the Stoic *heirmos* as a chain invites us to suppose that the Stoics understand the metaphor in the same way we do. The root of the error, however, lies not in the substitution of one metaphor for another (chain for string), but in our interpretation of the metaphor—be it string or chain. It has been argued that a crucial difference between the metaphor of the *heirmos* and that of the chain is that the former emphasizes the continuity of the conjoined causes, while the latter emphasizes their discreteness.[5] But the beads in the necklace are surely as distinct as the links in the chain. It is after all the necklace that is the *heirmos*, not the string that connects the beads.[6] The *heirmos* is a set of strung-together beads, just as the chain is a set of linked-together loops. They share the common feature of being interconnections of disparate items. This, I shall argue, is the operative feature in the Stoic metaphor; hence nothing important is lost in construing the *heirmos* as a chain.

[3] The etymology, also invoked in Arius Didymus in Eus. *Praep. Evang.* 15. 5 (*SVF* 2. 528), is of course false: εἱμαρμένη comes from the perfect passive participle of the verb μείρομαι (to apportion, to allot), while εἱρμός and the cognate verb εἴρω derive from the unrelated Indo-European root ϝερ, root also of the Latin *series*, as well as the Homeric Greek σειρή (chain). Chrysippus' invocation of the etymology shows that he too accepts the characterization of fate as an εἱρμός, even if, as Bobzien claims, he does not accept it as proper definition (Bobzien 1998: 50).

[4] Alex. *Fat.* 193. 5–8, 195. 13–16; Gellius 7. 2. 1–2 (quoted and discussed by Hankinson 1996: 192, 201–3); Eustathius also uses the term ἅλυσις to characterize the Stoic εἱρμός in a passage highly reminiscent of the passages in Alexander just cited (*Commentaria ad Homeri Iliadem* 2. 514. 25; cf. 515. 5 (van der Valk).

[5] Hankinson 1996: 192–3, 201–3.

[6] The continuity that Hankinson takes to be invoked by the metaphor of the *heirmos* is admittedly prominent in another Stoic metaphor, that of the rope. As the Stoic speaker in Cicero's *On Divination* explains, 'nothing that is going to be happens spontaneously (*subito*); rather, the passage of time is like the uncoiling of a rope (*quasi rudentis explicatio*), bringing about nothing new and unfolding nothing original' (Cic. *Div.* 1. 127 = LS 55O; tr. Long and Sedley; quoted and discussed by Hankinson 1996: 204). However, the rope is not invoked as an explanation of the *heirmos*; rather, it is presented as a metaphor for the passage of time from the past through the present and to the future. Now, *we* might suppose that such a sequence of events is a causal chain *par excellence*; however, the issue before us is not how we, but how the Stoics, conceive of a causal chain.

The Stoic chain of causes is, at the very least, a system of interconnected causes. Just what sort of interconnection the Stoics have in mind is the question this chapter addresses. I shall argue that a causal chain of the modern variety is not at all what the Stoics have in mind as an *heirmos* of causes, and that a better model for understanding the relation between causes in the Stoic chain is provided by the Stoic doctrine of cosmic sympathy, which posits a complex set of relations of mutual causal influence between the various bodies in the cosmos.

The metaphor of the causal chain is so central to our understanding of causality today that we tend to lose sight of the fact that it is a metaphor. As we understand it, a causal chain is a sequence of events, each of which is cause to its successor and effect of its predecessor—as when striking a match leads to a fire which burns down the house. This interpretation is not, however, licensed by the metaphor of the chain itself, for the links in a literal chain are no more related by temporal succession than the beads in a literal necklace are. We interpret the contemporaneous and symmetrical relation of connection and contiguity between the links in a literal chain as a temporal relation of succession and asymmetrical causal dependence between events. The naturalness of this interpretation is due not to the metaphor itself, but to our background assumptions about causality: in particular, to the view that causes are events and that causation is an asymmetrical relation between successive events.

To determine how the Stoics interpreted their own causal chain we must accordingly identify their background assumptions about causality, and their theoretical motivation for invoking the metaphor of the chain. The first of these tasks is the subject of the next section, where we will see that causes and effects, as the Stoics conceive of them, do not stand in the iterated relation of temporal succession characteristic of the modern conception of a causal chain. The second task will be the burden of the rest of the chapter, which argues that the relation between causes in the Stoic chain is that of mutual influence, and that the Stoic conception of fate as a chain of causes has its theoretical basis in their famous doctrine of 'cosmic sympathy'.

STOIC CAUSES

The fundamental causal relation recognized by the Stoics is that between the principles *logos* and matter. *Logos*, which they identify with god, pervades every body in the universe and quite literally holds them together:

[The Stoics] think there are two ultimate principles (ἀρχαί) of the universe (τῶν ὅλων), that which acts (τὸ ποιοῦν) and that which is acted upon (τὸ πάσχον). That which is acted upon is unqualified substance, that is matter; that which acts is the reason (λόγος) in it, that is, god. For this, since it is everlasting, constructs every single thing throughout all matter. (DL 7. 134 = LS 44B; tr. Long and Sedley; cf. Sextus, *AM* 9. 75 = LS 44C)

Thus any body can be analysed into its passive matter and the active *logos* that holds it together. This latter principle is what the Stoics call a cause:

[The Stoics] say that there are two things in nature from which everything is made—cause and matter. Matter lies passive, receptive of anything, but idle if nothing moves it. Cause, on the other hand, that is reason (*ratio*; sc. = λόγος), forms matter in whatever way it wishes, and produces from it all kinds of products. Thus for anything, there is that from which it is made, matter, and that by which it is made, cause. (Seneca, *Epistles* 65. 2 = LS 55E)

And indeed only such an active principle (*id quod facit*) is a cause (Seneca, *Ep.* 65. 4).

As we shall see in more detail later on, the Stoics maintain that the divine *logos* permeates the cosmos as a corporeal 'breath' (*pneuma*) (Stob. *Ecl.* 1. 79. 1–2 = LS 55M) which manifests itself to different degrees in different kinds of bodies. While in its simplest manifestation this corporeal *logos* is the 'tenor' (*hexis*) that holds bodies together, its effects are not limited to mere cohesion. The *logos* in living things is the nature (*phusis*) responsible for their growth and development, and in animals it is the soul (*psuchê*) that causes activities such as perception and locomotion (Galen, *Intr.* 14. 726. 7–11 = LS 47N; Philo, *Quod Deus Sit Immutabilis* 35–6 = LS 47Q; *Leg. Alleg.* 2. 22–3 = LS 47P; Sextus, *M*. 9. 81). Causes of this sort are σπερματικοί or 'seminal' *logoi*—so called because their effects typically unfold in an orderly sequence of events, as in the case of the unfolding morphogenesis of a plant from seed to maturity (Aetius 1. 7. 33 = LS 46A).[7] This sequence of events, however, is not one of cause and effect, on the Stoic paradigm. Rather, the entire sequence of events is the effect, while the cause is the nature or soul that persists through the sequence.[8] A cause, so conceived, is not succeeded by its effects, but exists simultaneously with them.

The Stoics extend this model of causation to cases where one unified body acts on another:

The Stoics say that every cause is a body which becomes the cause to a body of something incorporeal. For instance, the scalpel, a body, becomes the cause to the flesh, a body, of the incorporeal predicate 'being cut'. And again the fire, a body, becomes the cause to the wood, a body, of the incorporeal predicate, 'being burnt'. (Sextus, *M*. 9. 211 = LS 55B; tr. Long and Sedley)

In contrast with the original model, where the active and passive principles are aspects of a single unified body, the agent and patient in these changes are

[7] On the σπερματικοὶ λόγοι and the divine principle in matter, see also Gourinat (pp. 59–62 above).

[8] Hankinson rightly emphasizes the fact that, on the Stoic view, these unfolding sequences are unified and continuous processes (1996: 192–4). He takes the metaphor of the εἱρμός to capture this continuity, while that of the chain, he claims, emphasizes the discrete identity of the successive events. But this would seem to imply that the successive events in the process are the causes linked in the *heirmos*—which Hankinson himself rejects as contrary to Stoic doctrine (p. 200).

different unified bodies (for example, the scalpel and the knife), as a result of which, the cause need not exist simultaneously with its effect. The scalpel, for example, exists even before it cuts the flesh and it could continue to exist without ever making the cut. However, as in the case of the divine *logos* and the seminal *logoi*, the cause is not succeeded by the effect it produces. The scalpel does not pass out of existence upon cutting the flesh.

An objector keen to show that Stoics causes are succeeded by their effects might object that the scalpel's activity of cutting the flesh is succeeded by the existence of the wound. This activity, however, is not what the Stoics identify as the cause. One might suppose that, strictly speaking, it must be, given the Stoic view that all causes are manifestations of the active principle, *logos*.[9] But this is a mistake, for the kind of 'activity' essential to the active principle is contrasted with passivity, not with latency or potentiality.[10] The Stoic cause is distinctively 'active' in that it is the agent (τὸ ποιοῦν), as opposed to the patient (τὸ πάσχον), of change. Even if this cause is in *activity* (ἐνέργεια) while bringing about that change, so too is the body on which it exercises its agency (Clement, *Stromata* 8. 9. 26. 3 = LS 55C), and these two activities are simultaneous. We do not have here an example of a Stoic cause being succeeded by its effect.

Now, the Stoics do employ the notion of an 'antecedent cause' and explicitly invoke such causes in articulating the consequences of their thesis of fate:

> If all things come to be by fate, it does indeed follow that everything comes about by antecedent causes (*causis antepositis*)... (Cic. *Fat.* 41; cf. 9, 21, 23, 31; cf. Alex. *Fat.* 192. 8–11 = LS 55N2)

One might expect that these causes, at any rate, are succeeded by their effects. This expectation, however, is not borne out by the Stoics' own application of the notion. Chrysippus, for example, famously claims that the person who pushes a cylinder and thereby sets it in motion is the antecedent cause of its rolling (Cic

[9] Thus Salles 2005: 4–5. Bobzien 1998: 51 also argues, on different grounds, that the cause exists only at the time when the body is exercising its influence. Causes are not simply bodies, she claims, but 'bodies *while* and insofar as they actively produce... an effect in a body' (ibid.). While Bobzien is right to insist that Stoic causes are 'relatives' (e.g. the knife is a cause only in relation to the flesh and the effect it produces—cf. Sextus *M*. 9. 207, discussed by Bobzien 1998: 19), it does not follow from this that they exist only as long as their causal activity does. Father too is a relative notion, but fathers do not (typically) expire upon completion of their fathering activities. That X is a cause of Y even when it is not engaged in bringing about Y is an important presupposition of the Stoic view, if Frede is right that the Stoic doctrine of causes reflects their interest in ascribing responsibility (Frede 1980: 225).

[10] Frede 1980: 218–19 conflates the active nature ascribed to Stoic causes (that they ποιοῦσιν as opposed to πάσχουσιν) with their being in activity (ἐνέργεια). But the latter surely applies to the passive affection of bodies as well as to the agency of the bodies that affect them. Indeed, ἐνέργεια is explicitly classified as an incorporeal at Clement, *Stromata* 8. 9. 26. 3 (LS 55C), which would disqualify it from being a Stoic cause. To be sure, the Stoics do insist that a cause engages in activity (ἐνεργεῖν)—thus Frede cites Sextus, *PH* 3. 14 and Clement, *Strom*. 1. 17. 82. 3)—but I take this to be an additional requirement to that of agency (ποιεῖν). The distinctively 'active nature' of a Stoic cause consists in its being the agent, as opposed to the patient, of change.

Fat. 43).[11] This is to contrast the pusher's causal role with that of the cylinder's own 'force and nature' (*suapte vi et natura*)—presumably the *logos* internal to and constitutive of it. But while the push is antecedent to the rolling (or at any rate to most of the rolling[12]), it is the pusher who is identified as the cause, and he continues to exist throughout the cylinder's rolling. (It is presumably because the push is antecedent to (most of) the rolling that the pusher is classified as an antecedent cause.[13]) We will return to the topic of antecedent causes and their role in the chain of causes; the salient point for our present purposes is that, for the Stoics, not even antecedent causes need be succeeded by their effects.

On the modern conception of a causal chain, by contrast, causes *are* succeeded by their effects. This is one serious presumption against supposing that the Stoic conception of the chain of causes is the same as ours. Another objection to such an interpretation arises from the fact that in a causal chain of the modern variety the effects of causes function, in turn, as causes of further effects. This is at odds with two fundamental tenets of Stoic causal doctrine: first, that causes are bodies (Aetius, *Plac.* 1. 11. 5 = *SVF* 2. 340 = LS 55G) and, second, that the effects of causes are not bodies; in the canonical causal locution, 'Body A is the cause to Body B of incorporeal C' (Sextus *M* 9.211). From these two theses it follows that the effects of Stoic causes cannot themselves function as causes of further effects.[14] The modern notion of the causal chain, by contrast, presupposes that each link in the sequence is both cause of the next item and effect of the previous one. This reflects the modern understanding of causes as events that give rise to other events. On this conception of causality, the fundamental causal relation (famously problematized by Hume) is that of one event producing another. The Stoics, by contrast, conceive of causes as bodies that act on other bodies. The fundamental unanalysed causal relation, on this picture, is that between agent and patient: one body acting on another. An event may be the product of this interaction, but not being a body itself, it cannot act on any body to produce a further event.[15]

[11] The objection to which he is responding in making this example (Cic. *Fat.* 40) concerns antecedent causes, so it is safe to conclude that the example Chrysippus offers is of antecedent causation. Aulus Gellius' report of the example concurs in identifying the person who pushes (rather than the push) as the antecedent cause (*Noctes Atticae* 7. 2. 11).

[12] Presumably, both the hand and the cylinder move together during the push and then the latter continues on its way even after the former has stopped moving.

[13] Here I follow Bobzien (1998: 20–1).

[14] A point often noted: Long and Sedley 1987: i. 343; Sandbach 1989: 81 2; Hankinson 1996: 194; Bobzien 1998: 18, 50.

[15] Thus Frede writes, 'it is only in a very metaphorical sense that an event could be said to produce an effect' (1980: 218). Such considerations are presumably what the Stoics have in mind when they explain the incorporeal status of the effect by invoking its status as an activity (ἐνέργεια) of a body (Clement, *Strom.* 8. 9. 26. 3 = LS 55C, discussed above in n. 10). Interestingly, they do not thereby imply that the motion induced in a body is an incorporeal, if Galen is right that, for the Stoics, the motions of bodies are themselves bodies (*Qual. Inc.* 6 = *SVF* 2. 385).

This is not to say that, on the Stoic view, the relation between events and their successors (such as the activity of the scalpel and the resulting incision) is mysterious or inexplicable. We share with the Stoics the core idea that the scalpel cutting the flesh is an instance of causality. We differ from them, however, in the philosophical articulation we give to this idea—in particular, in how we use the vocabulary of cause and effect. While *we* might articulate the causal relation via 'the activity of the scalpel is the cause of the incision', the Stoics do so via 'the scalpel is the cause to the flesh of being cut'. While it is possible to translate the Stoic causal *locutions* into our own causal idiom *salva veritate*, we risk serious error if we interpret their causal *notions* (such as that of the *heirmos*) in the light of our own conception of causality. Authentically Stoic causes and effects cannot be related in the sort of iterated transitive relation envisaged by the modern interpretation. Whatever is a cause cannot be an effect, and whatever is an effect cannot be a cause.

Sextus is not our only source for this crucial piece of Stoic causal doctrine. Clement of Alexandria articulates explicitly what is implicit in the view that Sextus reports: a cause cannot have a cause:

> Causes are not causes *of* each other, but they are causes *to* each other. For the pre-existing condition of the spleen is the cause, not of the fever, but of the fever's coming about; and the pre-existing fever is the cause, not of the spleen, but of its condition's being intensified. In the same way . . . the stones in the vault are causes *to* each other of the predicate 'remaining', but they are not causes *of* each other. And the teacher and the pupil are causes *to* each other of the predicate 'making progress'.
>
> Causes are said to be causes *to each other* sometimes of the same [effect], as the merchant and retailer are the causes to each other of making a profit; but sometimes of different [effects], as in the case of the knife and the flesh; for the knife is the cause to the flesh of being cut, while the flesh is the cause to the knife of cutting. (Clement, *Strom.* 8. 9. 30. 1–30 = LS 55D; tr. Long and Sedley)

Clement's first example here might appear to license a modern interpretation of the Stoic causal chain. Even though Stoic causal notions do not permit us to say that the splenetic condition is the cause of the fever (both of these being bodies, in their view) they do permit us to say that the former is the cause of the latter's coming to be. This indicates that while the Stoics would object to causal locutions of the form

(II) Body A is the cause *of* body B,

they do not object to the paraphrase

(III) Body A is the cause of the coming-to-be of body B.

Thus even if, strictly speaking, Stoic causes, being bodies, do not have causes, they still can have causes for their suitably incorporeal coming-into-being. These observations might lead one to suppose that corresponding to every incorporeal effect is the coming-into-being of something suitably corporeal to function as a cause, and thus that causal notions of bona fide Stoic provenance can consistently be used to describe an iterated series of causes and effects: Body 1 (by acting on another body) causes the coming to be of Body 2, which (by acting on another body) causes the coming to be of Body 3, and so on.[16] Let us concede this for the sake of argument. At most it answers our second objection to construing the Stoic causal chain as an iterated sequence of cause and effect. It fails however, to provide any positive reason to suppose that this is what the Stoics have in mind as the *heirmos* of causes.

Indeed, if we look to our passage in Clement for guidance about how to interpret the Stoic causal chain,[17] rather than plunder it for materials out of which to construct an analogue of a modern causal chain, we find no evidence of a Stoic concern with iterated sequences of causes producing other causes. Instead we find a preoccupation with *causes acting on each other*—the disease that gives rise to a fever, which in turn affects the course of the disease; the stones in the arch that hold each other in place; the merchant and the retailer who execute a mutually profitable transaction. Such relations of mutual influence are in fact naturally captured by the metaphor of the εἱρμός, which we moderns are so quick to interpret as an asymmetrical temporal relation of succession. In a non-metaphorical necklace, each of the beads has an effect, more or less directly, on the others. Like the stones in an arch, the beads 'are causes *to* each other of the predicate "remaining"'. Might a system of reciprocal causal influence be what the Stoics have in mind as the εἱρμός of causes that they identify with fate?[18] Attention to the cosmological dimensions of the thesis of fate will give us good reason to suppose that they do.

STOIC COSMOLOGY

A fruitful route to discovering how the Stoics understood the claim that fate is an '*heirmos* of causes' is to pursue one of their alternative characterizations of fate, which identifies it with Zeus or god:[19] 'God, intelligence, fate, and Zeus are all one, and many other names are applied to him' (Diogenes Laertius 7.135; tr. LS 46B). One of these other names is *logos*:

They say that the very fate, nature and rationale (λόγος) in accordance with which the all is governed is god. It is present in all things which exist and happen, and in this way

[16] Long and Sedley 1987: i. 343 sketch a version of such a proposal.

[17] While Clement is not writing as a Stoic, he is evidently relying on Stoic notions. Indeed, his clarification in this passage of the sort of relations that obtain between causes addresses a question that naturally arises for the interpretation of the Stoic *heirmos*: in what relation do its constituent causes stand to one another?

[18] Bobzien identifies the 'interconnection' (ἐπιπλοκή) of causes with such a nexus (1998: 51, 95, 169, 219, 269), but distinguishes this from the εἱρμός, which she interprets as a linear sequence of causation (269). In response, I would point out that the metaphor of the *heirmos* does not demand a temporal interpretation (any more than the metaphor of interweaving does). Once we recognize this, it is easy to appreciate the *heirmos* as a metaphor for mutual interconnection.

[19] On fate as Zeus or god, see also Cicero, *ND* 1. 39 = LS 54B; Plut. *De Stoic. Rep.* 1056c = LS 55R.

uses the proper nature of all existing things for the government of the all. (Alex. *Fat.* 192. 25–8 = LS 55N4; tr. Long and Sedley)

Logos, we have seen, is the active causal principle that holds each thing together and is the cause of its activities. In an individual body it is the particular tenor (*hexis*), nature (*physis*), or soul (*psuchê*) that holds its material together and is responsible for its activities. According to Stoic cosmology, all such instances of *logos* are portions of a *pneuma* (breath) that pervades the universe (Aetius 1. 11. 5 = LS 55G).[20] Zeus, the *logos* of the world, is accordingly identified with this *pneuma*:[21]

The Stoics made god out to be intelligent, a designing fire which methodically proceeds towards creation of the world, and encompasses all the seminal principles (σπερματικοὶ λόγοι) according to which everything comes about according to fate, and a breath (πνεῦμα) pervading the whole world, which takes on different names owing to the alterations of the matter through which it passes. (Aetius 1. 7. 33 = LS 46A; tr. Long and Sedley)

The identification of Zeus with the cosmic *pneuma* means not simply that the divine *logos* is distributed throughout the cosmos into the individual tenors, natures, and souls of its constituents. The Stoics insist that the cosmos itself is a unified (ἡνωμένον) body.[22] And it is the *pneuma*, on a cosmic scale, that unifies it:

[Chrysippus] first assumes that the whole of substance is unified (ἡνῶσθαι) by a breath (πνεῦμα) which pervades it all, and by which the universe is sustained (συνέχεται) and stabilized (συμμένει) and made sympathetic with itself (συμπαθὲς . . . αὑτῷ). (Alex. *De Mixtione* 216. 14–16 = LS 48C; tr. Long and Sedley, slightly altered; cf. 223. 26–7, 227. 8–9 = *SVF* 2. 475)

Just as each individual body, plant, or animal in the cosmos has unity (ἑνότης) in virtue of its particular tenor, nature, or soul, the cosmos as a whole is unified (ἡνῶσθαι) by its all-pervading *pneuma*. In fact, the Stoics argue, the cosmic *pneuma* is a rational soul; the universe is a living rational animal with the divine *logos* as its governing principle.[23]

Since the Stoics identify fate, the chain of causes, with Zeus, and Zeus with the *pneuma* that unifies the cosmos, we might suppose that they identify fate with this *pneuma*. In fact, Chrysippus is reported to have written, in his work 'On the

[20] On the nature of the *pneuma*—about which there is some ambiguity in our sources—see Hahm 1985 as well as Cooper (Ch. 4 of this volume).

[21] See also Origen, *Cels.* 6. 71 (Borret ed. 358. 17–19), quoted by Bobzien 1998: 52 n. 95.

[22] Sextus, *M.* 9. 79–80; Alex. *Fat.* 191. 30–1, 192. 11–12 (LS 55N1, 2); DL 7. 140, 143; cf. also Proclus, *In Tim* (*SVF* 2. 533); Epictetus, *Diss.* 1. 14. 1; Alex. *Mixt.* 227. 8–9.

[23] Sextus, *M*.9.81–5 (*SVF* 2. 1013); cf. also Cic. *ND* 1. 37, 39; DL 7. 139, 142–3. Some of these texts refer to the cosmic *pneuma* not as soul but as 'nature' (*phusis*): Cleomedes, *Circul. Doctr.* 1. 1. 11 (Todd) = *SVF* 2. 546; [Plutarch], *De Fato* 11. 574d (*SVF* 2. 912). See also Alex. *Fat.* 192. 25–6; Cic. *ND* 2. 33 and *passim*. It seems particularly characteristic of Cicero's discussions of Stoic fate to identify the unifying force of the cosmos as *natura*.

Cosmos', that 'fate is a pneumatic power (δύναμιν πηευματικήν) governing the order of the whole' (Stob. 1. 79. 1 = *SVF* 2. 913) and Plotinus reports that those who posit 'a cause which penetrates all things, not only moving but also making each thing' call this principle 'fate' (εἱμαρμένη) (*Ennead* 3. 1. 2 = *SVF* 2. 946).[24] That the Stoics take the cosmos to be unified by fate is independently attested by Alexander of Aphrodisias, whose presentation of the Stoic thesis of fate makes it explicit that if the thesis is false 'the cosmos would be torn asunder and no longer be a unity, governed eternally by a single order and government' (Alex. *Fat.* 192. 11–12 = LS 55N2). The unifying function of fate is also reflected in the Stoics' frequent characterization of fate as an interweaving or interconnection:

Everything comes to be from a naturally linked binding together and interweaving (*omnia naturale conligatione conserte contexteque fiunt*). (Cic. *Fat.* 31)

[Chrysippus] says that what is fated is no different from what is necessitated . . . according to a conjoined interweaving (ἐπιπλοκή) of things. (Aetius, *Plac.* 1. 27. 2 = *SVF* 2. 916)

Such a characterization is naturally understood as an explanation of the metaphor of fate as an *heirmos*.[25] Plotinus evidently assumes so when he describes the (presumably Stoic) 'principle (ἀρχήν) that interweaves (ἐπιπλέκουσαν) and as it were strings together (συνείρουσαν) all things with each other' (*Enn.* 3. 1. 7 = *SVF* 2. 986). Aetius confirms that the items thus 'interwoven' are the individual causes and *logoi* encompassed by Zeus:

The primary fire [sc. Zeus] is as it were the seed which possesses the principles (λόγους) and causes (αἰτίας) of what has come to be, is coming to be, or will come to be. The interweaving (ἐπιπλοκή) and following of these is fate . . . (Aristocles in Eus. *Praep. Evang.* 15. 14. 2 = LS 46G; tr. Long and Sedley; cf. Plotinus, *Ennead* 3. 1. 4 = *SVF* 934; 3. 1. 2 = *SVF* 2. 946)

There is thus considerable overlap between the doctrine that fate is a chain of causes, and the doctrine that the divine *pneuma* connects all the causes in the cosmos into a unity. This gives us reason to expect that the connection between causes involved in the Stoic chain will be that by which *pneuma* unifies the various elements in the cosmos. So let us return to the *pneuma*, and see how it unifies the disparate causes that it encompasses.

SYMPATHY

Many texts indicate that the disparate bodies in the cosmos are unified in virtue of a relation that the Stoics dub 'sympathy' (συμπάθεια). While the doctrine

[24] On the relation between fate and *pneuma*, see Bobzien 1998: 45–7.

[25] Bobzien prefers to distinguish the metaphor of interweaving from that of the *heirmos*, attributing the former to Chrysippus and the latter to later Stoics (1998: 50). But once we abandon the assumption that the *heirmos* is a transitive temporal sequence (as Bobzien explicates it: pp. 51, 95; cf. 269), it is easy to see how it too is a kind of interconnection.

of cosmic sympathy has its most famous proponent in Posidonius, it goes back at least as far as Chrysippus. For example, in a passage we have already seen, Alexander of Aphrodisias attributes to Chrysippus the view that the divine *pneuma* unifying the cosmos makes it 'sympathetic' with itself (συμπαθὲς . . . αὑτῷ) (Alex. *De Mixtione* 216. 16 = LS 48C). A few pages later, when enumerating the Stoic doctrines that he says depend ultimately on their account of mixture, Alexander refers to their view of 'the unity (ἕνωσις) and mutual sympathy of the whole with itself (συμπάθεια πρὸς αὑτό)' (*Mixt.* 227. 8–9 = *SVF* 2. 475).

Other reports indicate that whether the cosmos is governed by a single nature is closely connected, by the Stoics, to the question of whether it is has internal sympathies:

> The universe is governed (διοικεῖσθαι) by nature (φύσει), agreeing with itself and having sympathy with itself (σύμπνουν καὶ συμπαθῆ αὐτὸν αὑτῷ ὄντα). (Ps.-Plutarch, *De Fato* 11. 574d = *SVF* 2. 912)

> If the substance of the whole were not naturally suffused throughout the whole, neither would the universe be able to be held together (συνέχεσθαι) and governed (διοικεῖσθαι) by nature (ὑπὸ φύσεως), nor would there be any sympathy (συμπάθεια) among its parts. (Cleomedes, *Caelestia* 1. 1. 69–71 (Todd) = *SVF* 2. 546)

Shortly before this passage, Cleomedes points to the 'sympathy' between its parts as one of the signs that the universe is governed by a single nature (*Caelestia* 1. 1. 11–13 (Todd) = *SVF* 2. 534).

Indeed, in reports of Stoic doctrine, the question of the unity of the cosmos and that of its internal sympathy regularly come in the same breath, as in Epictetus, *Diss.* 1. 14. 1–2, where the question, 'Is the whole a unity?' (ἡνῶσθαι τὰ πάντα) is immediately followed by 'Are terrestrial things in sympathy with the heavens?' The Stoic-leaning Philo of Alexandria cites approvingly those who unify the universe 'by showing the sympathy and community of its parts with each other'.[26] Cicero too points out that the proponents of cosmic unity appeal to natural sympathy in support of their claim (Cic. *De Div.* 2. 33–4). Marcus Aurelius, when discussing the affinity rational creatures should have for each other, indicates that unity (ἕνωσις) is to be achieved through sympathy (*Meditations* 9. 9. 2).

Sextus Empiricus lists several examples of such sympathies in his report of Stoic arguments for the unity of the cosmos:

> [78] Some bodies are unified (ἡνωμένα), others are formed of things joined together (ἐκ συναπτομένων), others of separate things (ἐκ διεστώτων). Unified bodies are those controlled by a single tenor (ἕξις), like plants and animals. Bodies made of parts joined together . . . are for example cables, turrets, and ships. Those made of separate things . . . are like armies and flocks and choruses.

[26] Philo, *De Migr. Abr.* 179: τῇ τῶν μερῶν πρὸς ἄλληλα κοινωνίᾳ καὶ συμπαθείᾳ/. Marcus Aurelius too mentions sympathy as a reason to attribute an organizing principle to the cosmos (*Meditations* 4. 27). Both passages are discussed by Laurand 2005: 522–3.

[79] Since the universe is a body, it must be either unified, or made from conjoined or separate parts. We can prove that it is not made of conjoined or separate parts from its various sympathies (ἐκ τῶν περὶ αὐτὸν συμπαθειῶν). For **according to the waxings and wanings of the moon, many animals on land and sea decline and grow, and the tide ebbs and flows in certain parts of the sea. In the same way, in accordance with certain risings and settings of the stars all sorts of changes take place in the surrounding atmosphere**. . . These things make it clear that the universe is a unified body.

[80] For in bodies formed of conjoined or separate things, the parts do not 'sympathize' with each other (οὐ συμπάσχει τὰ μέρη ἀλλήλοις). For example, if all the soldiers in an army perish, the sole survivor does not suffer anything passed on to him from the others (κατὰ διάδοσιν). But in the case of unified things there is a kind of sympathy (συμπάθειά τις); for example, when the finger is cut, the whole body shares its condition. So the universe is a unified body. (Sextus Empiricus, *M*. 9. 78–9 = *SVF* 2. 1013; tr. adapted from Bury)

The sympathies here cited in 9. 79 involve, at the very least, regular correlations between celestial and terrestrial phenomena. Other reports make it clear that the Stoics take these correlations to be causal. Seneca, writing as a Stoic, says so explicitly: the course of the tides is due to the influence of the moon (*Prov*. 1. 2. 4). Similarly, Cicero confirms that, for the Stoics, the tides are 'governed (*gubernantur*) by the motion of the moon' (Cic. *Div*. 2. 34; quoted in full below). And the Stoic spokesman Balbus in Cicero's *De Natura Deorum* 2. 50 indicates that the plant cycles are due to celestial cycles: 'Many things flow and issue from [the moon], which nourish animals and cause them to grow, and cause plants to flourish, and attain maturity'.[27]

Although the most famous putative examples of cosmic sympathy are Posidonius' invocations of celestial influence over the terrestrial (Augustine, *Civ. Dei* 5. 2), the scope of the causal sympathies invoked by the Stoics is much broader. For example, Cicero's criticisms of the Stoic belief in divination (whose efficacy they claimed rested on underlying causal sympathies: *Div*. 2. 124, 142) shows that some of the sympathies that they invoke are entirely within the sublunary realm. When discussing divination from entrails (in which, for example, fissures in the liver of a sacrificial animal might be taken to predict the discovery of buried treasure), Cicero writes:

What connection (*cognationem*) could these [e.g. fissures in the liver] have with the nature of things? Even if nature is joined together (*iuncta*) and continuous (*continens*; sc. συνεχομένη) by a single harmony (*uno consensu*), which is the view of the physicists, especially those who claim that everything is a unity (*qui omne quod esset unum esse*

[27] Here I disagree with Laurand, who claims that the bodies involved in cosmic sympathy do not act on each other, the divine *pneuma* being the only cause of the correlations (pp. 525–30). The divine *pneuma*, however, does not exert causal efficacy except through the causality of the constituents of the cosmos. The only exception would be the period of the conflagration when these bodies have been consumed (Origen, *Cels*. 4. 14 = LS 46H), but συμπάθεια is not a feature of this part of the cosmic cycle.

dixerunt), what possible connection (*coniunctum*) could the cosmos have to the discovery of treasure? If an increase in my fortune is indicated by the entrails, and nature brings this about, then the entrails must be connected to the universe (*coniuncta mundo*), and the nature of things must be in control of my fortune. Are not the physicists ashamed to claim such things?

Let us suppose there is some connection (*contagio*) between things in nature, which I am prepared to concede—for the Stoics collect many examples of these:

i. the livers of mice are supposed to get larger in the winter;
ii. the dry pennyroyal blooms and its seed pods burst on the shortest day;
iii. apple seeds enclosed within the fruit reverse their orientation [sc. on that day];
iv. when some strings of a lyre are struck, others resonate;
v. oysters and all shellfish increase and decrease in size with the waxing and waning of the moon;
vi. trees are opportunely felled when the moon is waxing since then the sap is dry.

[34] Why should I add the examples of

vii. the tides of springs and seas, whose ebb and flow are governed by the motion of the moon?

Innumerable examples can be given to illustrate the natural connection (*cognatio naturalis*) between distinct things. Nonetheless, these have no bearing on my present contention: if there is a fissure in some liver, does this indicate wealth? By what natural connection (*coniunctione naturae*) and as it were harmony and mutual agreement (*concentu atque consensu*), which the Greeks call συμπάθεια, can there be coordination (*convenire*) between the fissure in a liver and my small fortune, or between my small profit and heaven, the earth, and the nature of things? (Cicero, *Div.* 2. 33–4 = *SVF* 2. 1211)

While most of the examples of sympathies conceded here by Cicero are instances of celestial influence over the terrestrial, the fourth example (resonance of strings within a lyre) concerns causal influence between terrestrial entities, as does the contested example of divination from entrails—the alleged sympathy between the sacrificial entrails and the discovery of treasure. Similarly terrestrial (or at any rate sublunary) is Chrysippus' citation of the atmospheric influence on human bodily and intellectual temperament:

[7] . . . Let us reply to [Chrysippus] about the connection between things itself (*de ipsa rerum contagione*) . . . We can see how great a difference there is in the natures of different regions; some are healthy, others are plague-ridden, in some people are phlegmatic and as it were overflowing with moisture, in others they are dried up and parched; and there are many other things which differ greatly between one place and another. At Athens the air is thin, and for this reason the people of Attica too are thought to be more sharp-witted, while at Thebes it is dense, and for this reason the Thebans are stupid but strong. However, that thin air will not make anyone listen to Zeno or Arcesilaus or Theophrastus; nor will that thick air make anyone seek victory at Nemea rather than at the Isthmus . . . [8] What influence can the nature of the place have over our walking in Pompey's portico rather than in the Campus Martius, with you rather than with someone else, on the Ides rather than on the Kalends? (Cic. *Fat.* 7–8; tr. Sharples)

Further examples of putative terrestrial sympathies include cases where one item influences or causes another—for example, a sexual dream that has an effect on a man's bladder disease (Cic. *Div.* 2. 143), or the relation between a disease and it symptoms (*Div.* 2. 142–3). They also include cases where the phenomena in question are linked by a common cause or other mediating condition: the tree whose blooming signals the right time to plough (Cic. *Div.* 1. 16); the coordination of the tides in the Atlantic and Mediterranean, or in a harbour and its sea;[28] the heron's flight and the coming tempest that it signals (*Div.* 1. 14).

In addition to celestial-terrestrial sympathies and entirely terrestrial sympathies, some Stoic examples of putative sympathies concern causal influences entirely within the celestial world—for example, the *concentus*[29] of the planets with the sun and moon mentioned by the Stoic spokesman Balbus at *De Natura Deorum* 2. 119. Still others show that the influence of heavens on earth is reciprocated:

> The stars are of a fiery nature, and for that reason are nourished by the vapours of the land, sea, and waters, which the sun raises up out of the fields and waters that it warms. When they have been nourished and renewed by these, the stars and the whole aether pour them back down, and then draw them back again from the same source . . . (Cic. *ND* 2. 118)

The heavens produce rain, which moistens the earth, and in turn the vapours rising from the moistened earth nourish the stars.

Our textual evidence thus shows that the 'sympathies' cited by the Stoics encompass a considerable diversity of relations of causal influence, not necessarily direct, holding between a wide range of the elements in the universe. The broad scope of these sympathies should not be surprising, since, as we have seen, the Stoics insist that the sort of unity characteristic of the cosmos is an instance of the complex system of interdependence displayed by the different parts of a living organism (Sextus, *M.* 9. 81–5; cf. DL 7. 139, 142–3). Indeed, Plotinus reports a Stoic view according to which the fate that governs the universe is like the nature governing a plant, whose parts are in mutual interaction (*πρὸς ἄλληλα συμπλοκήν, ποίησίν τε καὶ πεῖσιν*) with each other (Plotinus *Enn.* 3. 1. 4. 5–10). In the same vein, the Stoic speaker in Cicero's *De Natura Deorum*, immediately after insisting that the unity of the cosmos approximates that of an organism (2. 82), goes on to give a detailed illustration of the complex system of mutual influence he has in mind:

> If those things that are sustained by roots in the earth live and flourish due to the art of nature, certainly the earth herself is sustained by the same force—she who, when impregnated with seeds gives birth in abundance to all things out of herself and nourishes and causes to grow the roots in her embrace. She herself is nourished in turn by the upper

[28] Priscianus Lydus, *Solutiones ad Chosoem* 6 (p. 69. 19–76. 20 Bywater ed.) = Posidonius Fragment 219 (Edelstein and Kidd).

[29] At *Div.* 2. 34, Cicero explicitly identifies the Latin *concentus* as the rendering of the Greek *συμπάθεια*.

and outer natures, while her exhalations nourish the air, the aether, and all the higher things. Thus if the earth is sustained by nature, the same principle holds for the rest of the universe: for if roots depend on the earth, animals are sustained by breathing in air, and the air itself is implicated in our seeing, hearing, and making sounds, since none of these things can be done without it. And it even moves along with us: whenever we go anywhere or make a movement it seems as if to part and make way for us. (Cic. *ND* 2. 83)

Just as the heavens and the earth influence each other, so do the air and the various constituents of the world. Cicero's source makes a point of invoking a wide range of mutual causal influence. The stars and other 'higher and outer' natures are part of a causal nexus with the earth, the air, and animals and plants; all directly or indirectly affect or are affected by the others. The celestial bodies affect air directly (2. 118) and, via air, they influence animal life, perceptions, and activity (2. 83). The animals and plants in turn affect air directly, and so indirectly affect the celestial bodies and the earth that is nourished by it. In this way the universe approximates the system of mutual influence in a natural organism.

What light does this conception of the universe shed on the doctrine that fate is a *heirmos* of causes? We have seen that, according to the Stoics, this complex system of mutual causal influence, which they dub 'sympathy', makes a unity of the disparate parts of the cosmos. In addition, they maintain that it is through such sympathies that *pneuma* unifies the cosmos, and that fate, like the divine *pneuma*, links together all the causes in the world. Therefore, it seems reasonable to conclude that precisely this system of mutual influence is the 'chain of causes' that the Stoics identify with fate. In fact, there is direct ancient testimony linking fate and causal sympathy. Pseudo-Plutarch tells us that one of the main points in Chrysippus' discussion of fate is the claim that 'the universe is governed (διοικεῖσθαι) by nature (φύσει), agreeing with itself and having *sympathy* with itself (σύμπνουν καὶ συμπαθῆ αὐτὸν αὑτῷ ὄντα)' ([Plutarch], *De Fato* 11. 574d = *SVF* 2. 912). The evidence therefore converges in support of the conclusion that the Stoic chain, or string, of causes is the complex system of mutual causal influence among the bodies in the cosmos.[30]

ANTECEDENT CAUSES, FATE, AND SYMPATHY

The precise scope of this system of interaction is not evident. Whether, for example, every body in the cosmos is supposed to be directly affected by every other (the maximal possibility) is unclear from our sources. The organic conception of the cosmos, for example, does not require it; within an organism there can be independent subsystems that are only indirectly related, via their

[30] Bobzien notes a connection between the nexus of causes and sympathy (1998: 169, 219) but does not develop the point.

causal relation with other parts. One dimension of the system of sympathy that can, however, be inferred with confidence, is that it extends seamlessly from the past into the present and the future.

Alexander's report of the Stoics makes this clear. After reporting their doctrine that the cosmos is unified by a principle that operates 'in the manner of an orderly chain' (κατὰ εἱρμόν τινα καὶ τάξιν: *Fat.* 192. 1), he elaborates:

[A] The prior things become causes to those that come to be after them (τοῖς μετὰ ταῦτα γινομένοις), and in this way all things are bound to each other (συνδεομένων ἀλλήλοις). [B] And it never happens that something comes to be in the world that does not have something else that follows it and is bound fast to it as to a cause. [C] Nor is any of the things that come to be later (τῶν ἐπιγινομένων) able to be severed (ἀπολελύσθαι) from what came before, so as not to follow one of them (τινι ἐξ αὐτῶν ἀκολουθεῖν) as if bound fast to it (ὥσπερ συνδεόμενον). [D] Rather, for everything that comes to be there is something else that follows it (ἐπακολουθεῖν), being bound to it of necessity as to a cause. [E] And everything that comes to be has something prior to which it is bound as to a cause. (Alex. *Fat.* 192. 2–8 = LS 55N; tr. revised from LS)

This passage testifies to a Stoic preoccupation with the fact that the chain of causes is temporally extended. An unwary reader might be tempted, especially by all the invocations of 'following' (ἀκολουθεῖν), to suppose that Alexander is here describing a sequence of events each of which is cause of its successor and effect of its predecessor.[31] However, although 'following' (ἀκολουθία) evidently is an important feature of the Stoic εἱρμός (Alexander, *Mantissa* 185. 1–5; Aristocles in Eusebius 14 = LS 46G; Gellius 7. 2. 3 = LS 55K), it need not indicate temporal succession. In its non-logical uses, ἀκολουθεῖν (or, less often, ἕπεσθαι)[32] describes the causal influence of one entity on another—as when I follow the nurse into the doctor's office. Plutarch describes Zeno as following (ἀκολουθεῖν) the Peripatetics in his account of happiness (*Comm. Not.* 23. 1, 1069–70; *SVF* 1. 183).[33] Indeed, in the common Stoic examples of cosmic sympathy, it is natural to say that the tides follow the moon.

The two causal theses expressed in our passage fit very well with the thesis of cosmic sympathy.[34] First of all, we are told, everything in the cosmos is a cause 'to' some later thing ([A], [B], and [D]). Second, everything has some earlier thing 'to which it is bound as to a cause' ([C], [E]). Alexander here employs the

[31] Indeed, Alexander cites temporally successive events (e.g. night following day) as counter-examples to the Stoic claims just quoted (*Fat.* 194. 27–195. 1). I agree with Hankinson (1996: 194–5) that this is a polemical misconstrual on Alexander's part.

[32] Logical uses: '"It is light" follows "It is day"', but not the other way around (DL 7. 73; *SVF* 2. 215); cf. Sextus, *M.* 8. 276.

[33] Following the laws, or being enslaved to a leader, are also unproblematic examples of 'following' (ἀκολουθεῖν), as is the influence of one part of the mind on another (e.g. appetite following reason). Following laws or principles: Arius Didymus in Eus. *Praep. Evang.* 15 (*SVF* 2. 528); Stob. *Ecl.* 2. 7 (*SVF* 3. 613); slave following master: Plotinus, *Enn.* 3. 1. 2 (*SVF* 2. 946); appetite following reason: Plut. *Vir. Mor.* 9. 449c (*SVF* 3. 384).

[34] Hankinson (1996: 195) agrees that the doctrine of causal sympathy is reflected in this passage.

causal dative ('x is the cause to y'), which in the canonical Stoic causal locution expresses the influence of one body over another. The two theses thus affirm that every body in the cosmos affects and is affected by at least one other body. The successive iterations of this relation of influence yield an extended connection of 'cause to cause'—precisely what the Stoic *heirmos* is supposed to be.

Alexander's report emphasizes that these causal connections occur across time: the causes in each case are in some sense 'prior' to their effects. Such temporal qualifications cannot mean that the influencing body is succeeded by the body it affects, since one body cannot affect another except at a time when they both exist. Presumably, therefore, these are intended as instances of what the Stoics call antecedent causation. As in the case of the rolling cylinder of Cicero, *Fat.* 41, discussed above, the 'antecedent cause' of the rolling is the person who sets the cylinder in motion by giving it a push (and it is presumably the temporal priority of the push to the rolling that underlies the classification of the pusher as 'antecedent cause').

Antecedent causality, so understood, is the influence of one cause on another. In the present example the causes in question are the person (whose soul gives rise to his various activities) and the cylinder (whose own 'force and nature' gives rise to its simple activity: rolling). Neither the cylinder will roll, nor the person push, Chrysippus tells us, without being first affected by an external cause (Cic. *Fat.* 42). The cylinder's actually rolling at a given time requires an interaction between these two distinct causes, one of them acting as a 'cause to' the other. Antecedent causation, therefore, is the mechanism whereby the activities of individual causes are integrated into the fabric of fate. To claim that something has an antecedent cause is to affirm that its cause is part of the causal nexus.[35]

Here it may be useful to distinguish between explaining an event and saying that it is fated. To say that the cylinder rolls because of its nature (or that the plant grows because of its nature) is simply to state the cause of the rolling (or growing). This is to explain it. To say that the rolling (or growing) is fated, by contrast, is to say that the cause cited in the explanation was itself acted on by an antecedent cause which precipitated its causal activity.

Having noted this connection between antecedent causation and the doctrine of causal influence, we should not find it problematic that Chrysippus is reported to have described fate as a system of antecedent causes.[36] This does not imply that he thinks some causes are exempt from the influence of fate. To see why not, we need only consider the first of the causal theses insisted on in Alexander's

[35] On the relation between fate and antecedent causes, I agree with Hankinson 1996: 196–9. Where I diverge from him is in concluding that antecedent causation, construed as an external body precipitating the activity of another body, constitutes the connection between causes in the Stoic εἱρμός. Hankinson, by contrast, locates the connection in the continuity of the processes generated by the *spermatikoi logoi*.

[36] Cic. *Fat.* 41; cf. Plut. *De Stoic. Rep.* 1056b–c = LS 55R; cf. Cic. *Top.* 59. For discussion, see Hankinson 1996: 198; Frede 1980: 240; Bobzien 1998: 301–14.

report above, which amounts to the claim that everything is an antecedent cause. (Indeed, the conjunction of the two theses in the passage amounts to the doctrine that everything both is and has an antecedent cause.) To quantify over antecedent causes, therefore, is to quantify over all causes. Even construed as a system of antecedent causation, fate leaves no cause outside its scope.

Let me conclude by sketching how the chain of causes applies in the case of human action. According to the thesis of fate, we have seen, such action arises from a nexus of causes. This is, of course, not to say that it is governed by the moon. On the contrary, the Stoics claim, our actions are governed by our souls. A person's character (a particular tension of her soul) causes her action in the paradigmatic way in which the *logos* of a body is the cause of its activities.[37] But any particular action also has an antecedent cause, the Stoics insist (and in so insisting subsume the action under the thesis of fate). The antecedent cause is an external body that makes an impression on the soul. It may, for example, be an untended valuable that elicits in an observer of a thievish disposition the impulse to steal. The soul, on this picture, is both agent (in causing the action) and patient (in being acted on by the external object). Thus we have an instance of the relation of 'cause to cause' that makes up the fabric of fate. To claim that the action is fated, on this view, is to insist that this cause (the character) does not operate in isolation from and unaffected by other causes. That is, even when we exercise our causality as agents, we are subject to the influence of other causes.

Since it is the influence of external causes that enmesh our actions in the nexus of fate, it is not surprising that ancient critics of Stoic fate allege that it has the unwelcome consequence that we are 'compelled by external causes'.[38] In contrast with modern expressions of incompatibilism, this worry does not focus on causal determination *per se*, but on external influence.[39] The Stoics' critics are not objecting that our actions would be *determined* by external factors (for it is the combination of the external and internal causes that determine actions).[40]

[37] Thus Zeno says that temperate activity (σωφρονεῖν) is caused by the virtue of temperance (σωφροσύνη) (Stob. *Ecl.* 1. 138. 20 = LS 55A3). Frede, by contrast (1980: 245–6), like Long and Sedley, construes σωφρονεῖν as the condition of 'being temperate'; thus he concludes that its relation to the cause, the virtue of temperance, is one of 'conceptual necessity'. Instead, I take σωφρονεῖν to refer to the activities that display temperance; on this reading, Zeno is invoking the standard causal role of *logos* (as it is manifested in the soul) producing an activity.

[38] Plut. *De Stoic. Rep.* 23. 1045b; cf. Alex. *Mantissa* 174. 28–30; *Fat.* 185. 18–19. Epicurus expresses a similar worry in the letter to Menoeceus, when he refers to the 'fate of the philosophers' as a 'mistress' (δεσπότις) to which we would be 'enslaved' (δουλεύειν) (DL 10. 133–4: cf. Plotinus, *Enn.* 3. 1. 2). Here I disagree with Boys-Stone 1996: 81.

[39] It is easy to slip into the assumption that the 'antecedent causes' in question causally determine the effect (thus Frede 1980: 234–5). But as Chrysippus' examples (Cic. *Fat.* 42–3) show, and as Hankinson has noted (1996: 199), the so-called antecedent causes are external bodies that precipitate the activity of the primary cause (as in the case of the person who pushes the cylinder and thus sets into motion).

[40] Boys-Stone 1996: 81, by contrast, takes the objection to be that external causes determine our actions, and thus infers that it is not legitimately raised against the Stoic thesis of fate. But he does so on the assumption that the Stoic chain of causes is a sequence of determining conditions.

Nor are they worried about alternative possibilities.[41] Rather, in keeping with the Stoic conception of causality as an interaction between agent and patient, they worry that human agents would be 'patients' relative to the agency of external forces. It is beyond the scope of this chapter to explore the Stoic response to this objection. For our present purposes, it suffices to note that the objection focuses on precisely the aspect of fate that, I have argued, is captured in the metaphor of the *heirmos*. For a cause to be part of the Stoic *heirmos* is for it to be acted on by other causes.

REFERENCES

Bobzien, S. (1998) *Determinism and Freedom in Stoic Philosophy* (Oxford: OUP).

—— (1999) 'Chrysippus' Theory of Causes', in K. Ierodiakonou (ed.), *Topics in Stoic Philosophy* (Oxford: OUP), 196–242.

Borret, M. (1967–76) *Origène, Contre Celse: Introduction, Texte Critique, Traduction et Notes* (Paris: Éditions du Cerf).

Boys-Stones, G. (1996) 'The Ἐπελευστική Δύναμις in Aristo's Psychology of Action', *Phronesis,* 41: 75–94.

Bury, R. G. (1968) *Sextus Empiricus,* with an English translation (4 vols. Loeb Classical Library. Cambridge, Mass.: Harvard UP).

Edelstein, L., and Kidd, I. G. (1972) *Posidonius* (Cambridge: CUP).

Frede, M. (1980) 'The Original Notion of Cause', in M. Schofield, M. Burnyeat, and J. Barnes (eds.), *Doubt and Dogmatism* (Oxford: Clarendon Press), 217–49.

Hahm, D. (1985) 'The Stoic Theory of Change', *Southern Journal of Philosophy,* 23 suppl.: 39–56.

Hankinson, R. J. (1996) 'Cicero's Rope', in K. A. Algra, P. W. van der Horst, and D. T. Runia (eds.), *Polyhistor: Studies in the History and Historiography of Ancient Philosophy* (Leiden: Brill), 185–205.

—— (1998) *Cause and Explanation in Ancient Greek Thought* (Oxford: OUP).

Laurand, V. (2005) 'La Sympathie universelle: Union et séparation', *Revue de Métaphysique et de Morale,* 4: 517–35.

Long, A. A., and Sedley, D. N. (1987) *The Hellenistic Philosophers* (2 vols. Cambridge: CUP).

Meyer, S. S. (1999) 'Fate, Fatalism, and Agency in Stoicism', *Social Philosophy and Policy,* 16(2): 250–73.

—— (2003) 'Review of Susanne Bobzien, Determinism and Freedom in Stoic Philosophy', *Philosophical Review,* 112(3): 405–9.

Pease, A. S. (1920–3) 'M. Tulli Ciceronis De Divinatione', *University of Illinois Studies in Language and Literature* (Repr. New York: Arno Press, 1979).

Salles, R. (2005) *The Stoics on Determinism and Compatibilism* (Aldershot: Ashgate).

[41] A point developed at length by Bobzien 1998: 276–90 and discussed in my review (Meyer 2003). Salles, by contrast, argues that the existence of alternative possibilities is at issue in the ancient debate (Salles 2005: 51–68, 78–81).

Sandbach, F. H. (1989). *The Stoics*, 2nd edn (London: Duckworth; Indianapolis: Hackett).

Sharples, R. W. (1983) *Alexander of Aphrodisias on Fate: Text, Translation and Commentary* (London: Duckworth).

——(1991) *Cicero, on Fate and Boethius, The Consolation of Philosophy,* ed. with an introduction, translations, and commentaries (Warminster: Aris & Phillips).

Todd, R. B. (ed.) (1990) *Cleomedes, Caelestia* (Leipzig: Teubner).

——and Bowen, A. C. (2004) *Cleomedes' Lectures on Astronomy: A Translation of The Heavens with an Introduction and Commentary* (Berkeley, Calif.: University of California Press).

Valk, M. van der (ed.) (1971–87) *Eustathii Thessalonicensis Commentarii Ad Homeri Iliadem Pertinentes* (Leiden: Brill).

PART II

ELEMENTS, COSMOGONY, AND CONFLAGRATION

4

Chrysippus on Physical Elements

John M. Cooper

My ultimate purpose here is to examine, discuss, and interpret a difficult excerpt in Stobaeus' fifth-century AD anthology. It alleges to report—uniquely, it appears—a distinction Chrysippus drew between three different applications of the term στοιχεῖον or element (i.e. physical element).[1] Stobaeus lists this passage as giving opinions specifically of Chrysippus 'about the elements out of substance' (περὶ τῶν ἐκ τῆς οὐσίας στοιχείων), though in holding them he says Chrysippus was following Zeno, the leader of his sect.[2] Hermann Diels identified this selection as an excerpt (his fr. 21) from Arius Didymus' late first-century BC *Epitome of Physical Doctrines*.[3] I print a translation below, with the text in an Appendix, as it is given in von Arnim. The text is not without its problems, and I indicate in footnotes to the text which of the principal editors' textual interventions I accept and follow in my translation. Whether this text presents a single, continuous excerpt from Arius Didymus, or instead some compilation of Stobaeus (or an earlier anthologist whose work Stobaeus employed) from dispersed passages of Arius (or even of Arius and another source),[4] I propose to treat it at face value, as giving a unified report on

[1] *Ioannis Stobaei Anthologii Libri Duo Priores*, ed. C. Wachsmuth (Berlin: Weidmann, 1884), book 1, ch. 16^c, i. 129–30.

[2] The phrase ἐκ τῆς οὐσίας here is unexpected. Usually a substance is said to be, or come to be, ἐκ its elements (as in the official Stoic definition of an element, Diogenes Laertius 7. 136, cited below p. 105); the usage here seems to reverse the expected order of derivation. I take it that here we are to understand the preposition as expressing the fact that in considering a substance's elements we engage in an act of abstraction; substances present themselves as constructed wholes, and it is only by analysis 'from' them that we arrived at a conception of their elements.

[3] H. Diels, *Doxographi Graeci* (Berlin: De Gruyter, 1965; orig. publ. 1879). The passage is included in H. von Arnim, *Stoicorum Veterum Fragmenta* (Leipzig: Teubner, 1903), ii, as fragment 413; A. A. Long and D. N. Sedley, *The Hellenistic Philosophers* (Cambridge: CUP, 1987), translate most of the passage as their text 47A.

[4] The passage does have some abrupt transitions, at λέγεσθαι (line 13), τριχῶς (14), and γεγονέναι (21) (for this numbering of the lines, see the Appendix). For a useful summary of what we know or can reasonably conjecture about Stobaeus' methods of composition, see D. E. Hahm, 'The Ethical Doxography of Arius Didymus', in W. Haase (ed.), *Aufstieg und Niedergang der Römischen Welt*, 36/4 (Berlin and New York: De Gruyter, 1990), at pp. 2938–43.

Chrysippus' cosmological-cosmogonical theory.[5] The philosophically rich and interesting cosmological ideas that, as we will see, are brought to light by careful interpretation of the passage so construed, lend powerful support to this way of taking it.

Let me begin with a bare summary of the passage. Arius reports that Chrysippus endorsed, first, the use of the term 'element' in application to the four traditional Greek basic materials, fire, air, water, and earth, which he takes to be the exclusive basis in the actual world of all other materials—compounds of all kinds—and for the material constitution of all material objects. As Arius paraphrases or quotes Chrysippus himself to say, on this first usage of the term, fire ranks alongside three other elements, all four of them elements equally and on a par. In a second usage, he says, there is only one element, to which he again gives the same name of 'fire' (πῦρ): in this second usage, fire is called an, or the, element *par excellence* (κατ' ἐξοχήν) and self-sufficiently (αὐτοτελῶς). Having mentioned and discussed briefly in the first part of the excerpt apparently just these two usages (I will come back later to examine his discussion), Arius announces, as if resumptively (line 14), 'So, according to Chrysippus, element is spoken of in three ways.' In concluding the excerpt we then see Arius indeed explicitly distinguishing three applications of the term 'element'. The first two in the order of presentation here, at the end of the passage, at least appear to be, and I think in fact are, in reverse order, the two I have just briefly characterized: one according to which the four traditional elements are elements, plus another according to which only fire (or something called fire) is an element. Scholars have, however, not found the third application (κατὰ τρίτον λόγον λέγεται στοιχεῖον εἶναι . . ., line 19), as explained by Arius, easy to grasp, and its relation to anything that has preceded in the excerpt are difficult to make out. One thing that Chrysippus' (or Arius') presentation seems to make clear, though, is that in this third usage, too, as in the second, there is only one element—which, however, is not specified.

In the only two significant discussions I have found of this passage both authors fail to see in it even any formulation of the third application.[6] Josiah Gould thinks what we get instead is a redescription of that 'fire' which is the sole

[5] It is included as the last entry in the 'chapter' (X Wachsmuth) headed as 'On Principles and Elements of the All', preceded by excerpts of a cosmological sort from cosmogonical poets, pre-Socratic philosophers, Plato, Aristotle, and Epicurus, among others.

[6] Josiah Gould, *The Philosophy of Chrysippus* (Leiden: Brill, 1970), 119–20; M. Lapidge, 'Ἀρχαί and Στοιχεῖα: A Problem in Stoic Cosmology', *Phronesis* 18 (1973), 240–78. To these one should add H. Diels, *Elementum* (Leipzig: Teubner, 1899), 38–9. Diels says that what I am calling the third usage is in fact identical with the second: he condescendingly supposes that this escaped Arius' notice in this hasty compilation. Diels thinks that at the end of the excerpt, in lines 21–4 of my text (see Appendix), Arius provides in a little appendix three further 'definitions' of an element (the MSS text is corrupt in at least two places). On my view (see below) this passage continues the exposition of my third usage, i.e. application. Diels's speaking throughout of Chrysippean 'definitions' has misled him. What is in question in the passage is uses, in the sense of applications, of the term; such 'definitions' as may be found in lines 21–4 are provided by way of explication of the third application. (See n. 37 below.)

element according to the second usage, while Michael Lapidge contents himself with saying that 'the third definition is missing from the text' (p. 271). Long and Sedley, who do seem to find in the text separate formulations of three usages, tell their readers in a note to their Greek text (ii. 278), that to each of the three usages listed at the end of the passage some separate section corresponds in the first part of the passage, where (on my reading of it, as outlined above) only two usages are mentioned and briefly discussed. As I will explain below (see the notes to my translation, pp. 107–9), their assignments of corresponding passages do not in every case make good sense, and result, in effect, in double-counting. However, I believe they are right to this extent, that in explaining the first two applications in the first part of the passage (down to line 14) Arius has intended to prepare the basis for identifying also the ground for Chrysippus' third application. In effect, even though, as I think, the first part of the passage only explicitly mentions two usages, Arius is justified in his apparently resumptive declaration in beginning the last part of the passage, 'So, according to Chrysippus, element is spoken in three ways': once we understand the third application, as he goes on to explain it, we can see that it has in fact been prepared for in what has preceded.

Before I can come to closer grip with the problems (including some textual ones) posed by this extremely interesting text, I need to prepare the ground; most of the chapter is taken up with these preparations. Specifically, I need to address two questions: first, the distinction, in Stoic theory, between elements (in any and all applications of the term) and first principles (ἀρχαί), and secondly, I need to consider Chrysippus' theory (which, I will argue, differs from Zeno's by being greatly more philosophically sophisticated) of the processes or stages by which the substance of the world (taken as a whole) gets turned (or rather, turns itself) from its original condition (that is, its condition after having undergone ἐκπύρωσις or conflagration—so to speak, before the process of world-formation begins[7]) into the condition it is in after the four material elements of fire, air, water, and earth have been generated (thus providing the complete material basis for the subsequent formation of the compound materials and the material objects that the world actually contains).

My argument will be that that substance, in that original condition, is what Chrysippus means to indicate in his third application of the term στοιχεῖον as *the* στοιχεῖον. This substance (as I will argue we can infer from our excerpt) is *not* anything that ought to be called 'fire' (as in fact Zeno had called it, with typical lack of philosophical sophistication).[8] Instead (as we know from a passage of

[7] 'So to speak before': not literally so, since on Stoic theory time is a dimension of change, and when ἐκπύρωσις has been completed all change has ceased and has not yet begun again. See ch. 51 in Long and Sedley. (I disagree with Long and Sedley when they infer, i. 311, from the fact that 'god is continuously active during all states of the Stoic universe' that time continues even 'after' the end of the world order and its rebeginning. Activity does not necessarily imply any change of or in anything.)

[8] But see n. 25 below.

Philo shortly to be considered) it is called by Chrysippus a 'flash' of light (αὐγή). The second application, then, I will argue—to something which does deserve to be called by the name of 'fire'—relates to the first stage in the process by which, according to both Zeno and Chrysippus (and the Stoic school generally), this original substance turns itself so as, eventually, to reach the point at which the four material elements that exist in the actual world are finally formed. This substance that deserves the name of 'fire', and is the unique element according to Chrysippus' second usage, is not, of course, the material element fire—not even the element fire in its purest instantiation, the fire which, according to the Stoics, the sun is made of. That, obviously, only comes into being at a later stage, in the formation of the physical elements making up everything in the actual world, of which it, in making up the sun, is the leading and controlling part. We can call it (one wishes Chrysippus had had such a linguistic device at his disposal) 'proto-fire'. I will explain later how proto-fire, as a certain substance—a flaming, fiery mass—differs from ordinary fire, even the purest instance of it.[9]

Thus we get, as Chrysippus according to Arius Didymus promised, three ways of speaking of elements: one according to which the fire, air, water, and earth of the actual world are elements on a par with one another, a second according to which proto-fire is the unique element, and a third according to which the unique element is the originary substance, the 'flash' of light, that is the only substance in existence when the world has been completely conflagrated (ἐκπυρωθέν), as Philo Judaeus puts it (see below at n. 23).

So that is a rapid summary of the interpretation of the content of this Stobaeus excerpt for which I will argue. I turn now to the preparations I need to make before examining the excerpt in detail. I should emphasize that in setting out the preparatory materials I am going to be guided by my decision to see how Chrysippus' cosmology and cosmogony look if we do take seriously and at face value what Arius reports about Chrysippus on elements in this passage. Let me begin, then, with the Stoic theory of first principles (ἀρχαί), as distinct from elements (στοιχεῖα). I will explain as I proceed the basis and the importance of the distinction between principles and elements, which, I believe, all the Stoics observed quite carefully.

It is well-known, I hope, by now that Zeno developed this theory of principles through a very close, critical reading of the Eleatic Visitor's discussion, with Theaetetus in Plato's *Sophist*, of a so-called battle between some philosophical 'gods' and some unphilosophical 'giants' over, as the Visitor puts it (246a1, 5), 'that which is' (τὸ ὄν) or 'being' (τὴν οὐσίαν). The Visitor wants to know what on earth *τὸ ὄν* or *ἡ οὐσία is*. What, in other words, can we say to make it clear to ourselves what that is, in virtue of which anything that is or has being, actually *is*, actually *has* being? The Visitor himself makes a suggestion about this

[9] In offering this interpretation below, I am working out the views I advanced but only sketched in n. 33 of 'Stoic Autonomy', in my *Knowledge, Nature, and the Good* (Princeton: PUP, 2004), 221.

(247d8–e1), as something the non-philosophical 'giants', who think that body (σῶμα) and being are the same thing (246b1), should accept: namely, that that which is, or has being, is whatever has the capacity or power (δύναμις) either to act (ποιεῖν) or be acted upon (παθεῖν). Plato's giants, believing that body and being are the same thing, are what we can call corporealists. Now, the most fundamental philosophical commitment of Zeno and all subsequent Stoics is likewise to corporealism (indeed the Stoics are corporealists in a much more thorough-going way than Plato's giants turn out to be).[10] (It is a very great error, which introduces much obfuscation, to describe the Stoics, as many do, instead as 'materialists,' as I will shortly explain.) Hence, as we can see from our reports of their doctrine of first principles (ἀρχαί), Zeno and all subsequent Stoics accept the Visitor's proposal. As Diogenes Laertius reports their doctrine (7. 134), 'they hold that there are two principles of the whole of things (τῶν ὅλων), that which acts and that which is acted on' (τὸ ποιοῦν καὶ τὸ πάσχον). For the Stoics, each of these is a body, i.e. each of them is (1) extended in three dimensions and (2) offers resistance, since that is what, for them, it is to be a body.[11] That they

[10] The Visitor browbeats the giants (anyhow, the ones who are willing to argue and discuss matters at all, and don't brutishly insist on their senses, in particular the sense of touch, as simply *showing* that being and body are the same) into admitting that there surely *are* souls, and that there are invisible and intangible qualities like virtues and vices too (246e9–247c2). As Theaetetus explains, the first of these admissions poses no problem for them: they can and do say that souls are bodies or bodily. But as to the second, Theaetetus thinks they are caught, being ashamed to say either that virtues and vices are not beings, or that they are bodies (b9–c2). The Visitor accepts this, but does remark that the *brutish* giants would not be ashamed to insist that if anything *does* have being, it must be something that can be 'squeezed by hand'. Zeno and subsequent Stoics in effect followed that clue, maintaining that qualities *are* bodies; Chrysippus notoriously held that virtues and vices can be *seen*, anyhow by the perfected human being, a 'wise' person. See Plutarch, *On Stoic Self-Contradictions* 1042e–f.

[11] For the corporeality of each of the principles, see Cicero, *Academica* 1. 39, and Long and Sedley's discussion, i. 374. See also Diogenes Laertius 7. 134 (reading with the MSS σώματα rather than editors' ἀσωμάτους, which would make DL say they are incorporeal; I share Michael Frede's reasons for saying that, even if maybe DL did write 'incorporeal', there can be no doubt that in fact the Stoics held the principles to be bodies: see 'La Théologie stoïcienne', in Jean-Baptiste Gourinat (ed.), *Les Stoïciens* (Paris: Vrin, 2005), 213–32, at pp. 215–16). For the Stoic definition of bodyhood translated in my text (τὸ τριχῇ διαστατὸν μετὰ ἀντιτυπίας) see Galen, *On Incorporeal Qualities* cited in *SVF* 2. 381, p. 127. 6–7, and Plotinus, *Enneads* 6. 1. 26. 21–3 (and compare Clement of Alexandria in *SVF* 2. 359, virtually quoting from Plato's *Sophist*). This seems to carry over into the Stoics' very different physical theory an idea from Epicurus, that being is differentiated from nothingness or emptiness by the mark that it offers resistance (see Lucretius, *De Rerum Natura* 4. 419–44, cited by Long and Sedley as their 5B). But resistance for Epicurus means inability of any other being to pass through any given being (a mark that belongs in full force to atoms), which, given Stoic theory of through-and-through intermixture of bodies (see below, at n. 14), it cannot mean for them. That is why, as I explain this Stoic definition in my main text, we need to understand it in some other way. Reasonable discomfort with the definition's apparently Epicurean origins seems to have led some Stoics, such as Apollodorus in the 2nd cent. BC—weirdly, as Plotinus' objection in the passage cited makes clear—to define body simply as what is extended in three dimensions (see DL 7. 135). But that is obviously unacceptable, because it would apparently make geometrical solids bodies too. On my understanding, the addition of μετὰ ἀντιτυπίας is precisely intended to bar such a consequence. One should also recall, as Gábor Betegh has reminded me, that the cosmos, and indeed all of being (whether or not under circumstances in which a cosmos even

offer resistance means that they *occupy* space; they are physically there, taking up room (even if—see the next paragraph—another body can occupy the same space). They do not, like a geometrical solid, merely extend through or provide limits to some range of space.[12] And each of the two principles is something that has being, is an ὄν—*their* being ὄντα is what is needed to make everything else, all that is constructed out of and by them, be ὄντα too.[13]

Each of these principles (as to their further identities I will say something in a moment) is a distinct body from the other. Though each is extended in three dimensions, and indeed extended in all the same space where the other is also extended, and each offers its 'resistance' to the other (and only to it), they remain perfectly distinct each from the other, as *two* bodies. This is possible, the Stoics think, because the nature of body is such that two bodies can be in all the same places, through and through interblended in a 'blend' of the two bodies. No matter into how small regions you might chop the blend up, both bodies will be present, both occupying the whole of that region. What shows that both bodies are there, everywhere there, is that the distinctive presence of each can be detected in any region, however small it might be, through its effects.[14] And in fact, according to the Stoics, these two bodies, the active principle and the passive or acted-on one, are completely blended with one another in the way just indicated. We see this when Diogenes continues in the same passage, 'that which is acted on is qualityless substance (ἄποιος οὐσία), i.e. matter, that which acts is reason in it, i.e. god'. Here we see, perhaps to our surprise, that although both principles are ὄντα, in Stoic terminology only one of the two, the passive principle, gets called οὐσία, being or substance. God or reason, although a distinct body from that other body, is *in it* (not the other way about—and it is never apart from it, either). What results from the combination of the two basic

exists), is demarcated from a surrounding void, something that, exactly as for Epicurus, is physical nothingness and emptiness, neither offering resistance (in the Epicurean sense) nor (in the Stoic sense) being occupied by anything.

12 The Stoics hold (against Plato in the *Timaeus*) that the matter that god works on (which is their passive principle) has no character of its own (not even a trace-like one) that could impose any 'necessities' that would limit god's creative power (god is the active principle). God can make of matter literally *anything* he wants; he can endue any of it with *any* quality he likes. Only his own perfect rationality imposes any restrictions as to what he *does* endue any of it with. (See Frede, 'La Théologie', 221–2.) So 'resistance' as a defining mark of body must not be interpreted to imply that matter, as a body, forces or limits god in any way. Because both god and matter *occupy* the same space, however, and are not, like geometrical figures, simply spread through it, each has something it must confront and engage with in mutually occupying the same space. That—nothing more—is how we must interpret the notion of 'resistance' here.

13 For their both being ὄντα see Plutarch, *On Common Notions* 1073e, ὄντος τὸ ποιεῖν τι καὶ πάσχειν.

14 On the Stoic theory of 'blends' (κράσεις) see esp. Alexander of Aphrodisias, *De Mixtione* 3–4 (excerpts from which are given in *SVF* 2. 473). Alexander is right to emphasize from the outset the application of this theory of blends on the cosmic scale: claims about water and wine when mixed, and other such phenomenal mixed stuffs, are trivialities of physical theory, compared with this absolutely fundamental application.

ὄντα, the principles, is a *substance* (οὐσία) because what undergoes formation in that combination, namely, qualityless substance, is itself 'already' a substance.[15] The combination yields a *material* substance because this qualityless substance is (also) matter. So this material substance is matter—*but* matter *with* god or reason in it. And crucially so, since reason, being in it, and indeed only *by* being in it (since nothing acts directly on anything from a distance), is what activates matter in every way that it does get activated, including all the ways that it gets qualified as the substance that it then is. Thus god is ὄν but not οὐσία; god is not (trans)formed when any substance comes to be, as the substance's matter *does* undergo formation, in particular not in the 'coming-to-be' of this first substance, or rather in its fundamental constitution. This first substance, the only one there is when the world has been conflagrated, is not just a body, as the principles also are; it is a *material* body, a material substance. The two principles—reason and qualityless matter, i.e. prime matter—are always combined with one another and thereby constitute this eternal *first* material substance.

Thus this one substance, existing already at the very 'beginning', is constituted by those two other bodies that are the principles; it is constituted from reason's being *in* and acting *on* prime matter. Hence wherever we find that one material substance we find each of the two bodies that are the principles: god or reason, with its essential, unlimited power of action, and qualityless substance or matter, with its essential, unlimited capacity to be acted upon. Moreover, the one material substance (the substance composed of matter with reason or god in it) is itself always and necessarily qualified in some way. How could it not be, having within it the active principle or reason? (Below, I will discuss in just what way it *is* qualified, simply insofar as it is this first substance—in advance, so to speak, of more particular ways in which it, or different expanses of it, become qualified in the course of the formation of the cosmos.) Yet, everywhere where that substance is, there, there is also qualityless substance, i.e. the passive principle. And reason too, of course. These are *three* bodies, all three occupying all of the same places: the one material substance, qualified however it may be by god or reason's being in the matter that helps to make it up; plus qualityless substance or prime matter; plus reason. The one material substance (that *material* body) is *constituted by* those two other bodies, by the one body's (god or reason's) being in the other body (prime matter).[16]

[15] See Calcidius, *In Timaeum* 291, cited by Frede, 'La Théologie', 219, and Frede's discussion.

[16] Michael Lapidge (and not he alone: see e.g. Robert B. Todd, 'Monism and Immanence: The Foundations of Stoic Physics', in John M. Rist (ed.), *The Stoics* (Berkeley and Los Angeles: University of California Press, 1978), 137–60, at pp. 139–40) shows a fundamental failure to understand the Stoic theory of principles when he insists in his otherwise illuminating article, 'A Problem in Stoic Cosmology', *Phronesis* 18 (1973), 241–3 *et passim*, that they are nothing more than 'aspects', 'nominally distinct but essentially one', of the one substance. As Long and Sedley correctly say, these are not mere 'conceptually distinguishable aspects of a single body' (*Hellenistic Philosophers*, i. 273); something that was only conceptually distinguishable (a λεκτόν) would be an incorporeal, and on Stoic theory that would mean that it lacked causal powers altogether.

Now, it is important to understand clearly that these latter two bodies (matter and god or reason), unlike the first, the one they constitute, are *non-material* bodies. Reason is a non-material body in the straightforward sense that, though triply extended and occupying space, and possessed essentially of powers of acting, *it* is no material thing. It is *in* matter, but there is no matter in *it*. Qualityless substance or matter is likewise triply extended and occupies space, but it differs from reason in that *it* is possessed essentially of powers to be acted upon. It is a body and is *matter*, to be sure, but it is not a *material body*. That is because, for the Stoics, all material bodies are made of matter as one, but only one, component. Reason is the other. Prime matter, which is what this matter is, does not itself *have* any matter as a component. Compound materials in the actual world, by contrast, are *material* bodies, precisely because they *do* have lower forms of material body in them, as what they are made out of; even each of the four elements is made out of *some* material body, as we shall see. This Stoic distinction between mere bodies (including the two principles)—by the Stoic definition, mere bodies are extended in three dimensions and offering resistance—and *material* bodies, which satisfy the further condition of having matter in them, is crucially important. It is the reason why the Stoics cannot correctly be called materialists, but ought instead, like Plato's giants, to be called corporealists. For the Stoics, there are bodies (namely, the two principles) that are not *material* bodies. Material bodies are, all of them without exception, constituted by the presence with one another of both principles, and by the effects of the one principle on the other.

Hence, in sum, the qualityless substance remains in full and actual existence, and remains qualityless, as an essential constituent actually present in the one originary material substance, to which it contributes its own essential function of the ability to be acted upon. Similarly the other principle, reason, also remains and remains active, equally as a constituent of that originary material substance, to which it contributes its own essential function of being able to act. Thus, the originary substance is (qua being infused by reason) able to act on itself (qua completely passive matter). The continued existence, and presence everywhere, of the active principle follows from the fact that each of the principles is a distinct body, however interblended with one another they may be. Here, one should notice that when the passive principle is called qualityless substance or matter it is, of course, as I have already noted, not being denied that it has some properties: it has the properties of extension in three dimensions, and of offering resistance, since those are the defining characteristics of bodies, as such. And it has whatever further character comes to it simply from having god's thought spread uniformly and undifferentiatedly everywhere through it (I return to this point below). What is denied to it is only any further qualification. And, in particular, of course, it is denied to have any colours or shapes or textures or consistencies or any other members of the traditional category of quality, not to mention that it lacks being so qualified as to constitute a plant or animal, with any of its various possible qualities. Thus Stoic prime matter differs quite notably from the prime matter that late Aristotelians attributed to Aristotle: that version of prime matter (a mistaken attribution to Aristotle, and in itself an incoherent idea) is of something that has no characteristics at all. Stoic prime matter does have actual properties, it simply lacks qualities and other qualifications added onto that primitive base. God or reason, of course, adds all those additional qualities to whatever materials have them, through specific, differential thoughts about how those particular materials are to be; so prime matter necessarily lacks them. Understood the Stoics' way, prime matter is not an incoherent notion.

So much, then for the Stoic distinction between principles and elements. Principles are mere bodies, elements are always (in whatever specific usage) material bodies. They are material bodies from which other, more complex, material bodies come to be.

Let me now turn to my second question, which concerns Chrysippus' theory of the processes or stages by which the substance of the world turns itself out of its original condition to the point where the ordinarily recognized elements, fire, air, water, and earth, begin to be generated. First we need to think a bit more about this substance as it is before it 'turns' in any way at all. Diogenes Laertius tells us (7. 137) that one way the Stoics (he does not specify which ones) used the term 'world' (*κόσμος*) was in reference to 'god himself, i.e. the uniquely qualified individual out of (*ἐκ*) all substance [i.e. all the bare matter that there is, taken as a whole], who is indestructible and ungenerated, being the craftsman of the world-ordering, at certain periods of time consuming all substance into himself and, in the other direction, generating it from himself'.[17] A bit earlier in his exposition of Stoic physical theory (7. 136), again without naming specific Stoics, and so presumably meaning most or all of them, he has said about god or reason that[18]

being in the beginning all by himself, he turns (*τρέπειν*) all substance through air into water, and just as sperm (*τὸ σπέρμα*) is encompassed in generative matter (*ἐν τῇ γονῇ*), so he, being the seminal thinking or reason (*σπερματικὸν λόγον*) of the world, stays behind as such in the moisture, making matter well-suited for his purposes in the following stages of generation. Next he generates first the four elements, fire, water, air, and earth.[19]

[17] See also Plutarch, *On Stoic Self-Contradictions* 1052c, who cites specific works of Chrysippus for the doctrine of god's periodically consuming all substance (all matter) into himself.

[18] Here is the Greek text: *κατ᾽ ἀρχὰς μὲν οὖν καθ᾽ αὑτὸν ὄντα τρέπειν τὴν πᾶσαν οὐσίαν δι᾽ ἀέρος εἰς ὕδωρ· καὶ ὥσπερ ἐν τῇ γονῇ τὸ σπέρμα περιέχεται, οὕτω καὶ τοῦτον σπερματικὸν λόγον ὄντα τοῦ κόσμου, τοιόνδε ὑπολείπεσθαι ἐν τῷ ὑγρῷ, εὐεργὸν αὑτῷ ποιοῦντα τὴν ὕλην πρὸς τὴν τῶν ἑξῆς γένεσιν· εἶτα ἀπογεννᾶν πρῶτον τὰ τέσσαρα στοιχεῖα πῦρ, ὕδωρ, ἀέρα, γῆν.*

[19] It is not clear what the *γονή* is that semen is said to be encompassed in. Long and Sedley seem right (ii. 272) to reject Lapidge's suggestion ('Stoic Cosmology', in J. M. Rist (ed.), *The Stoics* (Berkeley and Los Angeles: University of California Press 1978), 166) that it means the womb. (Lapidge seems to have been guided too strongly by his idea that for Zeno the moisture here in question was conceived as a female principle, to match Zeus's male semen-injecting role; the womb,

I discuss this passage further just below. First, though, what does it mean that in the beginning god is by himself, having consumed all substance into himself? It is presumably not intended that god or reason is the only body then in existence, having somehow consumed into itself all matter; i.e. it is presumably not intended that in the beginning reason has absorbed ἄποιος οὐσία or prime matter into itself, i.e. into the non-material body that it is. As we have seen, the two principles are spoken of in correlation: god or reason is a body *in* another body, namely, prime matter. In that sense, commentators are right that for the Stoics reason or god, and matter, never exist in separation from one another; that reason or god is always spread through matter, even in these pre-cosmic circumstances.[20] What god's being by himself must mean is that at this beginning point all there is in existence is god or reason, pervading prime matter. That is to say, all *qualified* substance (not all substance, i.e. not prime matter itself) has been absorbed into god or reason. That, in turn, is to say that although he or it in his active nature retains and keeps on thinking to himself all the thoughts

he imagines, is what semen gets injected into according to Zeno's presupposed theory of animal reproduction.) As Long and Sedley say, the use of this word for the womb is rare and, so understood, it does not correspond correctly to the moisture here referred to. Long and Sedley themselves, following David Hahm (*The Origins of Stoic Cosmology* (Columbus: Ohio State University Press, 1977), 60), render it by 'seminal fluid', with many parallels in Aristotle's biological works (see H. Bonitz, *Index Aristotelicus* (Berlin, 1970), 160). But in that case it is awkward to translate σπέρμα here, as they no doubt correctly do, as 'sperm': at least, *our* word 'sperm', in its principal use, refers to the seminal fluid, not to something *in* it (and the parallels in Aristotle actually show that he often uses σπέρμα and γονή interchangeably—in accordance with this usage of ours). If one does translate γονή by 'seminal fluid', then one ought to render σπέρμα by 'spermatozoa', not sperm, or at least one should indicate clearly that one means 'sperm' to be understood in that, its secondary meaning (*OED*). And in fact, Aristotle does sometimes speak of animal σπέρμα as something distinct from the γονή, i.e. the semen, that it or its power is found in. But I am not certain that Stoic reproductive theory would follow him in this. In any event, on Aristotle's theory while the sperm is the agent in generation, the seminal fluid itself is not the material out of which the new animal is made; but the Stoics' analogy here is from 'sperm' in some reproductive fluid to Zeus in the primordial moisture as matter for him to work upon, generating the four elements out of it. Hence the analogy intended may rely upon a reproductive theory that postulates some moisture provided by the female animal in her womb as what the sperm enters and gets encompassed by. (But see Hahm's discussion p. 61, relying on a heavily allegorized mythological account of Dio Chrysostom, *SVF* 2. 622, in which it seems that Zeus is in a seminal fluid of his own devising; see also his further discussion, pp. 68–76). Hence I opt for my less committal translation of γονή by 'generative material'. This could be either some wet stuff provided by the female, or (with Long and Sedley) the semen itself as vehicle for spermatozoa. (I think it does not matter greatly precisely what reproductive theory is being presupposed in the analogy: it is clear enough what is being said, analogously, about Zeus as seminal thinking being within the moisture—a moisture that, of course, he creates from himself. That is the main thing of interest here.)

[20] It is important however not to confuse this truth about the Stoic system with the idea that, for them, god is always spread through (some) *material* body, e.g. through 'artistic' or 'crafting' or 'designing fire' (πῦρ τεχνικόν) or through πνεῦμα (breath). God has 'designing fire', or else πνεῦμα, as his vehicle only within the cosmos, since designing fire is the element fire in its pure state, and πνεῦμα is a compound out of two material elements, fire and air. They therefore do not exist when the world has been completely conflagrated, at least on Chrysippus' philosophically careful and consistent version of the theory. (For Chrysippus' view, see lines 19–21 in my text in the Appendix.)

that in the actual world get put into effect in introducing all the qualifications of matter that constitute all the different sorts of substance that there actually are, he is not then using those thoughts to *act* in any differential way upon particular expanses of matter so as to endow substances with their particular characters; he is therefore not then affecting matter with any of those qualifications.

In what way, then, *is* reason affecting matter, as it must in some way be doing, by being in it, in constituting the originary substance? What is the character of the material substance that, as I have said, is composed by god's being present in prime matter at this pre-cosmic stage? Zeno seems not to have hesitated to declare roundly that this originary substance is fire. That was because of the fostering and creative power of heat coming from the sun in the actual world. The originary substance *is* the fostering and creative substance *par excellence*,[21] so it must have seemed to Zeno, as it has seemed to most interpreters of Stoic cosmogony, that this substance is a huge creative, fostering, flaming, fire. When Stoics said (as we read in Aetius, *SVF* 2. 1027) that 'god is a crafting fire (πῦρ τεχνικόν), proceeding methodically to the generation of the world, containing within it all the seminal thoughts in accordance with which the various different things come about', they were obviously speaking of that originary substance, and describing it as fire of the very same type (crafting, as opposed to consuming and destructive)[22] that constitutes the sun and the stars, according to Stoic theory. But to describe it so is plainly inconsistent with the very theory of cosmic generation that this substance is being called on, in this very passage, to originate. Even πῦρ τεχνικόν is a version (the purest) of one of the four elements, and those, on the Stoic theory, are late-comers in the order of generation, as we have seen. If such statements are to be interpreted consistently with the Stoic theory of cosmogenesis, all references to 'fire' in connection with the originary substance have to be understood proleptically, by anticipation of the material embodiment of god or reason *within* the cosmos, once formed. Within the cosmos god *is* πῦρ τεχνικόν, as one component of the divine 'breath' or πνεῦμα that is god or reason's immediate vehicle for controlling the world's constitution and behavior.

Philo gives us clear and persuasive evidence that Chrysippus was well aware of the absurdity involved in claiming that the originary substance actually *was* a huge quantity of the element fire, even crafting fire, and it seems that before him Cleanthes was too. Philo tells us that 'when conflagrated[23] . . . it is necessary that

[21] See Aristocles apud Eusebius (*SVF* 1. 98), in saying that for Zeno fire is the στοιχεῖον of τὰ ὄντα. This application by Zeno of the term στοιχεῖον to the originary substance serves to justify Chrysippus in distinguishing the third of his usages of the term from the other two that he marks off from it in our Stobaeus passage.

[22] Zeno distinguished these two types of fire; see Stobaeus in *SVF* 1. 120.

[23] The Greek here is τὸν κόσμον ἐκπυρωθέντα. This refers, as Philo's discussion makes abundantly clear, to what is in existence when the world has been completely consumed in fire, i.e. when the successive process of first elemental earth, then water, then air being consumed by and converted into fire, has been completed. Philo is not referring to what the world is like while being conflagrated, but, as the aorist itself makes clear, what (so to speak) 'it' is like once the process of

the world changes either into flame (φλόξ) or into a flash of light (αὐγή): into flame as Cleanthes thought or into a flash as Chrysippus did'.[24] Now, a flash, I take it, is the product of a fire (in this case, the fire through which god or reason, in the process of conflagration, consumes into himself all qualified substance). But it is not, and arguably does not require, once it has been produced, a continuing fire, consuming something, to sustain it—as Cleanthes' candidate, a flame, definitely does. Just think of all those stars we see that, in fact, according to contemporary astronomical theory, are by now completely dead, though their light continues to reach us.[25] Thus, I take it, Chrysippus has a serious, and good, philosophical point in rejecting not only Zeno's loose talk of the originary substance as some huge quantity of the element fire (i.e. one of the ordinarily

conflagration has been completed. As Philo makes clear, and insists on in his polemic against the Stoics, they must and do accept that every fire needs fuel, so that once all the elemental earth, water, and air have been consumed, the fire has to go out. It is with respect to that condition—as to what is in existence once the fire does go out, and conflagration is over with—that Philo is making this statement (and preparing his objections).

[24] See Philo, *On the Eternity of the World* (Περὶ ἀφθαρσίας κόσμου) 90 (= *SVF* 1. 511, LS 46M). One should, however, read the whole of the surrounding discussion (sects. 85–93). The author (presumably Philo, though his authorship has been disputed) begins (sect. 86) by postulating three forms that fire can take, that of or in a live coal (ἄνθραξ) plus Cleanthes' and Chrysippus' two candidates, flame and a flash of light. This analysis, implying that even a flash of light is a form of fire, might, but might not, also be Chrysippus': we have to bear in mind that the author's purpose in considering Stoic views is to refute them all. He has a preference either for the Aristotelian theory of the uncreatedness and indestructibility of the world or the Platonist view of its creation but non-destruction through the will of the creator to sustain it forever (see sects. 7–19). By insisting that even light is a form of fire, the author can attempt to refute the Stoic theory of conflagration-and-regeneration by pointing out that, since when the cosmos is completely conflagrated the fire of the conflagration has finally been extinguished, there is no fire, nothing fiery, left that could be the basis for regeneration, not even (he argues) light (sects. 85, 93). As I point out below (last three paragraphs of my main text), our Stobaeus passage very conspicuously does not report Chrysippus as equating with fire (of any form or at any stage) the flash of light that counts for him as the sole element according to his third usage. So in what follows I prefer to interpret Philo's insistence that a flash of light is one form of fire as polemically motivated. I suppose that Chrysippus himself did not so regard it (and reasonably so, as I will suggest). If, however, Chrysippus did authorize the classification of even the flash of light that remains after conflagration as a form of fire, the crucial point, for him (dismissed by Philo), will be that it is something one can legitimately call fire that however, uniquely, does not require for its continued existence any burning mass of anything (even though it required that for its coming to remain in existence as the sole substance, once the conflagration has been completed). See further n. 23 above.

[25] Obviously Chrysippus cannot be thinking of any such example. But even without a theory of light as something that is emitted by a source and travels in space, the basic idea that light may survive the extinction of its source is readily available to common sense, and to ancient Greek philosophers. Light seems clearly to be something, though caused by fire (its source, according to the Greeks), that has an independent existence. That is all Chrysippus need have been thinking, as he struggled to find a suitable way of describing the originary substance. Of course, he is not thinking of ordinary light (as we know it within the cosmos, about which he has theories of his own); that is presumably why he chose 'flash' instead of 'light' to express his view of what the originary substance is like. This 'flash' has always lain at the base of everything in the world, but, after conflagration (as the last act, so to speak of the final cosmic fire), it gets released, to survive on its own. A flash is all the more something that can be conceived in independence from its source.

recognized four elements) in its purest form, but even Cleanthes' talk of a bare flame, in favour of the theory that it is a flash.

This flash is, therefore, the sole and complete character, according to Chrysippus, of the material substance that results from the bare presence of the active principle within the passive principle, before the cosmic cycle begins. When the world has been completely conflagrated god or reason affects prime matter in such a way as to produce and sustain this flash (and nothing else at all is then in existence). That flash *is* the originary substance, out of which then, by subsequent 'turns' or 'turnings' (*τροπαί*), new substances get produced—in a prescribed order, as we have seen (see DL 7. 136, quoted above p. 101). The ultimate result of these turnings is that the world as it now actually exists, gets (re)constructed. As we will see shortly, when we come to a closer examination of the Stobaeus excerpt, this originary substance, this flash, counts as the sole element, according to the third of the three usages of the term *στοιχεῖον* that Arius Didymus tells us Chrysippus endorsed. It counts as an element, as opposed to a principle (*ἀρχή*), because it is something (a material substance) 'out of which' (*ἐξ οὗ*) by subsequent turns 'things that come to be, first come to be, and into which they are in the end reduced'. This is the definition of *στοιχεῖον* that Diogenes Laertius gives on the Stoics' behalf, in 7. 136. By contrast, nothing that comes to be, comes to be *out of* the two principles—god or reason, and qualityless matter. The principles *compose* the originary substance, the flash, but this substance did not *come to be* out of them. Since the principles are eternal and ungenerated, and are always interblended, the originary substance itself is eternal and ungenerated too. Hence, the flash of the originary substance is the earliest thing in the account of generation that Chrysippus provides for us that deserves the name of 'element', since it is the earliest thing that fits the Stoic definition of 'element' just cited: it is something out of which and from which subsequent things that come to be *first* come to be, and into which they finally get reduced. God and qualityless matter do not fit this definition. Hence, they are not elements, but only, as in fact Stoic theory specifies them, principles (*ἀρχαί*).[26]

What, then, in order, are the turns subsequently taken by this originary substance, and what new substances do they lead to? Diogenes Laertius gives us a complete-seeming capsule-account, but we have to piece it together from two brief passages, one in 7. 136, quoted above, the other, occurring a couple of pages later, in 7. 142. Each of these adds something important not included in the other. The passage in 136 tells us that at first god or reason was all alone, in the way that I have already explained. God then turned the entire substance (i.e.

[26] This analysis shows that M. Lapidge was wrong to think that the Stoics were forced to abandon the distinction between *ἀρχαί* and *στοιχεῖα* when Chrysippus introduced *πνεῦμα* as a cosmic force ('A Problem in Stoic Cosmology', p. 277). Only the two principles ever counted for any Stoic as *ἀρχαί*, and nothing Chrysippus said about *πνεῦμα* (which wasn't an element for him, only a compound out of two elements, fire and air) had any tendency to compromise this distinction.

the originary flash) 'through air into water'. The passage in 142 adds that this first series of turns was '*from fire* through air into moisture'.[27] (The variation from 'into water' in 136 to 'into moisture' in 142 is insignificant, I think,[28] but the addition in 142 of 'from fire' is a significant addition, as we will see.) The passage at 136 then goes on to tell us that god 'stays behind' in this moisture or water as the 'seminal thinking of the world'; that is, he stays behind thinking his plans for the specific construction, first, of the four elements out of which everything in the actual world is made, and then (as 142 adds), by mixing these elements together in varying proportions, of 'plants and animals and the other natural kinds'.[29] This work of construction begins (as 142 tells us) by his 'turning' this moisture so that out of different parts of it, in separated areas, earth gets produced by condensation, while first air and then fire get produced by rarefaction out of other parts; the remaining parts of this moisture, which are neither condensed nor rarefied, thus give rise to the fourth element, water.

Thus we get two successive sets of 'turnings'. First, the originary flash gets turned (or turns itself) from something called fire through something called air to something called moisture (in 142: or something called water, in 136). Then we get, out of that moisture by further turnings, the construction of the four material elements of the actual world. It is obvious that the fire, air, and water or moisture referred to in the first of these two sets of turnings cannot be any version

27 Here is the Greek text of the passage in 7. 142 (as printed by Long and Sedley, 46 C, except that I accept with Wachsmuth the reading ἐξαερωθῇ, 'turned into air', the reading of one MS, instead of ἐξαραιωθῇ, 'rarefied', read by other MSS), followed by a translation. *γίνεσθαι δὲ τὸν κόσμον ὅταν ἐκ πυρὸς ἡ οὐσία τραπῇ δι' ἀέρος εἰς ὑγρόν, εἶτα τὸ παχυμερὲς αὐτοῦ συστὰν ἀποτελεσθῇ γῆ, τὸ δὲ λεπτομερὲς ἐξαερωθῇ, καὶ τοῦτ' ἐπὶ πλέον λεπτυνθὲν πῦρ ἀπογεννήσῃ· εἶτα κατὰ μίξιν ἐκ τούτων φυτά τε καὶ ζῷα καὶ τὰ ἄλλα γένη*. 'The world comes into being when the substance is turned out of fire through air into moisture; next the thick part of the moisture condenses and gets finished off as earth, while the thin part is made into air, and the latter when it is thinned even more generates fire. Next, by mixture out of these, plants and animals and the other natural kinds.' One must compare with this the parallel passage in Stobaeus (see *SVF* 1. 102), alleged to be verbatim quotation from Zeno. The comparison supports my preferred reading of ἐξαερωθῇ, since without it DL will not be telling us directly, as *SVF* 1. 102 does, that some of the moisture is rarefied *into air*: *τοιαύτην δὲ δεήσει εἶναι ἐν περιόδῳ τὴν τοῦ ὅλου διακόσμησιν ἐκ τῆς οὐσίας, ὅταν ἐκ πυρὸς τροπὴ εἰς ὕδωρ δι' ἀέρος γένηται, τὸ μέν τι ὑφίστασθαι καὶ γῆν συνίστασθαι, [καὶ] ἐκ τοῦ λοιποῦ δὲ τὸ μὲν διαμένειν ὕδωρ, ἐκ δὲ τοῦ ἀτμιζομένου ἀέρα γίνεσθαι, ἐκ τινὸς δὲ τοῦ ἀέρος πῦρ ἐξάπτεσθαι*. . . . 'Out of the substance there will have to be the following sort of periodic cosmic ordering of the whole, when a turning happens out of fire into water through air: one part sinks and constitutes earth, and out of the rest of it one part remains water while out of the vaporized part air comes to be, and from some of the air fire gets ignited . . .' Note that *SVF* 1. 102 (claimed by Diels to come from Arius Didymus) says explicitly that some of the moisture remains as water, or remains and becomes water. DL leaves this unsaid, though it seems intended.

28 The passage in 136, having mentioned water as what substance is turned into, goes on to speak of it instead as 'the moist', *τῷ ὑγρῷ*.

29 In a passage reported by Plutarch (*Stoic Self-Contradictions* 1053b, shortly after citing the passage discussed in n. 40 below) from the first book of his *On Providence* Chrysippus describes the situation when god is (proto-)fire before turning into this moisture as one in which he is his soul alone; when in the moisture he is that soul now in and possessing a body, namely the moisture.

of the four elements out of which the world and its contents are subsequently constructed. If they are called 'fire', 'air', and 'water', these designations must be meant with some other reference. They are, as one could say, and I suggested at the outset, proto-fire, proto-air, and proto-water, in that they are only somewhat *like*, in being fiery, airy, and watery respectively (see n. 35 below), but lack the determinate structure of, the three actual material elements called fire, air, and water respectively. (Note that in the first set of turnings there is no proto-earth; anything even like earth only emerges when earth itself, the material element, gets generated from the moisture that is said to have the seminal thinking of god in it.)

On this account the originary substance, which is the first material form that god adopts, turns into something called fire (i.e. proto-fire), which is the second material form he/it takes. This proto-fire remains as the 'seminal thinking' in the moisture out of which then he, as that proto-fire, constructs the four ordinary elements.

We are now ready to turn to the Stobaeus excerpt giving Chrysippus' account of three ways of speaking of elements (*SVF* 2. 413).[30] Here is a translation (parenthetical numerals mark beginnings of lines in the Greek text in the Appendix):[31]

From Chrysippus. About the elements out of substance,[32] he declares himself more or less as follows, in keeping with the leader of his sect, Zeno—saying that there are four elements, fire air water and earth, out of which everything is constructed, both animals

[30] It is important to see that these are not three *senses* of the word 'element' (στοιχεῖον). On each of the three ways of speaking, or usages, what gets counted as an element is conceived as fitting the very same definition: as we have seen (DL 7.136) an element is defined by the Stoics as that from which a thing *first* comes to be, and into which it is *ultimately* reduced. What differentiates the three usages depends on what, on that usage, gets counted as the 'first' material body from which all other things come to be. In one way of considering the process of coming to be, this first level can be identified as the four simple bodies (earth, air, fire, and water); in another as the originary material body from which cosmogenesis begins; in yet another, as proto-fire, the very first stage in cosmogenesis. See my comments below in my main text for explications of how one can understand the three different frameworks for counting something as first or lowest material constituent of things.

[31] In the Appendix, as noted, I provide von Arnim's text exactly as he prints it, with no departures. In the footnotes there I indicate exhaustively, and explain, which of his and other editors' emendations I do or do not accept, and which manuscript variants I follow, in departure from von Arnim. The text translated here is therefore not that of von Arnim printed in the Appendix (for the reader's convenience), but one that includes the departures indicated in my notes to his text.

[32] In fact the elements Arius goes on to identify are 'out of' substance only in the first two usages that Arius is about to set forth for us, not the third. Elements on each of the first two usages are made ultimately out of the originary substance by its turnings. Below we learn that this substance itself is called (the) element, in the third of the usages, and it, of course, is not 'out of' substance. It is not 'out of' anything at all, as I have explained, since though composed of god and prime matter, it does not come to be from them. The terminology of 'out of substance' at the beginning of the passage evidently reflects the commonplace understanding of an element in application to material stuffs, whereas Chrysippus' innovatory recognition of the originary substance as entitled also to the name of 'element', possessing a reasonable justification though it does, also expands our understanding of what the notion of elementhood encompasses.

and plants and the world as a whole and all the things contained in it, and into these they get dissolved. But that which is called an element *par excellence* is so called because the rest are constructed by change out of it as origin (ἐξ αὐτοῦ πρώτου) and all are dispersed and get dissolved into it in the end, while this one does not (6) admit of dispersal or reduction into any other. On this way of speaking, fire is called 'element' self-sufficiently, since it is not <ranked> along with another.[33] But on the first way of speaking,[34] fire is constructive of things along with other <elements>: the first change that happens is the constructing change out of fire into air, the second from the latter into water (11), third the yet further change in a corresponding way of water, once constructed, into earth. And in the other direction, from that [i.e. earth] being dissolved and dispersed, the first dispersal that happens is into water, second out of water into air, and third and last into fire. (All that is

[33] Long and Sedley's analysis (ii. 278) makes this way of speaking (λόγος), which they correctly see is the one that has just been introduced and briefly discussed immediately before (in lines 4–6 of my text printed in the Appendix: 'But that which is called an element *par excellence* . . . dispersal or reduction into any other' in my translation), correspond to the first of the three usages listed in resumption by Arius in lines 14–21. But they seem to think that, in adding here the specification of fire as what, on their understanding of this usage, counts for Stoics as 'the' element, Chrysippus was introducing his third usage. At any rate, following helpful discussion with Ricardo Salles, that is now how I understand their annotated translation (their textual notes in vol. ii do not provide a suitable context for explaining how they intended Chrysippus to be understood, and their interpretative comment in vol. i is not clear on this point). For them, then, the 'second usage' (according to the initial order, which is the first in resumption) is a general characterization (a second 'account'), or sort of definition, of 'element', as 'whatever the other ordinarily recognized elements are constituted out of and ultimately get resolved into', and the third 'usage' is found in the specification of fire as the one that (according to Stoics) fits that description. Long and Sedley likewise seem to find, with respect to the specification of earth air fire and water as 'elements' according to the first of Chrysippus' usages (in the initial order), a further general characterization (or 'account') of 'element,' indicated in their translation (see the bracketed phrase '[i.e. (1) above]') as to be found in the first line and a half of the text. (In that case, I guess they must have in mind some general idea of an 'element' as what comes from 'substance' as the account according to which the four usual elements will get counted as elements; this seems to me not to make good sense.) Perhaps, though, '[i.e. above (1)]' is a slip for '(2)', and they mean to count the first application itself, assigning the term 'element' to the four usual ones, as the first 'usage' without having any separate 'account' for the term. But, whichever of these two things they intended, their interpretation mixes *applications* with definitions or general characterizations (I suppose it is possible that Chrysippus could be guilty of this, so I would not want to exclude their interpretation simply on this ground). But on the first option (the one they actually propose, with '(1)' rather than '(2)' in their bracket), if the second 'account' is registered as one of Chrysippus' usages, alongside its specification as another one of them, then the corresponding first account, leading to the identification of four elements, ought to count as an additional 'usage' for Chrysippus—resulting in four, not three usages. In fact, it seems far better to take Long and Sedley's second 'account' together with its specification with fire, as constituting a single one of Chrysippus' three 'usages', i.e. as I have insisted, applications, of the term 'element'. (See DL 7. 136 and my discussion above for the only *definition* the Stoics provide of 'element'; we do not hear elsewhere about an alternative 'account' or 'accounts' such as Long and Sedley postulate.) Hence this 'way of speaking' with its specification in terms of fire corresponds to the first of the three usages as listed in resumption (the second in the order of prior presentation). (See the next three nn.)

[34] The first way of speaking referred to here is manifestly the one mentioned first above, namely the one according to which fire, air, water, and earth are elements. This is also, again manifestly, the usage listed and discussed below, lines 15–19, as the second of the three usages Arius sets out in resumption.

fiery is called fire, what is airy air, the rest similarly.)[35] So, according to Chrysippus, element is spoken in three ways: in one way it is spoken of fire, because the rest are constructed by change out of it and get their reduction into it; in a second (16) way, in the way that the four elements, fire air water and earth, are spoken of (since the rest are constructed through some one or some ones or even all of these: through the four as the animals and all the things on the earth are compounds, through two as the moon is constructed through fire and air, through one as the sun, for it is through one only, since the sun is pure fire); in the third way of speaking,[36] that is spoken of as element which is originally constituted (20) in such a way as to provide generation from itself methodically up to a conclusion, and out of that [sc. that conclusion] to receive reduction into itself similarly methodically. And he said there were explanations about element of the following sort, that it is both that which is most easily moved through itself,[37] and the principle <and> reason and the eternal power possessing a nature such as to move itself downwards to turning and from turning upwards, everywhere cyclically, both consuming (24) everything into itself and in the other direction reconstituting it from itself in an orderly and methodical way.

As I mentioned at the outset, we find no difficulty in understanding the first of the three usages Chrysippus distinguishes, according to this passage: the usage according to which there are four elements of equal standing. These are the four material substances out of which all other materials and all material objects in the actual world are constructed, and into which at their destruction they are ultimately dissolved, while thereafter new materials or material objects get constructed from the same elemental bodies that were previously their material

[35] I take it to be the point of this parenthetical explanation to make clear that proto-fire, though (if you like) fiery, is not actually *fire*, and proto-air and -water are airy and watery but not air or water (the ordinary elements everyone is already familiar with)—nonetheless, these proto-substances do deserve to be called, as Chrysippus has just been doing, by those *names*, because of their respective special characters as fiery etc.

[36] For Long and Sedley the exposition of this usage (the third according to my count) in what follows is limited to lines 19–21 of my text in the Appendix—they do not print or translate, and say nothing about, lines 21–4. Hence, for them, this usage, which they identify as being constituted simply by a general characterization of an 'element' (see n. 33 above), is here only explicated by what they must regard as a paraphrase in terms of 'providing generation from itself methodically'. On my view, in my main text below, the exposition continues down to the end of the selection, so that quite a bit more is said in explication of this usage; and this phrase is by no means a paraphrase of that general characterization (which in fact, as I understand it, applies to the item that is an element according to the *first* of the usages described in resumption, not at all to what counts as 'element' according to the third application). The explication of the third usage discusses not proto-fire (or, as Long and Sedley would say, simply fire, the basic one of the four elements), but the originary substance, as these further remarks in fact make clear. It is a significant mark in favour of my interpretation of this passage over theirs that I include within my analysis the whole of the passage, and do not lop off the final four or so lines as some further separate remark, or set of remarks, of Chrysippus on the general topic of elements.

[37] Diels (*Elementum*, 39) sees in this the first of three definitions of what it is to be an element that Arius now appends to the previous discussion of three, in fact, ways of applying the term. It is plain, however, that the following material relates to the third of the three ways of applying the term as explained in lines 19–21 of my text. I therefore think it is better to take this phrase ('that which is most easily moved through itself') to go with that material, relating to the third way of applying the term 'element'.

constituents. These are fire, air, water, and earth. This usage is referred to and explained in lines 2–3, 9–11, and 15–19 of my Greek text.

However, now that we have examined closely Chrysippus' account of the two sets of turnings by which the originary substance gives rise, eventually, to these four elements, one sees at once that in 9–13 (in my translation, 'the first change . . . third and last into fire') we are told, not, as you might have expected, about the *second* set of turnings, whereby, beginning with the creation of the material element earth out of the moisture that has god in it as seminal thought, we then get formed in order the material elements air and fire, with water coming to be out of the remainder of the moisture that arose as the conclusion of the first set of turnings. Rather, we are given there a brief account of precisely that *first* set of turnings. What lines 9–13 give us is a brief run-through of the turnings from proto-fire through proto-air to proto-water, resulting (as the first step of the *second* set of turnings) in the condensation of some of the proto-water into *actual* earth—the first of the material elements of the world as it actually exists. This is followed, as we now know, but is not mentioned here, by the generation out of that moisture, through rarefaction, of *actual* air and *actual* fire, with *actual* water coming to be out of the remainder of the proto-water of the moisture. This *second* series of turnings, apart from the mention of earth, is not described or mentioned in lines 9–13. Instead, as I just said, it is the first set of turnings that we are told about here. This *first* set of turnings begins, as we see reflected here in lines 9–13, from proto-fire: there, we are told, 'the first change that happens is the constructing change out of fire into air'.[38] But, as we now know, proto-fire is itself preceded in existence by the originary substance, the flash: the *absolutely* first turning was from that originary substance.

Accordingly, we can now also see that the usage Chrysippus takes notice of second in order here, at the beginning of the passage, according to which something called 'fire' is the sole element, an element *par excellence* and self-sufficiently,[39] refers to the proto-fire of the *first* set of turnings. This raises a question that it will be worthwhile to pursue briefly. It is true that the *first* thing that comes to be when the originary substance begins to turn itself is proto-fire, and the Stoic definition of element emphasizes that on any usage an element should be what things *first* come from (in this case, the ordinary elements come to be first from proto-fire, in the series (proto-) fire–air–water. But how are we to understand (proto-)fire as the *sole* element, on this usage (the sole element from which the ordinary elements come to be)? That seems to imply that ordinary fire, air, water, and earth have proto-fire as their only constitutive element. One might ask, however, why one should count fire as the *sole* constitutive material

[38] So, as we can now see, the addition in DL 7. 142 of '*from fire* through air into moisture' to the mere 'through air into water' of 7. 136 is very significant. This indicates, as we see here in our Stobaeus text, that these changes begin from proto-fire.

[39] The second usage is referred to and explained in lines 4–9 and 14–15 of my Greek text: 'But that which is called . . . along with another' and 'in one way . . . reduction into it' in my translation.

of the things that come from it, according to this second usage? What about proto-air and (even more) proto-water, i.e. that moisture in which, we are told by Diogenes Laertius, the seminal thinking of the world stays behind, making matter suitable for its purposes in generation of the world through the generation, to begin with, of the actual four material elements? Ought they not also to be counted as elements according to this second usage—so that on this usage there would be three elements, not four as on the first usage, and not, as Arius tells us Chrysippus in fact claimed, just one?

In addressing this question we need first to take notice of the fact that the descriptions I have cited above from Diogenes Laertius of the first set of turnings (7. 136 and 142), resulting in the moisture from which then the generation of the four elements begins, are presented by him as giving the view not only of Chrysippus but of Zeno and indeed other Stoics. In 136, for example, Diogenes concludes the passage I quoted with, 'Zeno speaks about them [sc. the four elements] in *On the Whole*, and Chrysippus in the first book of his *Physics*, and Archedemus in a work *On Elements*.' To these authors in the passage I quoted from 142 about the generation and destruction of the cosmos, he adds references to works of Posidonius, Cleanthes, and Antipater.[40] Indeed, the parallel text to 142 (*SVF* 1. 102) that I cited in n. 27, also from Stobaeus, is prefaced with 'Zeno declared himself expressly as follows': there is no reference there to Chrysippus, or to any of these others, at all. Of course, Chrysippus wanted to follow Zeno in his own ordering of the first set of turnings, as well as in the second set. But we must be ready to interpret him in ways that in fact distinguish his view from Zeno's. He himself would maintain that his own view is the one that Zeno really had in mind all along. But we already know that he rejected any idea that the originary substance was *fire*, as suggested in 142, either in the sense of the ordinary element or in any other sense that implies a substance flaming up in any way; and we can expect other related 'precisifications'. In speaking of the turnings in 136 Diogenes only uses the term 'element' in connection with the usual four, and leaves one with the impression that the references to air and water in his exposition of the first set of turnings, leading to the further turnings that form the four, are just references to two of these same four elements. Presumably that in fact faithfully reflects Zeno's own innocence. He did not mind that, strictly speaking, there *was no* elemental air or water, not to mention any fire, in existence before the completion of the second set of turnings. As we have seen, Zeno did not distinguish, as Chrysippus took great care to do, between the originary substance as a flash, not some fire of the ordinary elemental kind, even the purest version. Accordingly Zeno may, as I have already suggested, have loosely and naïvely spoken of the originary substance as fire, which then turned through air

[40] It is worth noting that Plutarch, quoting or paraphrasing from the first book of Chrysippus' *Physics* (*Stoic Self-Contradictions* 1053a = *SVF* 2. 579), confirms that the details reported in 142 represent Chrysippus' own view.

into water or moisture, and remained inside the moisture as a seed, working on it to generate from it the element earth by condensation, and the elements air and fire by rarefaction, leaving the rest of it to be or become the element water.

Chrysippus knew that, as stated, this made no good sense. In his own version, then, he must have clarified as follows. Actually, as we have seen, the originary substance is a flash, not a fire; the first turn out of it, as world-formation gets under way, is to proto-fire, with further turnings through proto-air to proto-moisture, at which point the four elements come into being by condensation and rarefaction of the proto-moisture—an additional set of turnings. Wishing then to distinguish the first set of turnings from the second by marking a distinct way of applying the term 'element' from that according to which the four basic material bodies make up all the world's actual materials, he declared (in our Arius excerpt) that *only* the body (namely, god as proto-fire) produced at the first of these turns should be counted as an element in that second way of applying the term. Apparently his thought was as follows. Anything deserving the name 'element' must be some stuff out of which things are made, as I noted just now. The things that are to be made in these turnings are, ultimately, the four ordinarily recognized elements. Even if these result from the condensation and rarefaction of proto-water or moisture, it would be a mistake to say that they are made *out of* water or out of anything watery, as in any sense a stuff they are made of (in the strictest sense, they and everything else is made out of the originary substance, of course). Proto-water is only a preliminary, short-lived transition-point in the turning of proto-fire so that the four elements get generated. The same applies to proto-air. Zeno's insight had been that it is the power and energy of fire that underlies all the actual materials of the cosmos and all material objects. The truth of this insight requires that in our account of world-formation we make this power and energy, and only it, as proto-fire, the material basis for the four elements. It is god as proto-fire that is spread through the four ordinary elements as what, in their different material constitutions (depending on their varying densities), they are made out of.

We should conceive proto-fire, proto-air, and proto-water in the following way. The originary substance turns itself *first* by creating proto-fire, as a stuff that is by its material nature so structured as to possess the fostering and generative powers that are needed for subsequent stages of world-formation. The originary substance does this by, so to speak, overlaying over itself just those qualities that are needed for this task. It thereby 'becomes' something like fire by giving itself these further fiery and fostering characteristics, ones that it needs to have in order to complete Zeus's plan of world-creation. This new substance cannot be called 'fire' in any strict sense, as I have already observed. But it does deserve to be so designated because of its powers of generation and sustenance.[41]

[41] That is surely why it is precisely at this point in the exposition, at line 13, where proto-fire has been introduced for the first time, that Arius or Chrysippus points out that he is using the term 'fire' in this connection to refer to whatever is fire-like (and so, too, for 'air' and the others).

This 'fire' produces moisture (again by overlaying everywhere over itself watery characteristics), in which it 'stays behind' as 'the seminal thinking of the world' (see DL 7. 136, cited above)—in the first instance, as the seminal thought of the four elements that it is immediately going to produce, using its generative powers. It remains in all four of these elements, and must remain in them as what they are made out of, in order for *those* elements to have the distinctive powers of generation and construction (as with fire and air) or more or less passive constitutability (as with earth and water) that, on Stoic theory, must belong to them, if they in turn are to play their roles in world-constitution. Proto-air is a mere stage through which proto-fire must pass as it produces proto-water ('fire' has to go though the more dense stage marked by proto-air, before it can become proto-water). Proto-water is also itself only a temporary stage in proto-fire's (and the originary substance's) 'turning' downward to world-generation. Proto-fire needs proto-water as a residing point in which its own generative powers can reside, while carrying out its generative activities, through condensation of that material in some parts of it (leading to the existence of earth), and through rarefaction of other parts of it (leading to air and fire), with the creation of proper water from the remainder. Thus, as I said, proto-fire, and only proto-fire (among the three proto-stuffs) counts as *the* element, on this second usage of the term: it is all by itself what the usually recognized four elements of the material world are made of.

Let us attend now to the third of Chrysippus' ways of applying the term 'element'. This is referred to in lines 19–24 of my Greek text ('in the third way of speaking . . . in an orderly and methodical way' in my translation). Two striking facts about this usage and about what on it gets spoken of as an element deserve immediate notice. First, according to this passage, whereas in both the first two usages the element or elements include something called 'fire', in reference to the third there is, quite strikingly, no mention of fire. Second, we should note the use of the dative ὁδῷ, which I have translated 'methodically', which appears three times in the exposition of the third usage (and nowhere in the exposition of the other two usages). The first two times it modifies the 'generation' of unspecified things from, and their 'reduction' into, the element on this usage. In his exposition, earlier in our excerpt, of elements according to both the other usages Arius Didymus has indeed spoken similarly of 'construction' out of and 'dissolution' into elements, but in neither case do we find this interesting qualification, 'methodically'. Moreover, in the third appearance of ὁδῷ we read of 'the principle <and> reason and the eternal power . . . both consuming everything into itself and in the other direction reconstituting it from itself in an orderly and methodical way'. We have already examined passages where god is described when the world has been conflagrated as being all by himself, having consumed all substance (DL 7. 137, 136), and subsequently generating it all out of himself, and another passage (Aetius in *SVF* 2. 1027) which speaks of god, when the world has been conflagrated, as 'proceeding *methodically* to the generation of the world'. So there can be no doubt that in this third way of applying the term Chrysippus means to be assigning the title of element to god, as the originary substance, i.e. to prime matter as qualified by having god or reason spread everywhere through it. Arius has prepared the way for us to recognize this third usage in what he has said in lines 9–10 ('the first change that happens is the constructing change out of fire into air'). I have argued above that this is a reference to the first change out of proto-fire. If we bear in mind, as we ought to in reading about Chrysippus' views, that proto-fire is a substance existing as the first step in the reconstruction of the world, and is not the originary substance itself that exists when the world has been conflagrated, it must be evident that lying behind it stands that originary substance, the flash.

This third way of speaking of an element, according to which the originary substance, as the absolutely first stuff out of which things subsequent to it in cosmogenesis are made, is the sole element, makes very good sense. It is out of this, by the first of those methodical turnings, but (it appears) not yet a *constructing* change (a change κατὰ σύστασιν), that proto-fire comes to be.[42] Since proto-fire is what (with their different consistencies) elemental fire, air, water, and earth are made out of (by 'constructing' changes), one can therefore well say that, in a certain sense, this originary substance is the most basic material body there is, the one out of which, ultimately, everything in the world is constituted. As such, given the Stoic definition (DL 7. 136) of element as that 'out of which things that come to be, first come to be, and into which they are in the end reduced', we can entirely appropriately give it the title of element.

Here we must bear in mind the Stoic theory of through-and-through interblending of bodies. Just as both the two principles are everywhere where

[42] Arius speaks of the 'turns' both from proto-fire through proto-air to proto-water, and from proto-water to the four ordinary elements, as 'constructing changes' (see συνίστασθαι κατὰ μεταβολήν in lines 4–5, τῆς . . . κατὰ σύστασιν . . . μεταβολῆς in line 10, and συνίστασθαι κατὰ μεταβολὴν in line 15). But in connection with the first turn of all, from the originary substance to proto-fire, he does not speak in that way. He says only that the originary substance itself is 'constituted' (συνέστηκεν)—it is not *constructed*, since that implies a coming-into-being, while this substance is eternal—so as to provide generation methodically all the way to an end and dissolution methodically back again into it (lines 20–1), and that it 'moves itself downwards to turning' (line 23). He does not call the move a constructing change. That might suggest that proto-fire, the result of the first turn, is not to be thought of as constructed out of the originary substance, but to have some other relation to it, even though it comes from it and eventually gets reduced to it. Perhaps this difference in language is not significant. But maybe it is. The first turn (the first 'move') is to proto-fire. In subsequent moves, first proto-fire turns so as to engage in a constructive change that produces other proto-elements, ending with proto-water, and then proto-water gets changed so as to construct ordinary earth, air, fire, and water. But the move from the originary substance to proto-fire yields the first material actually *within* the cosmos now already in formation. Perhaps Chrysippus' idea is that all *constructing* changes start from there, because all processes of construction must be carried out within the cosmos, or within the cosmos-in-formation, from materials already on hand. Construction only begins once proto-fire is on hand. That means that, though proto-fire materially derives, as everything else does also, from the originary substance, it does not get constructed out of it, but comes into being rather as the necessary presupposition to all constructed forms of matter.

either of them is, so also the originary body is everywhere they are too, and, as the world gets generated and the ordinary four elements come into being, anywhere any element is, everywhere there, there will also be several distinct other bodies, including proto-fire as well as the originary substance. Though, as we see in our Stobaeus passage (lines 16–19), the sun is made out of only one element, fire, and the moon of two, fire and air, most material things have, in differing proportions, all four. So in the case of the latter, they have everywhere in them all four material elements in some or other proportions, plus proto-fire, plus the originary substance (all of which count as elements, on one of Chrysippus' usages or another), plus reason and prime matter (which, of course, are not elements at all according to any proper usage, but principles).[43]

[43] I thank Prof. Ricardo Salles for providing the incentive to write up these ideas, which were first presented at a conference on 'God and the Cosmos in Stoic Philosophy' organized at the Instituto de Investigaciones Filosoficas of the Universidad Nacional Autónoma de México, 3–5 July 2006. I have profited greatly from the thorough discussion of the paper on that occasion, and thank particularly David Hahm, Brad Inwood, Thomas Bénatouïl, and Professor Salles himself, for enabling me to take into account issues that they raised in discussion. In addition, I owe Prof. Salles thanks for his acute written comments, as editor of this book, on the penultimate version. Responding to them helped me to improve the chapter at a number of places. I also thank Prof. Katerina Ierodiakonou for arranging an unusual mid-summer colloquium at the Department of Methodology, History and Theory of the Sciences of the University of Athens later in the same month, at which I benefited from the opportunity to air my ideas before a large and highly discriminating group of Greek and foreign ancient philosophers and their students. I owe special thanks to Michael Frede for several excellent criticisms and suggestions, confronting which led to important improvements. Finally, I thank Prof. Panos Dimas and his colleagues and students at the University of Oslo for a stimulating and helpful discussion when I gave the paper there in Sept. 2006. Thanks are likewise due for further discussion after I delivered the paper in Budapest at the philosophy department of the Central European University in Jan. 2009. I particularly thank Prof. Dimas and Prof. Gábor Betegh for their queries and suggestions, which led to significant improvements in this final version.

APPENDIX

(*SVF* 2. 413) Stobaeus, *Eclog.* I, p. 129. 1–130. 20 Wachsmuth

Χρυσίππου. περὶ δὲ τῶν ἐκ τῆς οὐσίας στοιχείων τοιαῦτά τινα ἀποφαίνεται, τῷ τῆς αἱρέσεως ἡγεμόνι Ζήνωνι κατακολουθῶν, τέτταρα λέγων εἶναι στοιχεῖα ⟨πῦρ, ἀέρα, ὕδωρ, γῆν, ἐξ ὧν συνίστασθαι πάντα καὶ ζῷα⟩[a] καὶ φυτὰ καὶ τὸν ὅλον κόσμον καὶ τὰ ἐν αὐτῷ περιεχόμενα καὶ εἰς ταῦτα διαλύεσθαι. τὸ δὲ ⟨πῦρ καὶ⟩[b] κατ᾽ ἐξοχὴν στοιχεῖον λέγεσθαι διὰ τὸ ἐξ αὐτοῦ πρώτου τὰ λοιπά συνίστασθαι κατὰ μεταβολὴν καὶ εἰς αὐτὸ ἔσχατον πάντα χεόμενα διαλύεσθαι, τοῦτο δὲ μὴ ἐπιδέχεσθαι τὴν εἰς ἄλλο χύσιν ἢ ἀνάλυσιν· [συνίστασθαι δὲ ἐξ αὐτοῦ τὰ λοιπὰ καὶ χεόμενα εἰς τοῦτο ἔσχατον τελευτᾶν· παρὸ καὶ στοιχεῖον λέγεσθαι, ὃ πρῶτον ἕστηκεν οὕτως, ὥστε σύστασιν διδόναι ἀφ᾽ αὑτοῦ καὶ αὐτὸ τῶν λοιπῶν χύσιν καὶ διάλυσιν δέχεσθαι εἰς αὐτό.][c] κατὰ μὲν τὸν λόγον τοῦτον αὐτοτελῶς λεγομένου τοῦ πυρὸς στοιχείου· οὐ μετ᾽ ἄλλου γάρ·[d] κατὰ δὲ τὸν πρότερον καὶ μετ᾽ ἄλλων συστατικὸν εἶναι, πρώτης μὲν γιγνομένης τῆς ἐκ πυρὸς κατὰ σύστασιν εἰς ἀέρα μεταβολῆς, δευτέρας δ᾽ ἀπὸ τούτου εἰς ὕδωρ, τρίτης δ᾽ ἔτι μᾶλλον κατὰ τὸ ἀνάλογον συνισταμένου τοῦ ὕδατος εἰς γῆν. πάλιν δ᾽ ἀπὸ ταύτης διαλυομένης καὶ διαχεομένης πρώτη μὲν γίγνεται χύσις εἰς ὕδωρ, δευτέρα δ᾽ ἐξ ὕδατος εἰς ἀέρα, τρίτη δὲ καὶ ἐσχάτη εἰς πῦρ. Λέγεσθαι ⟨δὲ⟩[e] πῦρ τὸ πυρῶδες πᾶν καὶ ἀέρα τὸ ἀερῶδες καὶ ὁμοίως τὰ λοιπά.

[a] This addition is due to Diels, and I accept it (so do Long and Sedley). Something has obviously dropped out of the text here (the sentence makes no grammatical sense as it appears in the MSS); the full list of the usual four elements given below in lines 17–18, with the explication in 17–18 of how animals and everything else on the earth are composed from them, justify the details of Diels's addition.

[b] This is an addition of Usener (Heeren had emended with the addition of *πῦρ* without *καὶ*), accepted by Diels, which Long and Sedley argue is unnecessary, and also distorts the structure of the passage. I follow Long and Sedley in not accepting any emendation here. In this sentence Arius/Chrysippus is explaining the ground on which something would deserve to be called 'element' *par excellence*; below, lines 8–9, he then tells us that this thing is 'fire' (i.e. proto-fire, as I explain in my discussion in the main text). Usener's/Heeren's addition is premature.

[c] Wachsmuth detects here an intrusion into the text of a marginal note by a reader (he detects a similar one further on in Stobaeus, at p. 154. 24 ff.). I accept this deletion. It seems just to repeat in somewhat different words what has just been said in lines 4–6.

[d] The phrase *οὐ μετ᾽ ἄλλου γάρ* appears in the MSS after *πρότερον*, line 9, where it certainly makes no sense. It was first moved here by Heeren, and is so printed by Diels, Wachsmuth, von Arnim, and Long and Sedley. The genitive absolute in the phrase beginning *κατὰ μὲν τὸν λόγον τοῦτον*, with the matching *κατὰ δὲ τὸν πρότερον* in 9, which is followed by the infinitive *εἶναι* (with *τὸ πῦρ* to be understood as subject), seems anacoluthic; Diels at first wondered whether we oughtn't to read *γίγνεσθαι* for *γάρ* in the transposed phrase, since that would provide an infinitive to attach the genitive absolute to, so as to match the infinitive in the following clause and remove the anacoluthon. This seems to me attractive, though radical; but in his Addenda (p. 854) Diels points to what he seems to take to be the same matched genitive absolute and infinitival clauses at p. 144. 6 W ff. (in another Stoic report, identified by Diels as Arius Didymus fr. 19), and concludes that no emendation is needed. Reluctantly, I follow him in this.

[e] Some connective is required here by the grammar; this addition is Heeren's proposal, accepted by all subsequent editors. I accept it.

Τριχῶς δὴ λεγομένου κατὰ Χρύσιππον τοῦ στοιχείου, καθ' ἕνα μὲν τρόπον τοῦ πυρός, διὰ τὸ ἐξ αὐτοῦ τὰ λοιπὰ συνίστασθαι κατὰ μεταβολὴν καὶ εἰς αὐτὸ λαμβάνειν τὴν ἀνάλυσιν· καθ' ἕτερον δέ, καθὸ λέγεται τὰ τέσσαρα στοιχεῖα, πῦρ, ἀήρ, ὕδωρ, γῆ (ἐπεὶ διὰ τούτων τινὸς ἢ τινῶν ἢ καὶ πάντων τὰ λοιπὰ συνέστηκε, διὰ μὲν τῶν τεττάρων, ὡς τὰ ζῷα καὶ τὰ ἐπὶ γῆς πάντα συγκρίματα, διὰ δυοῖν δέ, ὡς ἡ σελήνη διὰ πυρὸς καὶ ἀέρος συνέστηκε, δι' ἑνὸς δὲ ὡς ὁ ἥλιος, διὰ πυρὸς γὰρ μόνου, ὁ γὰρ ἥλιος πῦρ ἐστιν εἰλικρινές), κατὰ τρίτον λόγον λέγεται στοιχεῖον***[f] εἶναι ὃ πρῶτον συνέστηκεν οὕτως, ὥστε γένεσιν διδόναι ἀφ' αὑτοῦ ὁδῷ μέχρι τέλους καὶ ἐξ ἐκείνου τὴν ἀνάλυσιν δέχεσθαι εἰς ἑαυτὸ τῇ ὁμοίᾳ ὁδῷ. Γεγονέναι δ' ἔφησε καὶ τοιαύτας ἀποδόσεις περὶ στοιχείου, ὡς ἔστι τό τε δι' αὑτοῦ εὐκινητότατον καὶ ἡ ἀρχὴ ⟨καὶ ὁ σπερματικὸς⟩[g] λόγος καὶ ἡ ἀΐδιος δύναμις φύσιν ἔχουσα τοιαύτην, ὥστε αὐτήν[h] τε κινεῖν κάτω πρὸς [γῆν][i] τὴν τροπὴν καὶ ἀπὸ τῆς τροπῆς ἄνω πάντῃ κύκλῳ, εἰς αὐτήν τε πάντα καταναλίσκουσα καὶ ἀφ' αὑτῆς πάλιν ἀποκαθιστᾶσα τεταγμένως καὶ ὁδῷ.

REFERENCES

Arnim, H. von (ed.) (1903–5) *Stoicorum Veterum Fragmenta* (Leipzig: Teubner), vols. 1 and 2.

Cooper, J. M. (2004) 'Stoic Autonomy', in J. M. Cooper, *Knowledge, Nature, and the Good* (Princeton, NJ: Princeton University Press), 204–44.

Diels, H. (1965/1879) *Doxographi Graeci* (Berlin: De Gruyter).

——(1899) *Elementum* (Leipzig: Teubner).

Frede, M. (2005) 'La Théologie Stoïcienne', in J.-B. Gourinat (ed.), *Les Stoïciens* (Paris: Vrin), 213–32.

Gould, J. (1970) *The Philosophy of Chrysippus* (Leiden: Brill).

Hahm, D. E. (1977) *The Origins of Stoic Cosmology* (Columbus: Ohio State University Press).

——(1990) 'The Ethical Doxography of Arius Didymus', in W. Haase (ed.), *Aufstieg und Niedergang der Römischen Welt*, vol. 36.4 (Berlin and New York: De Gruyter), 2938–43.

Lapidge, M. (1973) 'Archai and Stoicheia: A Problem in Stoic Cosmology', *Phronesis*, 18: 240–78.

Long, A. A. and Sedley, D. N. (1987) *The Hellenistic Philosophers* (Cambridge: Cambridge University Press).

Todd, R. B. (1978) 'Monism and Immanence: The Foundations of Stoic Physics', in J. M. Rist (ed.), *The Stoics* (Berkeley and Los Angeles: University of California Press), 137–60.

Wachsmuth, C. (ed.) (1884) *Ioannis Stobaei Anthologii Libri Duo Priores* (Berlin: Weidmann), vol. 1.

[f] Wachsmuth saw a lacuna here, and von Arnim agreed. I think that was a mistake. The MSS text makes good enough sense without emendation. I follow Long and Sedley here.

[g] Some addition is obviously necessary here; Usener proposed the one here printed (accepted by von Arnim). Meineke and Diels, more modestly and more reasonably, suggested simply adding ⟨*καὶ ὁ*⟩. I follow Meineke and Diels. (Long and Sedley print this Stobaeus text only down to *ὁδῷ* in my line 21, so they offer no help here; they omit the important last lines.)

[h] The MSS here read *γῆν*; the correction is due to Wachsmuth: this is reasonable and gives what ought to be the right sense. I accept this emendation.

[i] Hirzel first proposed deleting this word, and he is followed by Wachsmuth. I accept Hirzel's deletion. The text of this whole clause is transmitted in confused form in the MSS, some of which give *τροφὴν* and *τροφῆς* instead of *τροπὴν* and *τροπῆς*, no doubt in error.

5

Chrysippus on Conflagration and the Indestructibility of the Cosmos

Ricardo Salles

The present chapter deals with Chrysippus' claim that the cosmos 'should not be said to die' (οὐ ῥητέον ἀποθνήσκειν: Plut. *De Stoic Rep.* 1052C, *SVF* 2. 604, LS 46E, quoted below). As I shall argue, the claim is of great significance for our understanding of early Stoic cosmology because it reflects a clash within the school between two conceptions of the conflagration (ἐκπύρωσις): Chrysippus' own conception and that of Cleanthes. The two parties agree that there will be a conflagration. The cosmos will be totally consumed by fire, which will burn until it exhausts all the available combustible matter. Also, they both maintain that the conflagration is a periodical phenomenon. It will be followed by a new cosmogony and a reconstitution of the individual entities that existed earlier. Indeed, the whole cosmos will be restored to its previous condition, but will burn up again at some point causing a new conflagration, and so on *ad infinitum*. As joint proponents of the conflagration and the everlasting recurrence of everything, Cleanthes and Chrysippus differed from other major Stoics who denied, or at least questioned, that a conflagration will ever occur.[1] The important issue on which Cleanthes and Chrysippus departed from each other is that of whether the conflagration entails the periodical *destruction* of the cosmos. In other words, is the cosmos periodically destroyed at each conflagration? Cleanthes believes it is. Given that the fire of the conflagration, i.e. flame, is, by its very nature, destructive of its fuel, the conflagration will destroy the whole cosmos. This destruction is not permanent, because a new cosmogony will begin and an identical cosmos will

The earliest version of this chapter was delivered in Sept. 2005 in Mexico City at the Facultad de Filosofía of the Universidad Panamericana and the Facultad de Filosofía y Letras of the Universidad Nacional Autónoma de México (UNAM). A later version was presented in July 2006 at the conference 'Dios y el Cosmos en la Filosofía Estoica' organized at the UNAM. I am grateful to all these audiences for their comments.

[1] Notably Zeno of Tarsus, Diogenes of Babylon, Panaetius, and Boethus. See Cic. *ND* 2. 118, Eus. *Praep. Ev.* 15. 18. 2 (*SVF* 3 Zeno Tarsiensis 5) and Philo, *Aet.* 76–8 (*SVF* 3 Diogenes Babylonius 27).

be created. But, however momentary it may be, it is a destruction nevertheless. As we shall see in section 3, there is strong evidence for this view in Cleanthes. Chrysippus, however, opposed this idea. In his formulation, the cosmos will not periodically 'die' at the conflagration—and of course if it does not die, it will not be periodically destroyed either, because the cosmos is an animal and animals are destroyed only if they die.[2] The flame of the conflagration will certainly destroy complex bodies such as plants and animals. But this consumption is not destructive of the cosmos *as a whole* because, Chrysippus thinks, it is not destructive of its most basic constituents, namely, the four elements. I argue that the conflict between these two Stoics is ultimately rooted in a difference between their elemental theories on the question of how fire acts upon the other elements.

In section 1, a reconstruction is given of Chrysippus' arguments. Its strength resides in the fact that it is based on Cleanthean premisses—theses that Cleanthes either defends as tenets of his own cosmology or that are fully consistent with his cosmology. In consequence, Chrysippus' argument reveals a tension in Cleanthes' cosmology: the latter cannot consistently claim, as he actually does, that the cosmos will be *destroyed* at the conflagration. Sections 2 and 3 are devoted to exploring the theses in Cleanthes that seem to generate this inconsistency. In section 4, I explain how this conflict comes from a difference in the elemental theories of Cleanthes and Chrysippus. In the Appendix, I explain how Chrysippus' argument differs from two other Stoic arguments for the indestructibility of the cosmos.

1. CHRYSIPPUS' ARGUMENT IN PLUTARCH, *STOIC. REP.* 1052C

As is normally the case in *De Stoicorum Repugnantiis*, Plutarch quotes Chrysippus for polemical purposes. The present text is part of a longer passage devoted to establish how the conjunction of several theses defended by Chrysippus yields the incoherent idea that there is and there is not such a thing as 'nourishment for Zeus'. According to 1052D (in Cherniss's translation, slightly modified):

So he is in conflict with himself not only because in the former passages [from Chrysippus' *De Providentia* quoted at 1052B–C] he declares that, except for the cosmos and Zeus, there is nourishment for the rest of the gods, but in the latter [quoted at 1052C–D] he

[2] See Cleanthes *ap*. Cic. *ND* 2. 32 *animantem esse mundum* and Chrysippus *ap*. Plut. *De Stoic Rep*. 1053B (*SVF* 2. 605, LS 46F). See also DL 7. 139 (on Antipater) and 142 (on Chrysippus). I use the term 'animal' and not 'living thing' (which is also a standard translation of ζῷον) because in Stoicism not all living things are ensouled, e.g. plants. Cf. the evidence on Stoic plants in *SVF* 2. 708–13.

states that there is nourishment of the cosmos *also*, but even more because he says [at 1052D] that the cosmos grows by getting nourishment from itself.[3]

The argument for the immortality of the cosmos occurs in one of the passages from *De Providentia* that Plutarch uses to claim that for Chrysippus there *is* nourishment for Zeus:

ἐν δὲ τῷ πρώτῳ περὶ Προνοίας *τὸν Δία φησὶν αὔξεσθαι, μέχρι ἂν εἰς αὑτὸν ἅπαντα καταναλώσῃ· ἐπεὶ γὰρ ὁ θάνατος μέν ἐστι ψυχῆς χωρισμὸς ἀπὸ τοῦ σώματος, ἡ δὲ τοῦ κόσμου ψυχὴ οὐ χωρίζεται μὲν αὔξεται δὲ συνεχῶς μέχρι ἂν εἰς αὑτὴν ἐξαναλώσῃ τὴν ὕλην, οὐ ῥητέον ἀποθνήσκειν τὸν κόσμον.*

In the first book of *De Providentia* he says that Zeus goes on growing until all things have been consumed in his growth: 'For, since death is separation of soul from body and the soul of the cosmos is not separated but goes on growing continually until it has completely absorbed its matter, the cosmos should not be said to die'.

One assumption of the argument—implicit in the idea that the cosmos has soul and body—is that the cosmos is an animal. From there the argument seems to proceed as follows: (1) for any animal, death is the separation of its soul from its body, and (2) at the conflagration, the soul of the cosmos absorbs, or consumes,[4] its own body or matter (*ὕλη*), but (3) the consumption or absorption of something A by something B, and the separation of B from A are two different processes. Therefore, at the conflagration the cosmos does not die (and hence is not destroyed as Cleanthes maintains). The soundness of the argument requires a proof of the thesis upheld in premiss (3) that the processes of separation and consumption envisaged in the argument are sufficiently different from each other to guarantee that the latter, in contrast with the former, is compatible with the persistence of the animal as a whole. I leave this issue for section 4 and concentrate on premiss (1).

What is the soul of the cosmos? And what is its body? In Stoicism, the soul of the cosmos is fire in the form of heat. There is a specific reason for this. The heat that the stars irradiate exerts on the cosmos an activity analogous to the one exerted on an animal by its soul. It holds it together and promotes thereby its persistence.[5] In Stoic technical terminology fire as heat is the cohesive cause, or *συνεκτικόν αἴτιον*, of the cosmos in the same way as the soul of an animal is

[3] *οὐ μόνον οὖν ἐν ἐκείνοις τοὺς ἄλλους θεοὺς ἀποφαίνων τρεφομένους πλὴν τοῦ κόσμου καὶ τοῦ Διὸς ἐν τούτοις δὲ καὶ τὸν κόσμον λέγων τρέφεσθαι μάχεται πρὸς αὑτὸν ἀλλ' ἔτι μᾶλλον, ὅτι τὸν κόσμον αὔξεσθαί φησιν ἐξ αὑτοῦ τρεφόμενον.*

[4] One MS (g) has *εἰς αὑτὴν καταναλώσῃ* which is the verb used by Plutarch in the present passage immediately before the quotation. Cherniss and Long and Sedley read with the rest of the MSS *εἰς αὑτὴν ἐξαναλώσῃ* as printed above. In *LSJ s.v.* the two verbs are given roughly equivalent meanings. Cf. *ἀναλίσκω* in DL 7. 136 (*ἀναλίσκων εἰς ἑαυτὸν τὴν ἅπασαν οὐσίαν*), cited and discussed by Cooper in Ch. 4 (pp. 000–00).

[5] In Cicero's formulation at *ND* 2. 28: 'the cosmos itself owes its continued preservation for so long a time to the same or a similar substance', namely heat (*calor*): *mundum etiam ipsum simili parique natura in tanta diuturnitate servari*. For the idea that the activity of Stoic cohesive causes is temporally coextensive with their effect, see notably Clem. *Strom.* 8. 9. 33. 1–2 Stählin *et al.*

the cohesive cause of the animal. Now, if heat is the soul of the cosmos, what is its body or matter? In general, Stoic matter is what receives the action of a cohesive cause. At the highest level of abstraction, it is the ultimate recipient of the action of the active principle or god.[6] At a cosmological level, the recipients of this action are the entities that everything in the cosmos is made of and made from: earth, water, air, and fire both in the form of the individual portions of heat that we find in animals, and in the form of flame, which may have salutary and beneficial effects on animals.[7]

At least premisses (1) and (2) express Cleanthean views. The first premiss is attested for Cleantes in *ND* 2. 23 (quoted in the next section) and I shall consider the evidence for (2) and (3) in sections 3 and 4 respectively. I begin by looking at the reasons Cleanthes adduces for thinking that the heat is indeed the cohesive cause of cosmos and, hence, its soul.

2. HEAT AS THE COHESIVE CAUSE OF THE COSMOS

Although the thesis is accepted by a number of different Stoics, it is developed in greatest detail in *ND* 2. 23–32, which is generally regarded in current Stoic scholarship as an account of Cleanthes' cosmology.[8] The argument is given in sections 23–8. In what follows I quote (in Long and Sedley's translation) the parts of the text that are more directly relevant to the discussion.

It is a fact that all things which undergo nurture and growth contain within themselves a power of heat without which they could not be nurtured and grow. For everything which is hot and fiery is roused and activated by its own movements, but a thing which is nourished and grows has a definite and regular movement; as long as this remains in us, so long as sensation and life remain, but when heat has been chilled and extinguished,

(*SVF* 2. 351, LS 55I 1–2), Plut. *De Stoic. Rep.* 1053F (*SVF* 2. 449, LS 47M1) and Alexander of Aphrodisias, *De Mixtione* 223. 25–36 Todd (*SVF* 2. 441, LS 47L). For the notion of the soul of the cosmos as cohesive of the cosmos in the Old Academy see Sedley 2002: 63 and Kupreeva in Ch. 6 (pp. 000–00).

[6] See DL 7.134 (*SVF* 2. 299–300, LS 44B), Alex. *Mixt.* 224. 32–225. 3 (*SVF* 2. 310, LS 45H) and Sext. Emp. *M*. 9. 11–12 (*SVF* 2. 301). For the probable origin of this dualism in the Old Academy cf. Theophrastus *ap.* Simplicius, *In Ar. Phys.* 26. 11–13 and Cic. *Acad.* 1. 6. 24–7. 29 discussed by Sedley 2002: 42–3 and Gourinat in Ch. 2.

[7] See *ND* 2. 25–8. These four elements are those referred to at 2. 25 when the text says: 'All the parts of the cosmos (I will however only specify the most important) are supported and sustained by heat' (*Omnes igitur partes mundi (tangam autem maximas) calore fultae sustinentur*). Notice that in this theory both heat and flame are sustained by cosmic heat. Although flame is identified as something destructive at 2. 40–2 (I discuss this passage in detail in sect. 3), it is acknowledged in 2. 25–8 that it may eventually 'salutarem inpertit et vitalem calorem' (at 27). I believe that the two passages are fully consistent with each other provided that we bear in mind that flame is destructive of its *fuel* and not necessarily of that which receives the heat it expels. See sect. 3.

[8] On the Cleanthean character of these sections see Solmsen 1961; Hahm 1977: 272 n. 1; Long and Sedley 1987: ii. 279; and Besnier 1996: 154 (who offers at pp. 153–64 a detailed and very helpful analysis of the structure of the whole passage).

we ourselves die and are extinguished . . . Therefore every living thing, whether animal or vegetable, is alive on account of the heat enclosed within it. From this it must be understood that the element heat has within itself a vital power which pervades the whole cosmos . . . It follows from this that, since all parts of the cosmos are maintained by heat, the cosmos itself too has been preserved over so long a time by a comparable and like element—and all the more so because it must be understood that this hot and fiery entity is extended in every nature in such a way that it contains the power of reproduction and the cause of generation, since it is that by which all living things, including those whose roots are sustained by earth, must be brought to birth and grow.[9]

The heat in question ultimately proceeds from the sun. And what the argument seeks to prove is that this heat is what gives cohesion to the cosmos as a whole. The precise structure of the argument is so elusive that some scholars have surmised either that Cicero has omitted essential parts of it or that the transmitted text is lacunose.[10] One possible reconstruction would proceed as follows. (1) For any living thing, its cohesive cause lies in the inner heat that pervades it, and (2) heat pervades all parts of the cosmos and especially those that are constitutive of all the others. Now (3) if something A pervades each of the several parts of something B, A pervades the whole of B. Therefore, given (1) and (2), it follows that (4) heat is the cohesive cause of each of the several parts of the cosmos, and, given (3), (5) heat is the cohesive cause of the cosmos as a whole.[11]

Let us focus on (1) and (2). As I mentioned in the previous section, Stoic causal theory stipulates that the cohesive cause of something is what keeps it in existence. As soon as the activity of the cause stops, the existence itself of the entity stops. In other words, C is the cohesive cause of something S only if the activity of C and the endurance of S are temporally coextensive with each other. Therefore, the truth of (1) requires as a necessary condition that the endurance of a living thing be temporally coextensive with the presence of heat within it. This coextension is argued for in our passage through the observation that living things invariably die when their inner heat is extinguished and is no longer present in them.

[9] *Sic enim res se habet, ut omnia quae alantur et quae crescant contineant in se vim caloris, sine qua neque ali possent nec crescere. Nam omne quod est calidum et igneum cietur et agitur motu suo; quod autem alitur et crescit motu quodam utitur certo et aequabili; qui quam diu remanet in nobis tam diu sensus et vita remanet, refrigerato autem et extincto calore occidimus ipsi et extinguimur . . .. Omne igitur quod vivit, sive animal sive terra editum, id vivit propter inclusum in eo calorem. Ex quo intellegi debet eam caloris naturam vim habere in se vitalem per omnem mundum pertinentem . . . Ex quo concluditur, cum omnes mundi partes sustineantur calore, mundum etiam ipsum simili parique natura in tanta diuturnitate servari, eoque magis quod intellegi debet calidum illud atque igneum ita in omni fusum esse natura, ut in eo insit procreandi vis et causa gignendi, a quo et animantia omnia et ea, quorum stirpes terra continentur et nasci sit necesse et augescere.*

[10] For discussion, see Hahm 1977: 267–73.

[11] The thesis attested elsewhere for Cleanthes that the sun is the cosmos's ruling principle, or ἡγεμονικόν (Eusebius, *Praep. Ev.* 15. 15. 7 (*SVF* 1. 499, *DG* 465. 5–6) and DL 7. 139 (*SVF* 1. 499, *DG* 332. 23–5)) seems to be a logical consequence of combining (5) with the idea also that the sun's fire is the ultimate source of heat in the cosmos and that something's ἡγεμονικόν is its cohesive cause. These two ideas are present in Cleanthes. See notably *ND* 2. 29–30. For illuminating discussion of this question see Besnier 1996: 156 n. 1.

Most of the text in sections 2. 25–30 is devoted to establishing (2). As a matter of fact, the proof focuses explicitly on the most important constituents (or 'largest': *maxima*) in the cosmos. These are earth, water, air, and fire (both in the form of the individual portions of heat that we find in living things, and in the form of flame insofar as it is a source of heat). If they are *maxima* in the sense that they are what everything else in the cosmos is made out of, it becomes apparent why they are singled out for attention. If every non-elemental entity is made out of them, then, all non-living things, and not just the four elements and the living things, must be pervaded by heat. Let me quote the passage that deals with the case of water (section 26, in Rackham's translation slightly modified):

Atque aquae etiam admixtum esse calorem primum ipse liquor aquae declarat [effusio], quae neque conglaciaret frigoribus neque nive pruinaque concresceret, nisi eadem se admixto calore liquefacta et dilapsa diffunderet; itaque et aquilonibus reliquisque frigoribus adiectis durescit umor, et idem vicissim mollitur tepefactus et tabescit calore. Atque etiam maria agitata ventis ita tepescunt ut intellegi facile possit in tantis illis umoribus esse inclusum calorem; nec enim ille externus et adventicius habendus eat tepor, sed ex intumis maris partibus agitatione excitatus, quod nostris quoque corporibus contingit cum motu atque exercitatione recalescunt.

Likewise, that water contains an admixture of heat is shown first of all by its liquid nature; water would neither be frozen into ice by cold nor congealed into snow and hoar-frost unless it could also become fluid when liquefied and thawed by the admixture of heat; this is why the moisture both hardens when exposed to a north wind or a frost from some other quarter, and it also softens itself when warmed, and evaporates with heat. Also the sea when violently stirred by the wind becomes warm, so that it can readily be realized that this great body of fluid contains heat; for we must not suppose the warmth in question to be derived from some external source, but stirred up from the lowest depths of the sea by a violent motion, just as happens with our bodies when they are restored to warmth by movement and exercise.

The argument may be interpreted in the following way. One possible reason for thinking against Cleanthes that water does not contain heat is that water may become ice and frost. These are the coldest things in the cosmos and as such the thing from which they originate—water—could hardly contain heat. The argument, however, offers sufficient grounds for countering the objection. According to Cleanthes a mass of water freezes only when it is distributed evenly over a certain surface. But this even distribution requires that it be in a fluid or liquid state, and this in turn requires that it be acted upon by heat for only heat can keep it liquid. Now the heat in question, Cleanthes believes, is not extrinsic to the water. Two complementary reasons are given for this.

(i) Although a rise in the temperature of a mass of water may require the warmth of an external factor such as a certain type wind, the function of the external factor is merely to stir up the heat that the mass of water already contains and that is intrinsic to it.

(ii) The change from a liquid state to a solid one, i.e. freezing, may be determined by an external factor, but the change back to a liquid state is something that water does *itself* (*idem*) once the action of external factors diminishes or disappears.

The argument of Cleanthes, therefore, comes out as an ingenious piece of reasoning. It shows that the very fact adduced as evidence against the thesis that heat pervades water presupposes as a necessary condition the truth of the thesis itself. In order to become ice or frost, water must be in liquid state, but in order for it to reach that state it must contain a kind of heat that is intrinsic to it.

3. CLEANTHES' THEORY OF CONFLAGRATION: IS IT CONSISTENT?

I have shown why in Cleanthes heat is the cohesive cause of the cosmos and, hence, its soul. I now turn to his theory of conflagration and to why it implies that the soul of the cosmos consumes its own body.

In Cleanthean physics, the very fire that gives cohesion to the cosmos, heat, is also responsible for gradually desiccating it. This desiccation occurs because of the nature of the process by which the sun gives cohesion to the cosmos. The sun 'is nourished by the vapours exhaled from the ocean' (*ND* 2. 40; cf. 3. 37), and although most of this humidity goes back to the earth under the form of rain, some of it is inevitably consumed by the fire of the sun and the stars (*ND* 2. 118).[12] To quote this latter passage (and Rackham's translation):

sunt autem stellae natura flammeae, quocirca terrae maris aquarum vaporibus aluntur iis qui a sole ex agris tepefactis et ex aquis excitantur; quibus altae renovataeque stellae atque omnis aether refundunt eadem et rursum trahunt indidem, nihil ut fere intereat aut admodum paululum quod astrorum ignis et aetheris flamma consumat.

But the stars are of a fiery nature, and for this reason they are nourished by the vapours of the earth, the sea and the waters, which are raised up by the sun out of the fields which it warms and out of the waters; and when nourished and renewed by these vapours the stars and the whole aether shed them back again, and then once more draw them up from the same source, with the loss of none of their matter, or only of an extremely small part which is consumed by the fire of the stars and the flame of the aether.

This gradual desiccation is what is responsible for the conflagration, as the text indicates immediately afterwards:

Ex quo eventurum nostri putant id de quo Panaetium addubitare dicebant, ut ad extremum omnis mundus ignesceret, cum umore consumpto neque terra ali posset nec

[12] I deal with this issue in detail in Salles 2005. For discussion of the origin of this theory, see Algra 2004: 178 n. 16.

remearet aer, cuius ortus aqua omni exhausta esse non posset; ita relinqui nihil praeter ignem.

From what will happen, our people think what they said Panaetius questioned, namely, that the whole cosmos will ultimately be ignited insofar as, once the humidity has been consumed, neither will the earth be able to be nourished nor will the air be able to flow, being unable to rise upwards once all water has been exhausted; thus nothing will remain but fire. (My translation)

Although the main idea is clear—the conflagration results from the consumption of water by fire[13]—the details of the theory are obscure. In what follows I offer a speculative reconstruction. As a conjecture, we may surmise that, according to this theory, there is a certain analogy between humidity and the heat that the sun irradiates. Just as humidity travels to the sun and back again to the earth in a cycle, the heat that the sun irradiates does not get lost, but also travels back to the sun in an equally cyclical process. If we suppose further that heat travels back to the sun by means of air (presumably by means of winds), a conflagration will have to occur. For given that air cannot move unless it contains humidity (as the text clearly implies), then, once the humidity in the cosmos has been exhausted, the heat on earth, being unable to travel back to the sun, will progressively increase. At this stage heat ignites whatever is left. The conflagration proper would begin when the ignition occurs.[14] Fire, under the form of heat, has already consumed water, but now, under the form of *flame*, it will consume the other two remaining elements, air and earth (and a fortiori everything that contains them and so the whole cosmos). Thus, if we return to our initial question—would Cleanthes accept the premiss that at the conflagration fire consumes the elements?—the answer should be in the affirmative.

Now, it is clear that Cleanthes regards this consumption as destructive of the cosmos as a whole. There is strong evidence for this. First, the gradual consumption of water by the fire of the stars is presented as a *loss*, which suggests something more substantive than a mere qualitative change (water gradually 'disappears': *intereat*). Second, and more importantly, in section 41 of *ND* 2, Cicero points out that for Cleanthes 'this fire of ours that serves the needs of daily life is destructive and it itself consumes everything and wherever it spreads it routs and scatters everything' (*Atqui hic noster ignis, quem usus vitae requirit, confector est et consumptor omnium idemque, quocumque invasit, cuncta*

[13] Cf. Alex. Aphr. *In Meteor.* 61. 34–62. 11 (*SVF* 2. 594) who also attributes to the Stoics the idea that the desiccation caused by the sun is responsible for the destruction of the cosmos.

[14] This particular point is explicitly made in connection with Cleanthes in Stob. *Ecl.* 1. 153. 7–22 (*SVF* 1. 497). At the conflagration, the whole is 'ignited' (ἐκφλογισθέντος τοῦ παντὸς). See also Philo, *Aet.* 90–1: μεταβάλλειν εἰς φλόγα. This does not imply that the fire of the *sun* is itself flame (which it cannot be according to *ND* 2. 41–2; but cf. 2. 118). It merely means that the fire of the sun—whether it is heat or flame—causes the ignition of the cosmos and its transformation into flame.

disturbat ac dissipat). It is true that in this passage Cleanthes is referring to flame and not to heat, which, in sharp opposition to flame, is 'vital and salutary; it preserves everything, feeds it, fosters its growth, sustains it and bestows it perception' (*vitalis et salutaris omnia conservat, alit, auget, sustinet sensuque adficit*). At the conflagration, however, when heat has exhausted all the existing water, it *ignites*, i.e. turns into *flame*, whatever is left: *omnis mundus ignesceret*. Given the nature of flame, the flames thus produced destroy everything else. In this respect, the conflagration in Cleantes is destructive indeed of the whole cosmos.[15]

In the next section I explain how exactly consumption differs from separation: in contrast with the latter, the former involves a transformation of what is consumed by what consumes it. The contrast is sharp because a separation between two entities involves a differentiation between them, one that would be lost if one is transformed into the other. We may therefore conclude that, if it is true that the soul of the cosmos consumes its body at the conflagration, one cannot also maintain that the conflagration destroys the cosmos: as an animal, its destruction would require the *separation* of its soul and body, but no such separation occurs at the conflagration if, as Cleanthes accepts, the soul of the cosmos consumes its body.

This tension in Cleanthes seems to result ultimately from a conflict between two irreconcilable intuitions within his cosmology about the conflagration. One of them rests on a certain conception of the action of heat and flame upon the other elements. The former consume the latter and in this process the latter are destroyed by the former. In this way they cease to exist and are substituted by something else. On this view, the conflagration is destructive of the cosmos because it destroys its basic constituents. The other intuition rests on the idea of the cosmos as an animal and of death as the separation of soul and body. According to this idea, the death and the destruction of the cosmos would require, not the consumption of the body by fire, but rather the *removal* of fire from the body. On this intuition, the conflagration is not destructive of the cosmos because it does not involve what it takes for the death of the cosmos as a living thing to be effected.

How can the tension between these two intuitions be dissolved? One way would be by arguing that the transformation of the other elements into fire is not destructive of them. If this is granted there is no further reason to suppose that fire is destructive of the basic constituents of the cosmos, which was the reason why the tension above arose in the first place. As I shall argue in the next section, Chrysippus offered this new account.

[15] Given his identification of god with the cosmos (*ND* 2. 30: *deum esse mundum*), his thesis that the conflagration destroys the cosmos seems logically to necessitate the (absurd) conclusion that god himself is destructible. For discussion of this issue, see Long 1990: 284–6, discussed in Algra 2004: 177–87.

4. TRANSFORMATION IN CHRYSIPPUS' ELEMENTAL THEORY

As will be seen in this section, Chrysippus, based on an elemental theory which differs from that of Cleanthes, maintains that elemental transformation does not involve the destruction of anything, but is merely a change in the density of fire.

In the Plutarch passage that preserves Chrysippus' argument for the immortality of the cosmos (*De Stoic. Rep.* 1052C), the consumptive action of fire upon the elements is contrasted with the process of separation between soul and body that characterizes death. The difference intended is not stated explicitly in the passage itself. But one possibility is that consumption, in contrast with separation, involves a *transformation*. When A is consumed by B (but not when it separates from B) A is transformed into B. The thesis goes back to Cleanthes himself, who is reported by Plutarch in *CN* 1075D to have said that at the conflagration, when the sun consumes all else, it 'assimilates to itself and transforms into itself the moon and all the rest of the stars' (*ἐξομοιῶσαι πάντα* <sc. *τὴν σελήνην καὶ τὰ λοιπὰ ἄστρα*> *ἑαυτῷ καὶ μεταβαλεῖν εἰς ἑαυτόν*). In Chrysippus, the connection between consumption and transformation is implied in the link put forward in *De Stoic. Rep.* 1052C between consumption and nourishment. For nourishment is a process where the food is transformed into what is fed and in which the size of what is fed *increases* as a result of this transformation.[16] As will be seen, however, a crucial difference between Chrysippus and Cleanthes is that in Chrysippus this transformation does not involve destruction.

The transformation of air, water, and earth into fire is a central topic in Chrysippus' elemental theory (important echoes of which are found in some Presocratics, for example, Anaximenes (DK 13A5–7) and Heraclitus (DK 22A5)). According to this theory, fire is more basic than the other three elements. It is the element in the strictest sense because it is not only what air, water, and earth are made from, but also what they are made *of*. In particular, all three are fire with three different degrees of compression or condensation. In consequence, the transformation of any of them back into fire is nothing but a decompression of fire itself. In this process there is no substantial destruction of anything. This aspect of Chrysippus' elemental theory is reported in Stobaeus, *Ecl.* 1. 129. 7–11 and 18–24, which I do not interpret in the exactly the same way as Cooper does in Chapter 4 (pp. 107–15).[17] I quote the text as edited by Long and Sedley and use their translation.

τὸ δὲ [πῦρ καὶ] κατ' ἐξοχὴν στοιχεῖον λέγεσθαι διὰ τὸ ἐξ αὐτοῦ πρώτου τὰ λοιπὰ συνίστασθαι κατὰ μεταβολὴν καὶ εἰς αὐτὸ ἔσχατον πάντα χεόμενα διαλύεσθαι, τοῦτο

[16] This is also implicit in Plut. *De Stoic. Rep.* 1052C–E.

[17] Although I agree with some of the central ideas put forward by Cooper (as I explain in the Introduction), we do not interpret the usage of the term 'element' referred to in the first line of

δὲ μὴ ἐπιδέχεσθαι τὴν εἰς ἄλλο χύσιν ἢ ἀνάλυσιν . . . πρώτης μὲν γιγνομένης τῆς ἐκ πυρὸς κατὰ σύστασιν εἰς ἀέρα μεταβολῆς, δευτέρας δ' ἀπὸ τούτου εἰς ὕδωρ, τρίτης δ' ἔτι μᾶλλον κατὰ τὸ ἀνάλογον συνισταμένου τοῦ ὕδατος εἰς γῆν. πάλιν δ' ἀπὸ ταύτης διαλυομένης καὶ διαχεομένης πρώτη μὲν γίγνεται χύσις εἰς ὕδωρ, δευτέρα δ' ἐξ ὕδατος εἰς ἀέρα, τρίτη δὲ καὶ ἐσχάτη εἰς πῦρ.

The element *par excellence* is so called because the remainder are composed out of it in the first place by alteration and into it lastly everything is diffused and dissolved, but it does not admit of diffusion or resolution into something else . . . the first change to occur is the one from fire into air by condensation, and the second ensuing from this into water, and the third with water being still more compressed on the same principle, into earth. Reciprocally, from the dissolution and diffusion of earth, the first diffusion is into water, the second from water into air, the third and last into fire.

Notice that according to the text water and earth are, respectively, compressed air and water. The idea that they both are compressed *fire* is not explicitly stated (only air is explicitly referred to as such). But unless we had strong independent reasons for thinking that Chrysippus denied the transitivity of compression, water is ultimately compressed fire if it is compressed air and air is compressed fire. And if so, then, given that earth is compressed water, earth too must be ultimately compressed fire. This passage therefore seems to be good evidence that Chrysippus believed that earth, water, and air are indeed, *all three*, compressed fire.

No distinction is made in this text between types of fire. But we may assume that for Chrysippus heat and flame are types of fire, that they differ from each other by their density, and that heat is more dense than flame. The full list of stages in the process leading from the densest of all the elements (earth) to the thinnest type of fire (flame) would be the following:

earth → water → air → fire 1 (heat) to fire 2 (flame)[18]

The whole theory may explain why Chrysippean air, water, and earth are not destroyed at the conflagration. They are not because that into which each of them gets transformed is nothing other than what each of them really is, namely, fire. If we follow this idea from Chrysippus' elemental theory, the body of the cosmos is a compressed portion of its own fiery soul. The compression of portions of fire into air, water, and earth explains the differentiation between cosmic soul and

the present passage (τὸ κατ' ἐξοχὴν στοιχεῖον) in the same way. According to Cooper, this usage refers to a pre-cosmic fire ('proto-fire'), whereas in my view it refers to ordinary fire: it is ordinary fire that transforms itself by condensation into ordinary air, water, and earth (and back into itself through rarefaction) as is explained here. In other words, I interpret these lines as a description of Chrysippus' version of ordinary, not pre-cosmic, reciprocal elemental change.

[18] It is an open question whether in Chrysippus anything that tranforms itself into 'fire 2' must first transform itself into 'fire 1' (as earth, for instance, cannot be transformed into air directly, but must first be transformed into water). The answer is presumably 'no' in the particular case of the conflagration. If we believe that Chrysippus followed Cleanthes in maintaining that heat causes the ignition of earth, heat was not regarded by him as a step in their transformation into flame, but as the cause of this transformation.

body at the cosmogony. But this differentiation is lost at the conflagration when by decompression or rarefaction the elements are transformed back into fire in its rarest form, i.e. flame.[19]

This conception of the conflagration is based on an elemental theory that Chrysippus took and developed from Zeno.[20] A quite different conception of the elements, however, emerges from the reports we have of specifically Cleanthean views about the elements: Cic. *ND* 2. 23–8 (which we examined in section 2), and Stob. *Ecl.* 1. 153. 7–22 (= *SVF* 1. 497, which is the only report in which Cleanthes' position on cosmology is differentiated from those of other Stoics, notably Zeno, cf. 1. 152. 19–153. 6 = *SVF* 1. 102). There is no trace in either text that air, water, and earth are made *of* fire nor, a fortiori, that their reciprocal change is just a qualitative change in the density of fire. As a matter of fact, Stobaeus reports that the cosmogony in Cleanthes begins when a residual mass of fire that was left from the conflagration starts growing (*αὔξεσθαι*) from the outer layers of the cosmos towards its centre and begins a new cosmogony.[21] Given its very nature, this growth cannot mean the gradual *condensation* of fire that Chrysippus envisages in his conception of the change of fire into the other elements nor, in John Cooper's interpretation of the cosmogony in Chrysippus (in Chapter 4), the process by which 'proto-water' generates the four elements at the cosmogony. As for Cic. *ND* 2. 41, the emphasis in Cleanthes' theory is laid on the notion that air, water, and earth contain fire. It acts upon them from inside to give them cohesion, inevitably consuming them in the long run. In this theory the four elements seem to be different and equally basic substances that are not reducible to or transformable into each other. Notice that Cleanthes' idea that the four elements are different substances acting upon each other is logically compatible with Chrysippus' idea that one of them is more basic than the other three in the sense that the latter are compressed versions of the former. To evince this compatibility think of the cohesive action of fire upon water. From a Chrysippean perspective, this action could be viewed as the action of a mass of uncompressed fire acting upon a mass of compressed fire. However, none of this is suggested in the sources we have for Cleanthes, and it is possible that he did not perceive or develop this compatibility between the Zenonian-Chrysippean model and his own.[22]

[19] This flame (which I identify with the fire of the conflagration in Chrysippus) will be extinguished once the other elements have totally transformed themselves into it. Whatever is left from this extinction will be the originary substance out of which the new cosmos will be generated. See the Introduction and Ch. 4.

[20] As Stobaeus clearly indicates at *Ecl.* 1. 129. 3–4, at the beginning of the section to which the present passage belongs.

[21] For the notion of residual fire, see also Philo, *Aet.* 89: *ἔφασαν ὅτι μετὰ τὴν ἐκπύρωσιν, ἐπειδὰν ὁ νέος κόσμος μέλλῃ δημιουργεῖσθαι, σύμπαν μὲν τὸ πῦρ οὐ σβέννυται, ποσὴ δέ τις αὐτοῦ μοῖρα ὑπολείπεται* (no individual Stoic is mentioned by Philo in connection with this view).

[22] Interestingly, the idea that something can give cohesion to itself, which is implicit in the example I just mentioned, *is* attested for Chrysippus. See Plut. *Comm. Not.* 1085 C–E (*SVF* 2. 444).

5. RECAPITULATION

It is no novelty that Chrysippus transformed the Stoic concept of conflagration into something positive, in which the notions of conflagration and of destruction are no longer associated with each other.[23] But I have argued that this evolution is best understood as part of an argument for the immortality of the cosmos that evinces a tension in Cleanthes' cosmology. Cleanthes cannot maintain, as he does, (i) that the conflagration destroys the cosmos, (ii) that fire is its soul, *and* (iii) that at the conflagration its soul does not separate from its body. These three theses form an inconsistent triad because the conjunction of (ii) and (iii) implies the non-destruction of the cosmos at the conflagration, which is the negation of (i). The philosophical basis of Chrysippus' conception of the indestructibility of the cosmos is to be found in his elemental theory. Its indestructibility requires the indestructibility of its basic material constituents, and these constituents are not destroyed when they are consumed by the fire of the conflagration. The reason is that this consumption is, in reality, a transformation of the basic constituents of the cosmos into fire. And this transformation is nothing but a qualitative change of fire. In this process, it simply moves from a state of compression to one of total decompression. Cleanthes and Chrysippus seem to describe the conflagration in very similar terms: according to the former, at the conflagration 'nothing will remain but fire' (*relinqui nihil praeter ignem*: Cic. *ND* 2. 118) and, according to Chrysippus, at the conflagration the cosmos will be 'fiery through and through' (διόλου πυρώδες: Plut. *De Stoic. Rep.* 1053B, *SVF* 2. 605, *LS* 46F). The difference is that in Cleanthes, given his conception of fire, this stage of the conflagration is destructive of everything, whereas for Chrysippus it clearly is not.

To end, one last question is in order. How is it possible for the cosmos itself to persist during the time when only its soul is in existence as is the case in the conflagration? According to Chrysippus' theory, once the body of the cosmos has turned into its soul, its soul will exist in a disembodied state until a new cosmogony begins. And yet Chrysippus maintains that the *cosmos* itself should not be said to die. But how can it? This issue is not addressed in any of the sources we have considered so far. The inference from 'the disembodied soul of the cosmos exists' to 'the cosmos itself exists' seems to require the further thesis that the identity of the cosmos is fixed by its soul and neither by its body nor by the combination of its soul and its body. There might be evidence for this thesis

[23] The evidence relating to the positiveness of the conflagration in Chrysippus is presented and amply discussed in Mansfeld 1979: 174–83. See also Mansfeld 1981: 304–9.

in Stoicism,[24] but I shall bypass the problem and leave its discussion for another occasion.

[24] Eusebius (*Praep. Ev.* 15. 20. 6, *SVF* 2. 809, LS 53W), according to whom the 'Stoics' believe that 'we ourselves remain when our souls have separated from our body' (διαμένομεν ἡμεῖς ψυχαὶ γενόμενοι τοῦ σώματος χωρισθέντες). The thesis that our identity is fixed by our soul, alone or mainly, is part of a tradition that goes back to ps.-Plato, *Alcibiades* (129D–131B) and Aristotle, *NE* 9 and 10 (1168b31–1169a3 and 1177b26–1178a22, with commentary by Cooper 1986: 144–80).

APPENDIX

There are at least two other Stoic arguments for the indestructibility of the cosmos that deserve consideration. In what follows, I provide a brief analysis of them and point out the substantive differences that separate them from each other and from Chrysippus' argument in Plutarch.[25]

In his commentary on the *Physics* (*In Ar. Phys.* 1121. 12–15 Diels = *SVF* 2. 576), Simplicius attributes to the Stoics in general a view that, according to Simplicius, also characterizes some Presocratics.

γενητὸν δὲ καὶ φθαρτὸν τὸν ἕνα κόσμον ποιοῦσιν, ὅσοι ἀεὶ μέν φασιν εἶναι κόσμον, οὐ μὴν τὸν αὐτὸν ἀεί, ἀλλὰ ἄλλοτε ἄλλον γινόμενον κατά τινας χρόνων περιόδους, ὡς Ἀναξιμένης τε καὶ Ἡράκλειτος καὶ Διογένης καὶ ὕστερον οἱ ἀπὸ τῆς Στοᾶς.

Those who affirm that the cosmos is everlasting, but that it is not always the same but different at different moments according to certain periods of time, make the one cosmos generable and destructible, as Anaximenes, Heraclitus, Diogenes and, later, those from the Stoa <maintained>.

The idea I wish to focus on is that, in a sense, the cosmos is both everlasting and destructible (and generable). Assuming that in this latter case the destruction is caused by the conflagration, it follows from the theory reported by Simplicius that, in different senses, the conflagration both is and is not destructive of the cosmos. The explanation, according to the passage, is that the cosmos 'is not always the same but different at different times'. Depending on how we construe the non-identity in question—numerical or qualitative—the theory may be interpreted in at least two different ways. On the numerical reading, once the conflagration that will destroy the present cosmos subsides, a new cosmos will be generated. This new cosmos will be numerically different from the present one, but qualitatively identical to it. In other words, the present cosmos and the new one will be different tokens of exactly the same type. On this view, the cosmos is everlasting in type even though each of its tokens is destructible. Alternatively, on the qualitative reading, the new cosmos will be numerically the same cosmos as this one, but with some qualitative differences. They are token-identical and this single token changes some of its qualities from one cosmic cycle to another. On this alternative view, the cosmos as a token is everlasting although some of its qualities are lost at the conflagration and replaced by different ones in the new cycle.

The text of Simplicius is ambiguous as to which of these two readings matches the meaning originally intended by the Stoics. But however we interpret the text, the reasons this theory adduces for the indestructibility of the cosmos are totally different from those adduced by Chrysippus in Plutarch's *De Stoic. Rep.* 1052C. Moreover, Chrysippus would

[25] In addition to these two texts, see also Zeno *ap.* Philargurios in *SVF* 1. 108 and Chrysippus *ap.* Eusebius in *Praep. Ev.* 15. 18. 2 (*SVF* 2. 596, LS 46K), which I do not discuss in the present chapter but which seem to report this same argument.

actually reject this theory on either interpretation. For there is evidence elsewhere that for him the cosmos is both numerically *and* qualitatively the same throughout the infinite series of cosmic cycles.[26] So although the present theory offers an argument for the indestructibility of the cosmos, the argument should be sharply distinguished from the one in the Plutarch. Furthermore, it should not be attributed to Chrysippus.

The second argument to be considered clearly belongs to Chrysippus. It is not explicitly stated anywhere, at least not in full. But it may be extracted from a report in Diogenes Laertius of the Stoic distinction between three different senses of the term 'cosmos' (7. 137–8 = *SVF* 2. 526, *LS* 45F). Diogenes attributes the distinction to the Stoics in general, but it is attested elsewhere for Chrysippus by name (Stob. *Ecl.* 1. 184. 8–12, *SVF* 2. 527, *DG* 465. 14–17), which strongly suggests that it is an innovation of his. (There is also evidence that it was later borrowed by Posidonius; cf. *DL* 7. 138 = EK 14.) Here is the Diogenes text:

Λέγουσι δὲ κόσμον τριχῶς· αὐτόν τε τὸν θεὸν τὸν ἐκ τῆς ἁπάσης οὐσίας ἰδίως ποιόν, ὃς δὴ ἄφθαρτός ἐστι καὶ ἀγένητος, δημιουργὸς ὢν τῆς διακοσμήσεως, κατὰ χρόνων ποιὰς περιόδους ἀναλίσκων εἰς ἑαυτὸν τὴν ἅπασαν οὐσίαν καὶ πάλιν ἐξ ἑαυτοῦ γεννῶν· καὶ αὐτὴν δὲ τὴν διακόσμησιν [τῶν ἀστέρων][27] κόσμον εἶναι λέγουσι· καὶ τρίτον τὸ συνεστηκὸς ἐξ ἀμφοῖν.

They speak of cosmos in three ways: <in one> it is god himself who is the peculiarly qualified individual consisting of all substance, indestructible and ingenerable, being the demiurge of the cosmic order and consuming at set periods of time the whole substance into himself and reproducing it again from himself; they also say that cosmos is the cosmic order itself of the cosmos; and thirdly it is what is composed out of both.

An argument for the cosmos' indestructibility is implicit in this passage: (1) even though the cosmic order is periodically destroyed by a conflagration, there is a sense in which the cosmos is not the cosmic order itself but that which generates it and into which it is dissolved, namely god; but (2) god is indestructible; therefore, (3) there is a sense in which the cosmos itself is indestructible.

The present argument is similar in some important respects to the one in Plutarch. In particular, they both use the same apparently technical term *ἀνάλωσις* to refer to how the fire acts upon the cosmos. As in Plutarch (*ἐξανάλωσις*, *κατανάλωσις*), it is a process by which fire *consumes* it (*ἀναλίσκω*). And like the argument in Simplicius, the present argument exploits the crucial idea that the term 'cosmos' may be used in different senses. In one of them, it carries the connotation of indestructibility (and ingenerability). But the present argument is clearly not the same as either of the other two. Unlike the one in Simplicius, it identifies the special sense in question, not by reference to the type–token distinction, but by reference to god: it is because the cosmos is, in a sense, identical to *god* that it is, in a sense, indestructible. This is also an important difference between the present argument and the one in Plutarch, which argues for indestructibility of the cosmos without appealing to the idea of god. The key notion in the argument in Plutarch is that of fire as soul of the cosmos as opposed to its matter or body. The identification of fire as soul is made, not through the idea of god, but through an analogy between the cosmos and animals: in both cases, it is argued, fire is cohesive.

[26] For numerical identity see Alex. Aphr., *In Ar. A. Pr.* 180. 33–6 Wallies (*SVF* 2. 624). For qualitative identity see Nemesius, *De Natura Hominis* 111. 20–112. 3 Morani (*SVF* 2. 625).

[27] Rightly deleted by Long and Sedley (ii. 268) as an interpolation.

In sum, Simplicius, Diogenes Laertius, and Plutarch should not be read as reporting the same argument. In all three the intended conclusion is the same, namely, that the cosmos is not destroyed at the conflagration. But in each case the conclusion is argued for on different grounds: physical in Plutarch (elemental change), logical in Simplicius (type–token distinction), and theological in Diogenes (god–cosmos identity). Moreover the argument in Simplicius could hardly be Chrysippean.

REFERENCES

Algra, K. (2004) 'Eternity and the Concept of God in Early Stoicism', in G. van Riel and C. Macé (eds.), *Platonic Ideas and Concept Formation in Ancient and Medieval Thought* (Leuven: Leuven UP).

Besnier, B. (1996) 'La Nature dans le livre II du *De natura deorum* de Cicéron', in C. Lévy (ed.), *Le Concept de nature à Rome: La Physique* (Paris: Presses de l'École Normale Supérieure).

Cooper, J. (1986) *Reason and Human Good in Aristotle* (Indianapolis, Ind.: Hackett).

Hahm, D. (1977) *The Origins of Stoic Cosmology* (Columbus, Ohio: Ohio State UP).

Long, A. A. (1990) 'Scepticism about Gods in Hellenistic Philosophy', in M. Griffith and D. Mastronarde (eds.), *Cabinet of the Muses: Essays on Classical and Comparative Literature in Honor of Thomas G. Rosenmeyer* (Atlanta, Ga.: Scholars Press).

—— and Sedley, D. N. (1987) *The Hellenistic Philosophers* (2 vols. Cambridge: CUP).

Mansfeld, J. (1979) 'Providence and the Destruction of the Universe in Early Stoic Thought: With Some Remarks on the "Mysteries of Philosophy"', in M. J. Vermaseren (ed.), *Studies in Hellenistic Religions* (Leiden: Brill).

—— (1981) 'Bad World and Demiurge: A "Gnostic" Motif from Parmenides and Empedocles to Lucretius and Philo', in R. van den Broek and M. J. Vermaseren (eds.), *Studies in Gnosticism and Hellenistic Religions* (Leiden: Brill).

—— (1999) 'Theology', in K. Algra *et al.* (eds.), *Cambridge History of Hellenistic Philosophy* (Cambridge: CUP).

Salles, R. (2005) 'Ekpúrosis and the Goodness of God in Cleanthes', *Phronesis*, 50(1): 56–78.

Sedley, D. (2002) 'The Origins of Stoic God', in D. Frede and A. Laks (eds.), *Traditions of Theology* (Leiden: Brill).

Solmsen, F. 1961: 'Cleanthes or Posidonius? The Basis of Stoic Physics', *Medelingen der Koninklijke Nederlandse Akademie van Wetenschappen, Afdeeling Letterkunde* 24.

6

Stoic Themes in Peripatetic Sources?

Inna Kupreeva

The question of relation between Aristotelian and Stoic philosophy of nature is complex and controversial. There is a notorious problem of the use of Aristotle by the early Stoics. Numerous affinities of approach suggest that Stoics did make use of Aristotelian legacy when formulating their own theories,[1] but the evidence, such as references to Aristotle in the Stoic works, is not forthcoming, and healthy scepticism seems to be indicated.[2] A different but related set of problems arises in connection with post-Aristotelian Peripatos, particularly Hellenistic and post-Hellenistic, for which we have less historical and literary evidence than for the schools of the Imperial period. In doxographical outlines, Stoic and Aristotelian positions are often presented as close to each other.[3] Some parallels in terminology and arguments have been pointed out by scholars.[4] We have evidence that some members of the Stoic school are influenced by Aristotelian physical doctrines.[5] On the other hand, the debates between the two schools in late antiquity are also well-documented. We know of anti-Stoic polemic by Alexander of Aphrodisias, chair of Aristotelian philosophy in Athens at the end of the second century AD (his criticism of Stoic theory of mixture, determinism, corporealist metaphysics and psychology, theory of elements, conflagration);[6] of anti-Aristotelian arguments about void by the Stoic Cleomedes.[7]

In this chapter, I am going to give only a brief preliminary survey of several texts which might suggest a close affinity between the Peripatetic and Stoic doctrines of principles. The goal is to see to what extent these parallels may amount to a genuinely shared position on any of the issues under examination. The discussion is organized by source rather than by topic. I start with the Antiochean account of the physics of 'Old Academy' presented in Cicero's *Acad.* 1. 24–9;

[1] This has been well shown in the seminal work by D. E. Hahm (1977).

[2] Famously argued by F. H. Sandbach (1985).

[3] Cf. Galen, *Propr. Plac.* 15. 1 Nutton; *Nat. Fac.* 1. 5. 12; Alexander, *De Prov.* 3. 15 Thillet.

[4] See below at pp. 140–1, 150 n. 90, and 154–55.

[5] *Mixt.* 3. 216. 9–15 Bruns.

[6] There are several detailed studies of these debates, cf. Todd 1976; Sharples 1983; Moraux 2001. On Alexander's school being located in Athens, see data in Chaniotis 2004.

[7] Bowen and Todd 2004.

continue in section 2 with 'physical' fragments of Peripatetic Critolaus, trying to see whether they could provide any ground for Antiochus' incorporation of Peripatetic physics. The third section deals with the 'suspect' cases of Stoicizing in the Peripatetic tradition: Xenarchus' arguments against the fifth body and the use of the concept 'prime matter' by Boethus of Sidon and Nicolaus of Damascus. In the fourth section, I look at some parallel discussions in Alexander of Aphrodisias.

1

Perhaps one of the earliest known statements of 'harmony' between Stoa and Peripatos is the one attributed to Antiochus of Ascalon (*c.*130–68 BC), who reinvents the system of the 'Old Academy' as shared by the three schools, including Peripatetics. The passage from Cicero's *Acad.* 1. 6. 24–9 which contains the 'physical' section of an epitome of Antiochus' system (presented in the dialogue by Varro) is well-known and has been discussed in recent literature.[8] The peculiarity of this epitome is that it forms a part of doxographical background to Antiochus' argument against the 'sceptical' Academy. The argument is complex and has as its main battlefield 'logic' (i.e. epistemology).[9] However, the physical section plays an important role as well, because a shared cosmological framework would make the adoption of Stoic epistemological approach more natural. There is evidence that Cicero did make a special effort to present the Antiochean side of the argument accurately.[10]

According to Varro, who speaks on behalf of Antiochus in Cicero's dialogue, this system traces its origin back to the Old Academy under its third scholarch Polemo, who taught Zeno of Citium. Zeno does 'institute a reform' of the system; but the description of this reform does not suggest any destructive purpose.[11] In fact, the physical doctrine of 'Polemo' seems to be least affected by changes, compared to ethics and logic. In *Acad.* 1. 39, Varro mentions only two points of this doctrine to which Zeno took exception. The first has to do with the existence of the fifth element; the second is Zeno's corporealist thesis: an incorporeal thing cannot have causal efficacy (both will be discussed shortly). The attribution to Polemo has been recently examined in detail by D. N. Sedley who concluded largely in its favour on the basis of his analysis of multiple parallels with the physics of the *Timaeus*.[12] I am going to focus on the features of this text which

[8] Görler 1990; Sedley 2002; Gourinat in this volume (Ch. 3).

[9] Antiochus' own polemical context is set by his debate against Philo's version of the history of the Academy in the 'Roman Books' (for which see Brittain 2001: 169–254).

[10] *Att.* 13. 19. 5, quoted and discussed in Griffin 1997: 16; see also Görler 1990: 123–6; Brittain 2006: pp. xxxv–xxxviii.

[11] *Acad.* 1. 35: *sed Zeno cum Arcesilam anteiret aetate valdeque subtiliter dissereret et peracute moveretur, corrigere conatus est disciplinam.*

[12] Sedley 2002; cf. Frede 2005: 217–19

might be imported by Cicero's Antiochus from Stoic or Peripatetic sources. (I reproduce the text in full.)

(1) 24 Of nature they spoke thus: they divided it into two aspects, one being active, another as it were yielding itself to it as the one which something could be produced. They held that power is in the active aspect, and something called matter in the one acted upon; yet so that each one is in the other, since neither matter can exist without being held together by some power, nor power without any matter (for there is no thing that would not be constrained to being somewhere). But that which is made up of both they already called body and as it were some quality< . . . >.[13]

(2) 26< . . . > So, some of those qualities are primary, others derived from them. The primary ones are of one kind and simple, and the derivative ones various and as it were of many sorts. Thus air (for we also use this word as Latin) and fire and water and earth are the first; from these are derived the forms of things with soul and those that are generated from the earth. Therefore the former [qualities] are called principles and (to translate from the Greek) elements; among these, air and fire have a power of motion and production, and the others, viz. water and earth, [a power] of receiving and as it were suffering. Aristotle thought that there was a fifth kind, from which stars and minds are made, unique and unlike those four I have mentioned.[14]

(3) 27 But they regard as underlying everything that has no appearance and lacks any of these mentioned qualities (let us by frequent usage make this word more familiar and common), something called matter, from which everything is moulded and wrought, which can receive everything and undergo all kinds of change in every part, and even pass away, not into nothing but into its parts which can be cut and divided infinitely, since in nature there is no minimum which cannot be divided, but all things that are moving move by infinitely divisible intervals.[15]

(4) 28 And since that power which we named 'quality' moves in this way and since it so turns this way and that, they believe that this very matter as a whole in its entirety moves along, and that those are produced which they call 'the qualified'; from these in the whole of nature which is cohesive and continuous with all its parts a single cosmos is produced, outside which there is no part of matter and no body, but all the things that are in the world

[13] 24 *de natura autem . . . ita dicebant ut eam dividerent in res duas, ut altera esset efficiens, altera autem quasi huic se praebens, ex qua efficeretur aliquid. in eo quod efficeret vim esse censebant, in eo autem quod efficeretur materiam quandam; in utroque tamen utrumque, neque enim materiam ipsam cohaerere potuisse si nulla vi contineretur, neque vim sine aliqua materia (nihil est enim quod non alicubi esse cogatur). sed quod ex utroque, id iam corpus et quasi qualitatem quandam nominabant* < . . . >.

[14] 26 < . . . > *Earum igitur qualitatum sunt aliae principes, aliae ex his ortae. Principes sunt unius modi et simplices; ex his autem variae ortae sunt et quasi multiformes. Itaque aer (hoc quoque utimur iam pro Latino) et ignis et aqua et terra prima sunt; ex his autem ortae animantium formae earumque rerum quae gignuntur e terra. Ergo illa initia et (ut e Graeco vertam) elementa dicuntur; e quibus aer et ignis movendi vim habent et efficiendi, reliquiae partes accipiendi et quasi patiendi, aquam dico et terram. Quintum genus, e quo essent astra mentesque, singulare eorumque quattuor quae supra dixi dissimile Aristoteles quoddam esse rebatur.*

[15] 27 *Sed subiectam putant omnibus sine ulla specie atque carentem omni illa qualitate (faciamus enim tractando usitatius hoc verbum et tritius) materiam quondam, e qua omnia expressa atque efficta sunt, quae omnia accipere posit omnibusque modis mutari atque ex omni parte, atque etiam interire, non in nihilum sed in suas partes, quae infinite secari ac dividi possint, cum sit nihil omnino in rerum natura minimum quod dividi nequeat; quae autem moveantur, omnia intervallis moveri, quae intervalla item infinite divide possint.*

are its parts, which are held by the sentient nature in which resides perfect reasoning, the same and eternal (for there is nothing more powerful, which could cause it to pass away);[16]

(5) 29 They say that this power is the soul of the world, being mind and perfect wisdom, which they call god, and as it were the providence for all things that are subordinate to it, providing first and foremost for heavenly bodies, then on earth things that concern the humans. They also call it necessity because nothing can be otherwise than has been established by it in the middle of as though fated and immutable nexus of eternal order. Sometimes too they call it chance because it produces many things that are unforeseen and not expected by us because of the obscurity of their causes and our ignorance of the same.[17]

1. A parallel for two principles, active (involving 'power') and passive (involving 'matter'), is indeed found already in Theophrastus' summary of Plato's natural philosophy.[18] But the terminology of the passage is also close to Stoic.[19] The word 'power' (*vis* = δύναμις) is used to refer to the active principle in several late reports.[20] We are not told explicitly whether the principles are corporeal. We can infer that they are not, both from Zeno's 'exception' and from the way the notion of the composite is introduced.[21] On the other hand, they are said not to exist without each other, i.e. outside a corporeal substance: this might strike some Platonist thinkers as a rather weak sense of incorporeality.[22] The choice of the word 'power' may go back to Plato, *Sophist* 246A, where δύναμις is used to secure a necessary compromise for a discussion between the 'Gods' and the 'Giants'.[23] Since Antiochus' unification

[16] 28 *Et cum ita moveatur illa vis quam qualitatem esse diximus et cum sic ultro citroque versetur, et materiam ipsam totam penitus commutari putant et illa effici quae appellant qualia, e quibus in omni natura cohaerente et continuata cum omnibus suis partibus unum effectum esse mundum, extra quem nulla pars materiae sit nullumque corpus, partes autem esse mundi omnia quae insint in eo quae natura sentiente teneantur, in qua ratio perfecta insit quae sit eadem sempiterna (nihil enim valentius esse a quo intereat);*

[17] 29 *quam vim animum esse dicunt mundi, eandemque esse mentem sapientiamque perfectam, quem deum appellant, omniumque rerum quae sint ei subiectae quasi prudentiam quondam, procurantem caelestia maxime, deinde in terris ea quae pertineant ad homines; quam interdum eandem necessitatem appellant, quia nihil aliter possit atque ab ea constitutum sit inter quasi fatalem et immutabilem continuationem ordinis sempiterni; non numquam quidem eandem fortunam, quod efficiat multa improvisa ac necopinata nobis propter obscuritatem ignorationemque causarum.*

[18] fr. 230 FHSG (= Simplicius, *In Phys.* 25. 5–15 Diels); Sharples 1995: 67–73, cf. Sedley 2002: 63.

[19] Cf. DL 7. 134.

[20] *SVF* 2. 311 (= Sextus, *M.* 9. 75); 2. 444 (= Plut. *Comm. Not.* 1085C); 2. 1044 (= Alex. *Mixt.* 225. 18 Br.). cf. also the usage in [Aristotle], *De Mundo* 6 and discussion in Moraux 1984: 37–48.

[21] If we take the force of *iam* to be adversative with respect to the preceding clause.

[22] In fact, Brittain 2006: p. xxxii, suggests that Antiochus probably accepted both Zeno's corrections, including the one to do with corporeality of the first principles.

[23] *Soph.* 247D8–E4: λέγω δὴ τὸ καὶ ὁποιανοῦν τινα κεκτημένον δύναμιν εἴτ' εἰς τὸ ποιεῖν ἕτερον ὁτιοῦν πεφυκὸς εἴτ' εἰς τὸ παθεῖν καὶ σμικρότατον ὑπὸ τοῦ φαυλοτάτου, κἂν εἰ μόνον εἰς ἅπαξ, πᾶν τοῦτο ὄντως εἶναι· τίθεμαι γάρ ὅρον ὁρίζειν τὰ ὄντα ὡς ἔστιν οὐκ ἄλλο τι πλὴν δύναμις. For discussion, see Brunschwig 1988: 64–76 and John Cooper in Ch. 4 (pp. 99–101) above. There is no question of an exact parallel between the ontology of the *Sophist* passage and Antiochus; only a way of establishing a terminological agreement between the two antagonistic positions (I am grateful to Ricardo Salles for querying this point).

project ultimately caters for both 'Gods' and 'Giants', the usage may be intentional.[24]

The unity of the cosmos which is held together by its active principle defined as god is stated a little later (in (4) and (5)), but the philosophical account of this unity is not specified (three different types are possible: Stoic monism, Platonic dualism, and Aristotelian 'pluralism' of principles). It is important to note that god is identified with the active principle which is an aspect of nature. This suits the Stoics with their immanent theology, but probably not Aristotle who speaks of the first unmoved mover as a transcendent principle.[25]

2. There is no indication of a specific method by which the first qualities are derived from the active and passive principles. The four elements are set in correspondence with four elemental qualities: this is close to the Stoic (possibly also Platonic), but not Aristotelian, method of derivation.[26] The manner of generation of complex qualities from simple ones is also not specified: all we learn is that complex forms are somehow constituted by simple ones (*ex his ortae . . . gignuntur*). Obviously, at least three different stories could be told here by the three schools (Stoic total pervasion and tension, Platonic mingling of elemental solids in the *Timaeus*, Aristotelian mixture in *GC* 1. 10) but the differences are not spelt out. We can infer from subsequent discussion that the process must be teleologically ordered.

The theory of the 'fifth element' is attributed to Aristotle alone (and not to [Plato], Speusippus, Philip of Opus).[27] This attribution presents a problem for understanding the force of Zeno's rejection of the fifth element: if it is an amendment to Polemo (or the 'Old Academy'), then it is not clear why Aristotle is named as its only proponent. If the amendment is addressed only to Aristotle, it is not clear why it appears in the context of the 'Old Academy'. Moreover, the version of the doctrine of aether in this passage, according to which it is a material of stars and minds, does not have enough ground in Aristotle's extant texts.[28] One possibility would be to take this reference to Aristotle as anachronistic, signalling the difference between the Antiochean consensus and

[24] This can be compared with the notorious textual problem of DL 7. 134 for which the *Suda* reads 'incorporeal' (ἀσωμάτους) instead of 'bodies' (σώματα) of the main MSS (cf. Todd 1978; Frede 2005; Cooper pp. 97–101 above).

[25] For detailed discussion of the place of prime mover in the cosmos, see Sharples 2002: 4–12. In *Acad.* 2. 121, Cicero speaking as a Philonian sceptic, points out that Strato the Peripatetic denies his god the role of creator.

[26] Cf. DL 7. 136–7 (= *SVF* 2. 580); cf. Galen, *De Elem. ex Hipp.* No trace of Plato's geometrical construction of the elements (alluded to by Varro in *Acad.* 1. 6) is found. For a parallel with *Tim.* 49D–E, see Sedley 2002: 58 and n. 36. According to Aristotle, each element is constituted by two qualities (*GC* 2. 4), but one of the two is regarded as 'dominant'.

[27] Cf. [Plato], *Epin.* 981B–C; Speusippus fr. 122 Isnardi (= [Iambl.] *Theolog. Arithm.*); [Plato] ap. Xenocraten (frgg. 264–6 Isnardi Parente (= Simpl. *In Phys.*)). One might wonder if the reference to Zeno's predecessors (*superiores*) at *Acad.* 1. 39 includes anyone other than Aristotle.

[28] This whole section was often taken to be a fragment of Aristotle's lost dialogue *De Philosophia* (Fr. 27 Ross). cf. Untersteiner 1963: 266; Moraux 1963: 1222; Gigon does not include this text in

near-contemporary Aristotelianism. As we shall see shortly, a theory of aether is attributed in the sources to Critolaus and his followers. Cicero's knowledge of Aristotelian psychology is based not just on the lost exoteric works, but also on post-Aristotelian Peripatetic writings and reports which might inform his perspective on the current doctrines of the school.[29] Antiochus' own position on the 'fifth substance' is not stated, but he probably was with the Stoic–Platonic mainstream.[30] Zeno's denial of the fifth element is presented here and elsewhere as a minor technical issue.[31] This is not the way it will be taken in later Aristotelian tradition.[32]

3. The description of matter as formless and qualityless substrate subject to change is found in Stoic sources.[33] The idea that prime matter undergoes change in its parts is in better agreement with Plato's picture of receptacle and Stoic theory of matter than with Aristotle's.[34] On the few occasions when Aristotle may be thought to employ the concept of prime matter, he uses it not in a collective sense, but relatively to a particular substance or process of change.[35] The thesis of infinite divisibility is well attested as Stoic, and could be given an Aristotelian reading (*possint* in *quae intervalla item infinite dividi possint* should be referring to potentiality in a technical sense).[36]

4. The active principle is said to move 'in this way', which probably refers to its ability to pervade the whole of matter. *Ultro citroque* could be taken in a neutral, most general sense of moving in every direction; or as referring to two reciprocal aspects of pneumatic motion.[37] This description of active principle passing through matter would certainly suit the Stoics.

The ending of the paragraph does have more affinities with the *Timaeus*: the Demiurge uses all materials to create the cosmos and the cosmos is eternal.[38] The thesis of eternity should not have been passed by Zeno without an amendment, as Cicero points out elsewhere in the dialogue.[39] But given the overall 'harmonizing'

his edn. of fragments. For further arguments against an 'all-inclusive' treatment of *De Philosophia*, see Hahm 1982, cf. Furley 1989*a*: 204–11. See p. 149–50 below.

29 Cf. Görler 1989; Furley 1989*a*; Sharples 2001: 169–73.

30 Cf. Brittain 2006: p. xxxii.

31 *Fin.* 4. 12–13.

32 Cf. Alex. *Mixt.* 10 and below, pp. 151–6 and 161–4.

33 ἄποιος DL 7. 134 (*SVF* Z 1. 85, 493 (Cl.), 2. 300; 3 Arch. 12); Sext. *M.* 9. 11 (= 2. 301); Plot. 2. 4. 1 (= 2. 320), Simpl. *In Phys.* 227. 23 (= 2. 326); ἀσχημάτιστος Orig. *Orat.* (= 2. 318); Galen, *Meth. Med.* 2. 7, 10. 155 K. (= 2. 322); 2. 326. For Platonic background, see Sedley 2002: 55–6; for interpretation of Stoic concept, see Frede 2005: 219–22; Cooper, pp. 96–102 above; and Gourinat Ch. 3.

34 *Tim.* 50A-C, [Ar. Did.] fr. 27. 463. 5–13 Diels.

35 *Metaph.* 7. 3, 1029a10–30; *GC* 2. 1, 329a24–32; cf. Charles 2004. For recent discussions, see Algra 2004; Broadie 2004; Charles 2004.

36 See LS 50, cf. *SVF* 482–91. It is attacked at length as Stoic in [Galen], *De Qualitatibus Incorporeis* 109–61 Giusta (cf. however, Sedley 2002: 67). The theory of motion 'by infinitely divisible intervals' has some affinities with Aristotelian theory as developed by Strato. Cf. fr. 82 Wehrli, Sorabji 1983: 377–9.

37 So Brittain 2006: 98 n. 25; Rackham translates: 'vibrates to and fro'. Cf. *SVF* 2. 451 (= Nemes. *Nat. Hom.* 18. 6 Morani). Cf. however, Sedley 2002: 67 n. 59.

38 *Tim.* 32C–33B, 41B.

39 *Acad.* 2. 119; cf. Donini 1982: 79; Furley 1989*a*: 203.

goals of this summary, it is possible that Antiochus was prepared to treat the Stoic theory as essentially the same as the other two on this point in that, according to the Stoics, the world is perishable in the end of each big annual cycle, but the sequence of cycles itself is unlimited.[40] Once again, the text suppresses the differences between the treatments of the question of eternity in three schools which could have been ripe already before the time of Cicero's philosophical *Lehrjahre*.[41]

5. The doctrine of providence is at home, in different ways, with Stoics and Platonists.[42] The case of Peripatetics is more complicated. Aristotle does not make any pronouncements about providence, but accepts natural teleology. His successors Theophrastus and Strato are sceptical about natural teleology; Strato is on record denying providence, as Cicero points out in *Acad.* 42, speaking as his own character on behalf of the 'Philonean' Academy to the Antiochean Lucullus.[43] A version of the doctrine of providence, according to which there is providence in the heavenly, but not in the sublunary realm, is attested as Aristotle's in the source possibly dependent on Critolaus.[44] A version which involves sublunary providence at the level of species (but not individuals) does not seem to appear until Alexander of Aphrodisias.[45] The use of both 'necessity' and 'chance' as synonyms for 'providence' could suggest a misleadingly easy way of incorporating Aristotelian doctrine, according to which strict necessity rules in the heavens, and chance and some sort of necessity (hypothetical) in the sublunary realm, but perhaps none of the operations can be rightfully described as providence in a traditional sense. The text of Cicero's *Academics* makes it clear that Antiochus selects a particular version of Aristotelianism for his historical concordance, leaving aside those features of the school doctrine which do not fit his historiographical project. In doing so, Antiochus uses the expository method familiar also from other ancient writers who in their polemical works, in order to clarify and strengthen their position on a certain point, provide a doxographical preamble or interlude presenting their position as a part of a broad intellectual consensus. The Stoics and Aristotelians are often treated as allies in polemical writings against Epicureans, on the subjects of teleology, role of gods, and the nature of mind.[46] In this kind of polemical context, the existing well-known differences between 'allied' schools become less important

[40] Cf. Mansfeld 1979; Long 1985: 26–31; Salles, pp. 126–9 above.

[41] Cf. Furley 1989*a*: 201–4. [42] See Reydams-Schils 1999.

[43] *Acad.* 2. 120–1; cf. Theophrastus, *Metaphysics* 10a22–11a12; Strato, fr. 32, 33, 35 (Wehrli). Discussion in Berryman 1996.

[44] Sharples 2002: 23; Mansfeld 1992: 134–52, with Critolaus fr. 15 Wehrli cited below.

[45] Sharples 2002: 18–36; edns. of *De Providentia* with translations by H.-J. Ruland (1976), S. Fazzo and M. Zonta (1999), P. Thillet (2003).

[46] This technique is used by Galen (see Kupreeva 2004; Tieleman 2007) and Alexander of Aphrodisias (see Todd 1976; Sharples 1983; Mansfeld 1988). Cicero resorts to it as well, for instance, in *De Natura Deorum* (see Furley 1989*a*: 203–4). On the use of doxographical techniques by ancient authors in general, see Mansfeld 1990.

than their basic agreement against the common opponents. On the whole the system of the 'Old Academy' outlined in this text seems to be better suited to Stoics than to Aristotle. This may be due to a genuine affinity between Zeno and Polemo or to the editorial activity of Antiochus (described by Cicero as someone *qui appellabatur Academicus, erat quidem, si perpauca mutavisset, germanissimus Stoicus*).[47]

Several points of Antiochean consensus would need a clarification with regard to the 'standard' Aristotelian doctrine. The active principle is described as immanent rather than transcendent. The theory of elements is different from the standard Aristotelian view (where each element is made up of two qualities). The concept of matter as an incorporeal passive principle without qualities is not explicitly used by Aristotle.[48] The force of corporeality or incorporeality as an attribute of this principle (which cannot exist on its own on both Aristotelian and Stoic accounts) is something that may deserve investigation. The unity of the universe seems to be construed in terms of physical cohesion provided by the motion of the active principle; if we try to apply this model to the Aristotelian cosmos, the only candidate for the role of the active principle will be the first body (of the heavens), whose motion although not pervading the cosmos, still brings about all the sublunary processes. The question of the eternity of the world (another point on which the three schools could have come up with at least three different accounts) is not highlighted. 'Providence' clearly is not a word from Aristotle's philosophical vocabulary.[49] But on this version of 'concordance' between the three schools, Aristotelians are committed to a belief in providence. There is a question whether this version of Aristotelianism is just a construct of tendentious doxography, or whether any developments in the current school doctrine might give support to any of these peculiarities.

2

In this section I consider several cosmological fragments of Critolaus of Phaselis, the scholarch of Lyceum active probably about two generations prior to Antiochus.[50] Epiphanius has the following report about Critolaus:

> Aristotle, the son of Nicomachus, was by origin a Macedonian from Stagira, according to some sources, a Thracian, according to others. He said that there are two principles, god and matter, and things above the moon fall in the scope of divine providence, while those under the moon subsist unprovided and are carried on by some irrational

[47] Cic. *Acad.* 2. 132; cf. Görler 1990.

[48] It can be argued that it does not necessarily conflict with Aristotle's system, but that is a different issue (Algra 2004: 92–4).

[49] The few occurrences of *meiromai* in the corpus are non-technical.

[50] For chronology, see Olivier 1895: 6; Wehrli 1969^{2}: 63–4; Dorandi 1999: 37. On the place of Critolaus in the history of Hellenistic Peripatos, see Sedley 1989: 117–18.

drive, as they chance. He says that there are two worlds, the upper and the lower, the upper being imperishable, the lower perishable. And he says that the soul is body's continuous activity (*endelekheia*). . . . And Critolaus the Phaselite believes the same as Aristotle.[51]

It is not clear that the summary goes back to Critolaus, but this is a plausible conjecture.[52] The statement of the main tenet does not differ considerably from the Antiochean summary. We find the two principles described as 'god and matter' where Cicero's Varro had the seat of power (god) and matter.[53] The *endelekheia* could be an error for *entelekheia*. It is impossible to tell whether it is just a scribal error or a reading in the original (whatever that was),[54] but it is the best evidence we have for the Peripatetic origin of this passage. The division of the universe into two different parts can be interpreted as Aristotle's distinction between the sublunary region of coming to be and perishing and the indestructible world of heavens. We have seen in the previous section that, if the Antiochean summary does indeed cover Peripatetics, they are committed to *some* theory of providence, perhaps developed under pressure from rival schools. Epiphanius concurs with Cicero/Antiochus on this point, apparently disagreeing about the substance of the theory.

The most detailed report of Critolaus' cosmology comes from Philo's *De Aeternitate Mundi,* which has preserved some of his arguments for the eternity of the world.[55] These arguments are of interest to us because they present the Peripatetic school position as distinct from the positions of both Plato's *Timaeus* and the Stoics.[56] They possibly give us a glimpse of the intellectual milieu in which this position is being established, although the exact amount of Critolaus' material in these passages is hard to assess.[57] The first argument printed by Wehrli is as follows:

And Critolaus, one of those who worshipped the Muses, an adherent of Peripatetic philosophy, used the following reasons in support of the doctrine of the eternity of the cosmos: 'If the cosmos has come to be, the earth too must have come to be; but if the

[51] Fr. 15 Wehrli = Epiph. *Adv. Haereses* 3. 31 (592. 9–14, 19 Diels).

[52] Wehrli 1969^2; Sharples 2002.

[53] Wehrli says that the report may be a result of Epiphanius' not distinguishing between Peripatetics and Pythagoreans. According to him, this is supposed to account for the bipartite division of the cosmos. For criticism of Epiphanius as a source, see Sharples 1998: 104 n. 296.

[54] Wehrli ad loc. suspects the text of Epiphanius (perhaps rightly). On the other hand, cf. Cicero, *Tusc.* 1. 22 (see J. E. King's note ad loc.). I am grateful to Ricardo Salles for querying this point.

[55] On the problem of authorship and structure of the text, see Runia 1981: 105–12.

[56] On the role of the three school positions in the structure of Philo's treatise, see Runia 1981: 112–21.

[57] Wehrli prints just two arguments; Olivier 1895: 16–18, lists five (or maybe four, depending on construal); Arnaldez (1969) also indicates that reference to the third person (Critolaus) is made at the beginning of ch. 74, which possibly allows us to add some material from chs. 74–5. Philo's report is possibly contaminated: scholars point out that sometimes it is difficult to decide whether the rhetorical prose belongs to Critolaus or the author. Nonetheless, it is possible to track some elements of these texts which exhibit consistency of doctrine.

earth is generated, then so surely is also the humankind; but human being is ungenerated, [being] of the eternal kind, as will be shown. Hence the cosmos too is eternal.'[58]

The proof of the eternity of humankind which follows upon this fragment is stated as a refutation of the view according to which the human race is not eternal but has come to existence.[59] The source of this opposing view is not entirely clear,[60] but a structural parallel with the argument against the eternity of the world refuted by Theophrastus is worth noting: 'if the world is eternal, then humankind must be eternal; but in fact it is not, as is clear from the recent origin of crafts'.[61] Theophrastus' argument is reported by Philo as refuting the impossible assumptions of the opponents. The text following Critolaus' argument is focused on the impossible consequence which follows from the opposite view, namely that humans (or living organisms) are earth-born.[62] In the end this proof invokes the Aristotelian (and Platonic) thesis that permanence of biological species through reproduction is the only way for human individuals to achieve immortality.[63] The continuity with Theophrastus' argument reported in the same source is remarkable. The overall position established by the proof does not seriously differ from that of Aristotle's.

The second of the arguments cited by Wehrli is as follows:

Critolaus, continuing his attack, used also the following argument: 'That which is its own cause of health is free from illness; also that which is its own cause of being awake is awake; if so, then also that which is the cause of its own existence is eternal; but the cosmos is the cause of its own existence, given that it is the cause of existence for everything else; hence, the cosmos is eternal.'[64]

[58] Fr. 13 Wehrli (Philo *aet.* 55): Κριτόλαος δὲ *τῶν κεχορευκότων** Μούσαις, *τῆς Περιπατητικῆς ἐραστὴς φιλοσοφίας, τῷ περὶ τῆς ἀιδιότητος κόσμου δόγματι συνειπὼν ἐχρήσατο τοιαύταις πίστεσιν· εἰ γέγονεν ὁ κόσμος, ἀνάγκη καὶ τὴν γῆν γεγονέναι· εἰ δὲ ἡ γῆ γενητή, πάντως καὶ τὸ ἀνθρώπων γένος· ἄνθρωπος δὲ ἀγένητον ἐξ ἀιδίου τοῦ γένους ὑφεστώτος ὥσπερ ἐπιδειχθήσεται. ἀίδιος ἄρα καὶ ὁ κόσμος*. Fort. *συγκεχορευκότων* Olivier.

[59] Chs. 56–69.

[60] Arnaldez 1969 ad loc. suggests Stoics, referring to Sextus, *M*. 9. 28 (who reports the argument by recent Stoics according to which the first men were earth-born; the context is the origin of the human notion of god).

[61] Chs. 130–1 (argument), 145–9 (Theophrastus' refutation) are reprinted as parts of Theophrastus' fr. 184 FHSG.

[62] The argument would refute Lucretius 5. 795–836; cf. Sedley 1998: 166–85, on possible links between Theophrastus and Lucretius.

[63] Ch. 69 Cohn is concluded with: *μένει γὰρ εἰς ἀεί, φθειρομένων τῶν ἐν εἴδει, τεράστιον ὡς ἀληθῶς καὶ θεῖον ἔργον*. There is no certainty that this material comes from Critolaus himself (and not an intermediary source). Some language may be owed to Philo (cf. Arnaldez on ch. 63), but the argument proper construes well with the brief formulation of fr. 13 and is immediately followed by the next argument directly attributed to Critolaus (= fr. 12 Wehrli). Sharples 1997: 160, rightly compares the reasoning to Aristotle's in *GC* 2. 10, but notably the motion of heavens is not invoked by Critolaus as a cause.

[64] Fr. 12 Wehrli: *ἐπαγωνιζόμενος δ' ὁ* Κριτόλαος *ἐχρῆτο καὶ τοιούτῳ λόγῳ· τὸ αἴτιον αὑτῷ τοῦ ὑγιαίνειν ἄνοσόν ἐστιν· ἀλλὰ καὶ τὸ αἴτιον αὑτῷ τοῦ ἀγρυπνεῖν ἄγρυπνον ἐστιν· εἰ δὲ τοῦτο, καὶ τὸ αἴτιον αὑτῷ τοῦ ὑπάρχειν ἀίδιόν ἐστιν· αἴτιος δ' ὁ κόσμος αὑτῷ τοῦ ὑπάρχειν, εἴ γε καὶ τοῖς ἄλλοις ἅπασιν· ἀίδιος ἄρα ὁ κόσμος ἐστίν.*

The statement of the argument is too short to give us a precise idea of its philosophical background.[65] Disease is invoked earlier in the treatise, as an example of the kind of destruction brought about by an internal cause.[66] Philo also cites *Timaeus* 32B5–C1, where Plato explains that the balanced elemental constitution of the cosmos accounts for its being free from disease and old age.[67] It is possible that Critolaus' second argument is meant to be an *ad hominem* strengthening of Plato's reasoning in the *Timaeus*, showing that the latter is strong enough to prove the eternity of the cosmos and not just its indestructibility. Whether Critolaus himself is the author of this plan is impossible to tell.[68]

Furthermore, the specific background of the argument is not immediately clear. It seems to be based on the analysis of several key concepts:

(1) If x is a cause of its own state P then it is in the state P.

(2) If x is the cause of its own existence, then it exists.

(3) The cosmos is [always] the cause of its own existence.

(3′) For it is the cause of existence of anything else.

(3″) i.e. it is the cause of existence of any of its states.

(4) Therefore the cosmos [always] exists, i.e. is eternal.

The point illustrated by the examples seems to be that in order for a thing to cause its own state it must be in that state. An example of the opposite case would be the state brought about by an external cause, in which case a thing would not need to be in that state (e.g. when the state of health is brought about by medicines; or being awake by an alarm clock). In the case of the world and its states, however, it seems that there is an assumption spelled out in 3′ and 3″, that the world as a whole is always a cause to any process; therefore it is the cause of any of its own states, and the cause of its own existence.[69] In the context of this argument the assumption does sound question-begging in that it is not clear what warrants the permanent supply of processes or states on whose existence the existence of the world seems to be made dependent.[70] Perhaps it could be taken as a reply to the argument for the destructibility of the world from the destructibility of its parts, in the sense that any process within the world, even the process of destruction, does have the world as its cause (and therefore is always local not global destruction). It is difficult to be certain also about the opponents of Critolaus. In his case, Stoics are not ruled out.[71] On the other hand, his

[65] Olivier 1895: 17 suggests that Philo abbreviated his source. [66] Ch. 20.

[67] Ch. 25. The term ἄγρυπνον is probably supposed to refer to the state of activity.

[68] It could be Philo's (cf. Runia 1981: 83–4), or Theophrastus' (cf. Sedley 1998: 172).

[69] A parallel with *Tim.* 32C–33B suggests itself.

[70] Olivier 1895: 17 calls this an 'ontological argument'; Arnaldez 1969 cites *Phaedo* 105C as parallel.

[71] As they are in the case of Theophrastus (cf. Runia 1981; Sedley 1998). cf. Olivier 1895: 17–19.

contemporary and near-contemporary Stoics, according to reports, do not show strong commitment to the theory of conflagration.[72]

All this, of course, does not tell us much about the details of Critolaus' analysis of the cosmic structure. We can only assume that Critolaus recognized the main tenets of Aristotle's cosmological doctrine, but there is not much information about his position on multiple controversial issues of interpretation (type and mechanism of causation exercised by the first unmoved mover, heavenly bodies and their relation to the unmoved principle).[73]

Fr. 16 attributes to Critolaus and Diodorus of Tyre the view that '[god] is the intellect [coming] from the impassive aether'.[74] The preposition ἀπό, 'from', is problematic. Olivier took it as meaning that the intellect, i.e. Aristotle's god, has material constitution.[75] But it could refer to the origin of the activity of intellect in the upper cosmos in a more general way. It does seem to suggest an idea of direction from the sphere of fixed stars downwards, but the report is too brief, so any details would have to be speculative. The only thing that seems uncontroversial is the reference to Aristotle's thought-god of *Metaph.* 12. 9.

The question of cosmic structure is closely related to the question of providence. It has been argued, on the strength of fr. 15 Wehrli, that Critolaus might be the first thinker in the post-Aristotelian tradition credited by the sources with the view that there is providence in the upper cosmos but not in sublunary region.[76] This view appears to be more restrictive than the 'Antiochean' version, according to which divine providence somehow still operates in the sublunary world, albeit on a diminished scale. A similar statement of 'diminished' providence in sublunary world is found in a number of other Peripatetic sources. An example would be [Aristotle], *De Mundo* 6, with the highest god operating through intermediaries,[77] like the Persian King, not personally involved in any affairs of the state but still exercising his power through appointed representatives.[78]

[72] Diogenes of Babylon (*SVF* III Diog, 27); Boethus of Sidon (*SVF* III Boeth. 7).

[73] Cf. Sharples 2002: 14. Note also the criticism of Chrysippus' example of Dion/Theon as stated by Philo: the example is reformulated, assilimating Dion to the whole cosmos, Theon (part of Dion without a leg) to his 'ruling' part (soul). Chrysippus' thought experiment (of amputating Dion's leg) will then result in the whole cosmos surviving without its ruling part (Philo, *Aet.* 50–1).

[74] Fr. 16 Wehrli (Stob. *Ecl.* 1. 1. 29b = 'Aët'. 1. 7. 21): *Κριτόλαος καὶ Διόδωρος ὁ Τύριος νοῦν ἀπ' αἰθέρος ἀπαθοῦς*. Cf. Sharples 2002: 14; Mansfeld 1992: 134–9.

[75] Olivier 1895: 44–6. Olivier considers the possibility of Stoic influence on the Peripatetics earlier than Critolaus, but thinks that Critolaus himself made no concessions to the Stoic doctrine in his physics (as opposed to ethics).

[76] The main evidence for attribution of this doctrine to Critolaus is Epiphanius' report above; but this view is attributed to Aristotle in a number of late doxographical sources (discussions in Mueller 1994; Sharples 2002). As we shall see below, Alexander of Aphrodisias is concerned with justification of some versions of this view.

[77] *De Mundo* 397b25–30. Cf. Alex. *Quaest.* 48. 5–8 Bruns: *εἴη τε ἂν οὕτω πρῶτον ἀπολαῦον τὸ πῦρ τῆς τοιαύτης ἐκ τοῦ θείου δυνάμεως, ἔπειτα τοῖς μετ' αὐτὸ διαδιδὸν αὐτήν, ὡς πάντα τὰ σώματα τῇ τοιαύτῃ διαδόσει μεταλαμβάνειν αὐτῆς, τὰ μὲν πλεῖον, τὰ δὲ ἔλαττον.*

[78] *De Mundo* 398a6–b28. See discussions in Moraux 1984: 37–48; Sharples 2002: 25–6; Fazzo 2002: 185.

The political character of analogy makes difficult a 'strong' interpretation of separation, or 'transcendence', of the divine being so construed. In fact, the use of military metaphor by Aristotle in *Metaphysics* 12. 10 (1075a13–15) raises similar problems. On the other hand, we can compare the political metaphors used by anti-Stoic authors to criticize Stoic theological doctrine: corporeal nature of the active principle and the doctrine of total pervasion bring about comparisons of Zeus with Proteus and all kinds of lower natures.[79]

The sparseness of our sources makes it particularly difficult to establish the relation between these two versions of the doctrine of providence in the post-Aristotelian Peripatos. In the fr. 37, Critolaus uses the god of the universe as a political metaphor. The fragment is based on two reports. Fr. 37a comes from Plutarch's discussion (in *Precepts of Statecraft*) of whether a public figure should be involved in all the minutiae of his office directly or whether he should occupy himself only with most important things and maintain dignified distance from everything else. Plutarch reports the view of Critolaus:

> But there are others who think the conduct of Pericles was more dignified and splendid, one of whom is Critolaus the Peripatetic, who claims that just as the 'Salaminia' and the 'Paralus', ships at Athens, were not sent out to sea for every service, but only for necessary and important missions, so the statesman should employ himself for the most momentous and important matters, as does the King of the Universe,
>
> For God great things doth take in hand,
> But small things passing by he leaves to chance,
>
> according to Euripides. (*Precepts*, 811C10–D7, tr. Fowler)

The force of the analogy is reverse to that of *Metaph.* 12. 10 and *De Mundo*: the main context here is political, and the highest god is cited as an analogy of a politician. The wording, particularly of Euripides' quotation, might seem to support Epiphanius' report. The second version of this fragment, from the *Life of Pericles*, seems to endorse a less restricted account of providence:

> And he [Pericles] avoiding constant presence and insolence, approached people as though at intervals, neither making pronouncements over every issue nor always coming forward to speak to the gathering, but giving himself, says Critolaus, as a warship 'Salaminia', to big tasks, and attending to others by sending as speakers his friends and comrades. (*Pericles* 7 = Critolaus fr. 37b Wehrli)

Here the 'divine' example is missing: it could mean either that it is not coming from Critolaus in fr. 37a or that it is omitted by Plutarch for stylistic reasons in this text.

[79] Cf. [Galen], *De Qualitatibus Incorporeis*: *εἰ δὲ μεταποιήσας αὑτὸν ὁ Ζεὺς ἰσαρίθμως οἷς ἔφην συμβεβηκόσι τροπὰς ἴσχει ποιοτήτων μυρίας, χείρων ἐστὶ τοῦ μυθολογουμένου Πρωτέως· ὁ μὲν γὰρ εἰς ὀλίγας φύσεις ἑαυτὸν μετεποίει καὶ μετεμόρφου καὶ ταύτας οὐκ ἀπρεπεῖς, . . . ὁ δὲ οὐκ ἔστιν ὅ τι μὴ καὶ τῶν αἰσχίστων γίγνεται· διὰ μὲν ἀνδρὸς ἄφρονος ἄφρων, διὰ δὲ αἰσχροῦ γνώμης αἰσχοποιός, διὰ δὲ ζῴων ἀλόγων ἀποθηριούμενος, διὰ δὲ λίθων καὶ ξύλων ἄψυχος γενόμενος, διὰ δὲ κοπρίου βδελυρὰ φύσις, ἔτι δ' ἀχρεῖον.* (20 Giusta).

If both this fragment and fr. 15 go back to Critolaus,[80] it might mean that a 'dualist' formulation of providence theory reported in Epiphanius' summary is compatible with 'diminished providence' illustrated by these analogies of human agency (statesman, general). That might imply some (perhaps very low) degree of involvement with sublunary region.[81] We don't know if Critolaus might have a good way of reconciling the two possible approaches. But such an attempt (if ever made) cannot be expected to be a simple adjustment to the Stoic doctrine of providence as far as possible within the Aristotelian framework.[82] The use of political or military metaphor seems to indicate a focus shift in Peripatetic discussion of providence from the criticism of teleology to the proof of causal priority of the first principle in the sublunary region.[83] The formulation in Epiphanius may reflect the former tendency.[84] Similar interpretation of the Aristotelian position is reported in some other sources; it is easy to see how it can be used for polemical purposes.[85] The second approach could have been taken on board by polemical doxography of a different type, for instance, by Antiochus. It seems clear, however, that the source of the tension lies within the Aristotelian system itself.[86]

As we have seen, Varro in *Acad.* 1. 26 attributes to Aristotle the doctrine according to which the fifth element is the material from which stars and minds are made.[87] The extant works of Aristotle do not have any explicit statement of such theory.[88] On the other hand, the Stoics do believe that the soul is made of a variety of fire which is the same as the fine substance of the heavenly region.[89]

[80] Ian Mueller raised a doubt concerning the origin of the Euripides quotation (1994: 155 n. 42). In fact, the same quotation from Euripides (with a minor change) occurs also in Plutarch's *De Cohibenda Ira* 464A, in a very similar moral context. It is hard to be certain about Plutarch's source: on each occurrence it seems to be from memory (as is also the quotation from Critolaus). But it is by no means impossible that Critolaus should quote Euripides (see Olivier 1895: 27–8 on related topics in Critolaus and a possible source of this fragment). The permanence of political context of quotation might also suggest the same source.

[81] *Pace* Sharples 2002: 23 n. 109.

[82] Moraux in his analysis of *De Mundo* 6 points out that the use of Stoic language and imagery does not mean that the author attempts to 'Stoicize' the Peripatetic theory: on the contrary, the task is to emphasize the distinct features of Aristotelian doctrine (Moraux 1984: 38–9, 78–9). The 'strong' formulations preserved in the doxographical summaries might be a result of similar emphasis in the original works.

[83] Cf. particularly the second argument for the eternity of the world discussed above (pp. 144–5).

[84] Perhaps in conjunction with a strong division between the heavenly and sublunary worlds emphasized by Aristotle in some key texts (*Metaph.* 12. 8; *GC* 2. 11).

[85] Cf. Atticus fr. 3 for comparison with Epicurus. [86] Cf. Sharples 2002: 23–4.

[87] *Acad.* 1. 26 above (for more references and discussion see Moraux 1963: 1213–22).

[88] See discussion of *On Philosophy* in Hahm 1982 and Furley 1989*a*: 204–11; cf. below p. 152 n. 103.

[89] *SVF* 2. 1021 (= DL 7. 147), 1. 124 (= Cens. *De Die Nat.* 4. 10); 1. 126 (= Varro, *Ling. Lat.* 5. 59), 2. 423 (= Aug. *De Civ. Dei* 8. 5); Cleanthes (*SVF* 1. 504); Mansfeld 1992: 139–40, cf. Cooper, pp. 103–7 above. Mansfeld 1992: 139 n. 17 finds the report about Zeno's objection to the fifth element in *Acad.* 1. 39 odd because of the closeness of the thesis to his own theory. It seems that if the report is to be taken as valid, then the main target of the criticism was not the aethereal constitution of the soul, but the nature of Aristotle's aether, namely, the fact that it is

According to the report preserved by Tertullian, the Peripatetic of Critolaus' circle believed that the aether which surrounds the cosmos is in some way the source of existence for human soul:

I don't mean just those who fashion [the soul] out of the visible bodily components . . . as Critolaus and his Peripatetics from some fifth substance, if indeed it is also a body because it encloses the bodies, but also the Stoics who, although they claim almost in our words that soul is spirit, insofar as blowing and spirit are close, will nonetheless readily persuade us that soul is body.[90]

This is seconded by Macrobius: 'Critolaus the Peripatetic [says] that soul is made of the fifth substance.'[91]

The view that human souls are made of the heavenly substance could be regarded as a sign of 'Stoicizing' attitude.[92] But the possibility of interpreting this report as based on the texts of Aristotelian corpus has been pointed out as well. The traditional interpretation of the sometimes asterisked passage concerning νοῦς θύραθεν in *GA* 2. 3 took it to refer to the direct effect of the divine intellect on the human soul.[93] Along the same lines the difficult text about the source of right desire in *EE* 8. 2 is taken to refer to the divine element within us.[94] We may note that Critolaus and Peripatetics are set apart from Stoics by Tertullian who thinks that the former are more committed 'corporealists' than the latter, despite the more elusive corporeal nature of their divine body—because it is not a spirit within the body.[95]

According to Cicero in *De Divinatione*, Cratippus who was a student of Antiochus and left the Academy for the Peripatetic school taught that human soul in part has a divine origin.[96]

The human soul is in some degree derived and drawn from a source exterior to itself. Hence we understand that outside the human soul there is a divine soul from which the human

impassive and does not mix with other elements. This type of aether could not be the material of soul without contradiction within the Stoic theory, and Zeno might want to distance himself from this concept of the fifth element (distinct nature of πῦρ τεχνικόν in the Stoic doctrine of elements notwithstanding).

[90] *Nec illos dico solos qui eam de manifestis corporalibus effingunt . . . ut Critolaus et Peripatetici eius ex quinta nescio qua substantia, si et illa corpus, quia corpora includit, sed etiam Stoicos allego, qui spiritum praedicantes animam paene nobiscum, qua proxima inter se flatus et spiritus, tamen corpus animam facile persuadebunt.* (Tert. *An.* 1. 5 = Critolaus fr. 17 Wehrli).

[91] *Critolaus Peripateticus constare eam* (sc. *animam*) *de quinta essentia* (*In Somn. Scip.* 1. 14. 20 = Critolaus fr. 18 Wehrli)

[92] Cf. Mansfeld 1992: 139–40; n. 89 above.

[93] *GA* 2. 1, 731b24–32a1. For modern deflationary readings, see Freudenthal 1995: 37; Caston 1999: 215–16; on the material of semen, see Rashed 2007: 287–91.

[94] *EE* 8. 2, 1248a16–33, cf. Sharples 2002: 11–12; van der Eijk 2005: 32–41.

[95] This latter remark may or may not refer simply to terminology (πνεῦμα is a good word to use as far as Tertullian is concerned), but the whole report probably owes something to the medical tradition as well. For Tertullian's sources, see Waszink 1947: 22*–44*; most recently, Polito 2006: 316–21. This report does again raise a question about the source of *endelekheia* in Epiphanius and elsewhere.

[96] Moraux 1973: 223–56.

soul is sprung. Moreover, that part of human soul which is endowed with sensation, motion and desire is inseparable from bodily activity; while the part that thinks and reasons is most vigorous when it is most removed from the body. (*Div.* 1. 70, tr. Rackham)

The most natural parallel that suggests itself is, of course, the *Timaeus*, but otherwise there is little clarity about the status of the divine and human soul. There is no evidence of any role for the aethereal body, or of a more precise cosmological framework of this theory of soul. Much is unclear with regard to Cratippus' original position, so it would be premature to draw any conclusions about its compatibility with Peripatetic theories of his time. But on the basis of the cited passage we can at least say that the issue of compatibility does not seem to be decisive in the matter of Cratippus' new school allegiance: it appears from Cicero that he can believe in the world-soul and at the same time be a card-carrying Peripatetic.[97]

A very brief survey of Critolaus' fragments shows that some of the 'Stoicizing' features similar to those signalled in the Antiochean concordance in *Acad.* 1. 26–9 (two principles, commitment to a theory of providence, aethereal constitution of the soul) are present in the near-contemporary Peripatetic doctrines. However, it seems that the properly doctrinal background of these features originates in the traditional Peripatetic agenda and reflects the problems within the Aristotelian tradition. Critolaus' arguments for the eternity of the world are characteristic in this respect, showing the same commitment to this tenet as is shown by the earlier Peripatetics such as Theophrastus.

3

In this section I shall discuss Xenarchus' arguments against the fifth body and the use of the notion of prime matter in some Peripatetic sources after Andronicus (Boethus and Nicolaus of Damascus). Andronicus' name is usually connected with his edition of Aristotelian corpus.[98] Although the details of his editorial contribution are elusive and we are rightly warned by Barnes not to make Andronicus' work a single landmark in what may be a continuous and complex process, to some extent it is still a convenient landmark, in that it signals the engagement of this group of Peripatetics with exegetical and doctrinal problems on the basis of 'our' (or at least 'nostratic') text of Aristotelian corpus. There have been suggestions that some Peripatetics of this generation were influenced by Stoicism.[99]

[97] Moraux has argued against the earlier attempts to interpret Cratippus' theory of natural divination as based on Stoic sources: 1973: 242–56.

[98] On Andronicus and his edn., see Moraux 1973: 45–141; Barnes 1997: 21–66. For the suggestion that both Boethus and Xenarchus could be Andronicus' students, see Moraux 1973: 197.

[99] For Xenarchus, Moraux 1973: 211–12; Gottschalk 1987: 1120; for Boethus and Andronicus, e.g. Sorabji 1999: 103–5.

The Fifth Element

Xenarchus of Seleucia criticized Aristotle's theory of aether in his treatise *Against the Fifth Substance.*[100] He is sometimes described as being only 'notionally' Peripatetic, in fact almost a Stoic, at least somebody heavily influenced by Stoicism.[101] His objections to Aristotle, along with the replies by Alexander of Aphrodisias, are preserved by Simplicius. There are two sets of arguments on *Cael.* 1. 2 reported by Simplicius under two lemmata (268b11–269a18 and 269a18–32), partly overlapping, and two references elsewhere.[102] The following arguments are reported by simplicius:

(*a*) Aristotle's division of all simple lines into straight and circular is flawed; the cylindrical helix is also simple (*apud* Simplic. *in Cael.* 13. 22–8, 14. 14–21).

(*b*) It is not correct that each simple body has only one simple movement: Aristotle's rectilinear simple motions belong to the bodies when they are still on their way to becoming complete elements; once they reach their natural place and become complete elements, some of them (namely air and fire) can move in a circle (21. 33–22. 17 = 42. 10–14).

(*c*) Two versions: (*c*′) A composite body may have a simple motion, in accordance with the prevailing tendency (Alexander) (23. 24–6). (*c*″) There are infinite composite movements but no infinite composite bodies (Simplicius) (23. 11–15).

(*d*) Even on Aristotle's account, each of the middle elements (water and air) has two natural movements; so it is not impossible for fire to have two (namely one rectilinear upward, another circular, depending on its location) (23. 31–24. 7).

(*e*) The body moving in a circle cannot be simple, because its different parts move with a different speed (on the equator and on the poles) (24. 21–7 = 42. 8–10).

(*f*) Aristotle makes a methodological mistake when he tries to prove his physical thesis by means of mathematical demonstrations (*ad hominem*) (25. 11–13 = 42. 7–8).

(*g*) There is a problem with Aristotle's proof that circular motion is natural to heavenly body only (50. 18–24).

(*h*) The principle 'one thing has only one contrary' is invalid: in Aristotle's ethics, virtue has a pair of vices as its contraries (55. 25–31).

[100] An edn. and full study of the fragments are still outstanding. Discussion of transmission in Rescigno 2004: 73–80 is very helpful.

[101] See n. 99 above.

[102] Some of these arguments are discussed in Falcon 2001: 150–74; Hankinson 2002/3; Sorabji 2007 provides a general overview.

(*i*) Aristotle's definition of 'the light' as 'floating above all bodies' is unsound, because it contradicts the other definition ('moving upwards') (70. 20–2).

(*j*) It is possible that there should be void outside the cosmos: Chrysippus' definition may be corrected to avoid controversy (286. 2–6).

These arguments are based on a close study of Aristotle's text, and directed first of all at finding the incongruities in the technical arsenal of Aristotle's arguments.[103] How much of a positive theory would follow upon this destructive criticism is a different matter; I shall attempt a very brief survey of general options offered by the fragments.

Argument (*a*) draws on near-contemporary work in geometry (Apollonius of Perga), where the properties of cylindrical helix are discussed and where it is sometimes treated as a simple line.[104] There is no explicit attempt to revise the Aristotelian concept of natural motion,[105] but we would like to know whether argument (*i*), where Xenarchus criticizes the 'static' definition of the light because of the apparent conflict with the 'dynamic' one, leads to any substantive doctrinal amendment.[106] Argument (*b*), supported by (*d*) and perhaps (*h*), could be taken as an implicit outline of what Xenarchus sees as a plausible theory of elemental motions: the elements move with Aristotle's simple motions till they get their proper forms (i.e. reach their natural places), and then they either continue to move (with natural motions of a different kind, in the case of fire and air) or rest (as water and earth).[107] But in each case, Xenarchus' criticism of Aristotle's arguments and the alternative suggestions seem to be based on a close systematic reading of Aristotle, which assumes the overall consistency of all the doctrines within the corpus. In (*b*), Xenarchus invokes Aristotle's idea that the elements reach their actuality and thus their proper form when they reach their natural place.[108] The notion that air and fire are in circular motion could be based on Aristotle's theory of exhalation in the *Meteorology*, where the cause of this rotation, the circular motion of the adjacent heavenly body, is not described as unnatural.[109] In (*d*), again, Xenarchus invokes Aristotle's theory of natural motions of the 'middle' elements (air and water),[110] in order to undermine

[103] Notably, there are no criticisms of the 'psychological' argument mentioned by Cic. *Tusc.* 1. 22, and no references outside the scope of *De Caelo*, except for the name of the fifth element in the title.

[104] For references and discussion, see Hankinson 2002/3: 24; Rescigno 2004: 172–6. We may notice a parallel generalizing tendency in Chrysippus who reduces basic kinds of motions to two, straight and curved, and speaks of many more derived from these by combination (*SVF* 492 = Stob. *Ecl.* 165. 15).

[105] For instance, along the lines of Strato's theory of 'squeezing out', cf. Strato frgg. 50–3.

[106] For the discussion of the question of 'lightness' and 'weightlessness' in Stoic physics, see Furley 1989*b*; Hahm 1977: 111–35; Wolff 1988.

[107] Cf. Moraux 2001: 199 n. 10, 211–12; Rescigno 2004: 198–9.

[108] *Cael.* 4. 4, 311a1–10.

[109] *Meteor.* 1. 3, 340b32–41a12.

[110] According to this theory, stated in *Cael.* 4. 4, air, which is naturally light, i.e. upward moving, is heavier than fire, and water, which is naturally heavy, i.e. downward moving, is lighter than earth.

Aristotle's argument according to which each simple body has only one natural motion. The argument in (*h*) is remarkable because it seems to assume doctrinal consistency throughout the corpus, the assumption we find in later commentators such as Aspasius and Alexander. Argument (*g*) may seem to favour fire as the outermost elemental layer. Xenarchus' unspecified objection is cited in support of the difficulty formulated—and resolved—by Alexander in order to show that circular motion is natural for the aether.[111] Argument (*e*) is strong enough to criticize any concept of a simple and unmixed layer. Although the resulting picture of elemental layers might have parallels with the Stoic universe, this is not decisive: the distribution of the elements is the same in the Platonic cosmos, with provisions made for vagueness of boundaries.[112]

The most expressly 'Stoic' argument seems to be (*j*). It is directed against Aristotle's thesis in *Cael.* 1. 9, 279a11–17 that 'there is no place nor void nor time beyond the heaven'. The objection has to do with 'void' and the context might suggest that Xenarchus is arguing in support of the *Chrysippean* concept of void.[113] Simplicius quotes Alexander's criticism of Chrysippus' concept of extracosmic void. If the void exists it is either (1) finite or (2) infinite. (1) If finite, it is limited by some body, which the Stoics deny. So it must be infinite. (2) If infinite:

(i) from Chrysippus' own definition of the void ('void is an interval which, while capable of receiving a body, has not received it'),[114]

(ii) void and body are relatives, and therefore

(iii) the infinite void and infinite body also must be relatives (i.e. just as void is the space which can receive a body, infinite void is the space which can receive an infinite body).

(iv) Thus, if the relation between the void and the body is as described by the Stoic definition, then the infinite body should exist, which the Stoics themselves deny.[115]

Thus the 'middle' elements can be taken to have two opposite natural tendencies—one 'dominant' and the other 'recessive', as it were, which is displayed only relatively to the natural tendencies of their respective 'extremes' (i.e. fire and earth).

[111] The aporia is as follows: assume that the fiery layer is moving in a circle not by a natural motion; then fire's natural motion will be upwards; its contrary, downward motion, will be also against nature. So there will be two contraries to the natural (upward) motion, which is incoherent. Moraux 1973: 201, followed by Falcon 2001: 106–7, attribute the aporia to Xenarchus; Hankinson 2002/3: 50–1 and Rescigno 2004: 234–5 think that it is formulated by Alexander himself as a thought experiment. In any case, Xenarchus' other arguments (*b*), (*d*), and (*e*) show that he is prepared to challenge Aristotle by suggesting that the circular motion of fire can be construed as natural.

[112] Cf. Falcon 2001: 157. Also in Aristotle's *Meteorology* 1. 3, the layers of dry and moist exhalation are not strictly demarcated (e.g. 340b33–41a4).

[113] The discussion of the so-called 'Archytas'' argument at 284. 28–286. 27 Heiberg; Rescigno 2004: 470–502.

[114] Ap. Simpl. 285. 32–5: κενὸν δὲ τοῦτό φασι διάστημα, ὃ οἷόν τε ὂν σῶμα δέξασθαι μὴ δέδεκται.

[115] 285. 27–286. 2 Heiberg.

Then Xenarchus is quoted (most likely by Alexander) as trying to save the Chrysippean thesis:

Xenarchus transformed 'capable of receiving' into 'receptive', so as to resolve in this way the absurdity brought about by the use of relatives; but the transformation has not added anything. For 'receptive' is nothing but 'capable of receiving', and if so it still remains a relative. (286. 2–6 Heiberg)[116]

This argument is regarded by Moraux as a clear evidence of Xenarchus' Stoic leanings.[117] But it is not entirely clear whether Xenarchus' correction was originally intended to support Chrysippus. Algra indeed has suggested that it could be read as an objection to Aristotle's denial of existence of void outside the cosmos in *Cael.* 1. 9, 279a13–15.[118] In this case, the replacement of 'capable of receiving' with 'receptive' would address a different kind of problem, namely, that void is defined as capable of receiving a body while in fact, as Aristotle says, there is no body outside the cosmos. 'Capable of receiving' would be taken as an 'open' possibility which can be realized, for all we know about the cosmos. 'Receptive' would be a pure, or 'counterfactual' possibility, referring to the intrinsic receptive power of the void which could have been actualized had it not been for the *de facto* impossibility of such realization.[119] Thus, whatever the exact thesis that is being supported by Xenarchus, the force of his amendment consists in making sense of the definition of void, without making it dependent on the realization of the possibility which it claims. Now, even if Xenarchus is in fact defending the Stoic position, it is unlikely that he is trying to block the *reductio* from the assumption of the 'infinite body', as is suggested by Alexander, for that would be a defence based on granting as a 'logical possibility' the assumption ('infinite body') while the goal (in Alexander's report) seems to be to invalidate that assumption.[120] It seems more likely that Xenarchus' original target was on a more general level (e.g. definition of the void), and Alexander adduced this modification of the position he criticizes in his own argument in order to show that his *reductio* (from 'infinite body') is strong enough to hold out against this adjustment which might be effective as an objection to Aristotle's argument.

The objection (*f*) (Aristotle uses mathematical proofs to demonstrate physical causes) could be read, as Moraux points out, as an *ad hominem* argument in Aristotelian context (with regard to Aristotle's criticism of *μετάβασις εἰς*

[116] *Ξέναρχος δὲ τὸ οἷόν τε δέξασθαι μετέβαλεν εἰς τὸ δεκτικόν, ὡς οὕτως λύσων τὴν ἐκ τοῦ πρός τι ἐπιφερομένην ἀτοπίαν τῇ θέσει· οὐ μὴν πλέον τι ἡ μετάληψις ἐποίησε, τὸ γὰρ δεκτικὸν οὐδὲν ἄλλο ἐστὶν ἢ τὸ οἷόν τε δέξασθαι, τοιοῦτον δὲ ὂν μένει πρός τι ὄν.*

[117] Moraux 1973: 203, 209–10. [118] Algra 2000: 171 n. 25; cf. Rescigno 2004: 483.

[119] Algra 1995 distinguishes between the 'logical' and 'physical possibility'; Moraux 1973: 202 speaks of 'counterfactual' possibility as 'Stoic' possibility. For the logical background of this concept, see Bobzien 1998: 97–144.

[120] Chrysippus ap Stob. (*SVF* 2. 503, esp. 163. 5–12 von Arnim), cf. also the arguments in Cleomedes, *Meteor.* 1. 1 (104–12).

ἄλλο γένος in *An. Post.* 1. 7).[121] A similar methodological concern is probably in the background of the following report. According to Julian's *Oration upon the Mother of Gods*, Xenarchus explains the coming to be of form and matter by the motion of the 'fifth body' and criticizes Aristotle and Theophrastus for enquiring into the transcendent principles of natural processes.

But some sharp Peripatetic, such as Xenarchus, says: we see that the cause of these (i.e. of form and matter being held together) is the fifth circular body. And Aristotle made himself laughable by investigating and worrying about these matters, and similarly Theophrastus. At any rate, he ignored his own words. For just as having arrived at the incorporeal intelligible substance he stopped and did not worry about the cause, saying that this is so by nature, so in the case of the fifth substance he should have accepted it being so by nature not investigating the causes any further, but stopped with them and not even venture out to the intelligible, for it is nothing by its own nature, and besides has an empty supposition. For I do remember hearing[122] that Xenarchus says such things. (Julian, *Or.* 8 (5) 3. 17–34)

The reference to Theophrastus (fr. 158 FHSG) can be compared with Theophrastus' methodological remarks concerning the scope of scientific explanation in *Metaphysics* 9b16–10a21.[123] Theophrastus points out that the study of the first principles of nature cannot seek to explain every principle: that would eliminate the idea of explanation (because of the infinite regress). This project is too rich, as it were, to make sense.[124] On the other hand, the project of studying heavenly bodies in astronomy, limited to the study of the geometry of their motions and positions, is too limited: it is based on description and lacks the account which is explanatory in a proper sense. Theophrastus seems to suggest that a suitable methodological approach would seek to establish the principles on the basis of their proper activities: just as a student of biology studies the soul only to the extent to which it is the principle of various activities of living beings, so the student of nature as a whole should study the first principles to the extent to which they serve to explain the heavenly motions.[125] Xenarchus' reproach to Aristotle and Theophrastus seems to be that they do not apply the same methodological restriction to the theory of the 'fifth' heavenly element. In this fragment, it is remarkable that Xenarchus wants to come across as a consistent Peripatetic, in fact, more consistent in this particular issue than the founders of the school.

It seems that despite some apparent affinities with Stoicism (criticism of the fifth substance; criticism of Aristotle's argument against extracosmic void), there is no clear evidence that Xenarchus is committed to any distinctly Stoic doctrine.

[121] In this sense, Posidonius fr. 18 EK, with its distinction between physics and astronomy in their respective subject matters and methods, would be following the same overall methodology.

[122] On 'hearing' possibly referring to a private reading from Xenarchus, see Sharples 1998: 94 n. 257.

[123] For a good recent discussion, see Rashed 2007: 262–8.

[124] *Metaph.* 9b16–24.

[125] *Metaph.* 10a9–19. cf. also fr. 159 FHSG (= Proclus, *In Tim.* 35A).

Rather, he seems to be criticizing 'Aristotelem per Aristotelem', appealing in his criticisms primarily to the texts of Aristotle and Theophrastus and pointing up inconsistencies within the doctrine taken as a whole. His suggestions could be seen as attempts to amend the system rather than replace it with a different (e.g. Stoic) system. The report according to which he had Aristotelian definition of the soul might seem reassuring in this respect as well.[126]

The Prime Matter

The prime matter seems to be the most likely genuine bridge-concept between the two systems.[127] Yet, as we shall see, the integration of this concept into the Aristotelian exegesis was not straightforward. In Peripatetic physical doxography, the principle of matter appears as a counterpart to the principle of form.[128] Characteristic are two excerpts published by Stobaeus, printed by Diels as frgg. 2 and 3 of Arius Didymus, in which a summary of Aristotelian principles is presented:

Aristotle. And since nature in accordance with its concept is a certain principle of motion and rest, neither matter, by its proper definition, can move, nor form. For one is always formless, the other is form; and the former is not a body, although of bodily nature, and the latter completely incorporeal. And they say that matter is not a body not just because it seems to lack the bodily dimensions, but because, in accordance with its proper definition, it also misses out on many other characteristics of the body: shape, colour, heaviness, lightness, generally all quality and quantity. For were it to partake of these, it would be of some kind and some quantity; but since it does not partake of them, it would not be a body, but bodily, because it underlies all the qualities as a matrix; for in the same way as form separated from matter happens to be incorporeal, so too matter is not body when form is removed. For the existence of body requires that the two come together.[129]

The next fragment (which I do not cite in full) outlines the concept of form:

Now, form is different from figure to this extent, namely, that one is going through in depth, another is on the surface; and one is similar to white colour used in painting,

[126] Cf. Aëtius 388. 16–20 Diels. (*Ξέναρχος ὁ Περιπατητικὸς καί τινες ἕτεροι τῆς αὐτῆς αἱρέσεως τὴν κατὰ τὸ εἶδος τελειότητα καὶ ἐντελέχειαν καθ' ἑαυτὴν οὖσαν ἅμα καὶ μετὰ τοῦ σώματος συντεταγμένην.*) One would like to know what Xenarchus made of Aristotle's theory of intellect, but the only explicit report preserved is very elliptic. According to it, Xenarchus understood the potential intellect as somehow the same as the prime matter (*ap.* Philop. 15. 65–9 Verbeke). Moraux (1973: 208) suggested that he could not attribute to Aristotle such an absurdity and must have posited this polemically in the course of his critical discussion of Aristotle's doctrine (cf. Alex. *Mant.* 2. 106. 19–23).

[127] Cf. Alex. *Mixt.* 213. 15–214.6; Galen, *Nat. Fac.* 1. 12; *De Elem. Sec. Hipp.* 4. 3–8.

[128] The meaning of 'principle' itself becomes a subject of discussion. The epitome of Aristotle's philosophy compiled by Nicolaus of Damascus opens with a most painstaking enumeration of all the meanings of the 'principle' he found in the corpus.

[129] [Ar. Did.] Fr. 2. 448. 1–12 Diels (= Stob. *Ecl.* 1. 11. 4).

another gives form to the substance of milk; the same [principle] is called 'form', insofar as it both gives matter form and provides it with [inner] structure.[130]

'Form' here refers to the principle of hylomorphic constitution discussed by Aristotle in physics.[131] The way the contrast with shape is drawn deserves attention: form pervades matter 'in depth', as it were physically imparting to it all its qualities. This reminds us of the description of the manner of operation of the active principle in the Stoic system.[132]

The notion of prime matter as 'formless and unqualified' is not found in this form in any of Aristotle's writings, but apparently is adopted by all the authors of this period (and later by Alexander of Aphrodisias).

Simplicius' *Physics* commentary has preserved for us a discussion of the concepts of matter and substrate (*ὑποκείμενον*) by the early commentators.[133] The issue under discussion is the relation between privation (*στέρησις*) and the substrate. If the substrate exists only in potentiality, and what is in potentiality is in the state of privation, it seems that the substrate is none other than privation, by its proper account.[134] To avoid this conclusion, we need to assume that the substrate exists in actuality, while the lack of forms which it can receive belongs to it accidentally.[135] This must be a solution to the problem given by Alexander of Aphrodisias, who is cited as explaining that when the substrate has a privation as an accident it is the matter of something, while the substrate as such (i.e. within a hylomorphic composite) does not have a privation.[136] Simplicius cites Boethus who seems to distinguish between the substrate-before-the-change and the substrate-within-the-composite-after-the-change as follows:

It is called 'matter' on account of its being *qualityless and formless*; for matter seems to be named relatively to the thing which will be; but once it has received a form, it is no longer called 'matter', but 'the substrate'; for something is said to be the substrate with respect to that which already exists.[137]

Alexander points out that Boethus fails to distinguish between the 'qualityless and formless' matter proper and the substrate-before-the-change, or 'antecedent' matter, which contains privations of relevant forms incidentally. Notably, Alexander corrects Boethus' analysis while retaining the notion of 'formless and qualityless matter' which he says is different from the two outlined meanings of substrate.[138]

[130] [Ar. Did.] Fr. 3. 448. 16–19 Diels.

[131] The list of synonyms at the beginning of the fragment (*morphê, entelekheia, to ti ên einai, ousia hê kata logon*, and *energeia*) is similar in form to the lists of defined terms in Nicolaus of Damascus.

[132] I have discussed some related issues in Kupreeva 2003: 325–40.

[133] Simplic. *In Phys.* 1. 7, 190a13–31. [134] Simplic. *In Phys.* 211. 9–10.

[135] Simplic. *In Phys.* 211. 10–13. [136] Simplic. *In Phys.* 211. 13–15.

[137] *Ap* Simpl. *In Phys.* 211. 15–18 Diels: *ἄμορφος μὲν οὖσα καὶ ἀνείδεος ὕλη λέγεται, ἡ γὰρ ὕλη πρὸς τὸ ἐσόμενον ὠνομάσθαι δοκεῖ· ὅταν δὲ δέξηται τὸ εἶδος, οὐκέτι ὕλη ἀλλ' ὑποκείμενον λέγεται· ὑποκεῖσθαι γάρ τι λέγεται τῷ ἤδη ἐνόντι.*

[138] 211. 13–15. Marwan Rashed has recently come up with a different analysis of Boethus' position in this passage. According to him, Boethus' prime matter has physical existence in actuality

The prime matter, according to Alexander, can receive both form and privation (the view criticized by Simplicius, who does not reject the notion of prime matter nonetheless).[139]

Another instance where the concept of prime matter is incorporated into Aristotle's usage is found in the *Epitome* of Aristotle's philosophy by Nicolaus of Damascus.[140] In his summary of Aristotle's theory of nature (based on *Physics* 2.1), Nicolaus distinguishes the matter of a particular from 'formless and unqualified matter':

1 . . . [b]<F>or nature is said to be matter, viz. matter of two kinds:
2 [a]One of particulars. This is without order (ῥυθμός), [[b] this possibly is shape] and it is without form in comparison with anything else, to which it ascends, and in which is contained all artificial matter and every element, e.g. fire and water and the rest.
3 The other matter which is supreme is wholly unspecified and without form. (Tr. Drossaart-Lulofs)

Drossaart-Lulofs points out a parallel between this description of particular matter (where the Greek word ῥυθμός is transcribed in Syriac) with Aristotle's *Metaphysics* Δ4, 1015a7, where matter is described as ἀρρύθμιστος, and argues that Nicolaus interprets the ὅλως πρώτη ὕλη of 1015a8 as the 'supreme formless matter'. If he is right, it means that Nicolaus overlooked Aristotle's examples which make clear that 'first matter' is used in the sense of generically prior (thus bronze is the proximate matter of statue, but water is 'prior' generically, because it is the matter of all metals).[141] The list of meanings in *Metaphysics* Δ contains no references to the formless and qualityless prime matter, but Nicolaus cannot omit it in his classification, and so includes it without any (or on slightest) textual evidence.

We can see that Hellenistic prime matter, with its 'Stoic' epithets 'formless', 'qualityless', is accepted by virtually all post-Aristotelian Peripatetics, but they do show some amount of uncertainty about finding its analogues in Aristotelian texts. When eventually it is accepted as the matter of the four elements, it is also

(Rashed 2007: 199–205). This analysis is open to some difficulties: first, not even the Stoics made such a strong claim about their own corporeal prime matter, and secondly, it seems that the force of Alexander's correction is to establish proper logical distinctions between the concepts, two of which were conflated by Boethus (211. 18–19, assuming this remark is a summary of Alexander's position). We do not need to take this ambiguity as a sign of a particular theoretical commitment on behalf of Boethus (it is not clear in what way his assimilation of prime matter to the antecedent matter was supposed to work). Alexander's point (*pace* Rashed 2007: 205) is not just that the antecedent matter is always accompanied by a privation as an accident, but also that the principle of prime matter is to be distinguished from any particular antecedent matter in that the former (unlike the latter) can take on both the form and the lack of form for any form F: hence the emphasis on the distinction between the 'negation' and 'privation'. (Most probably Alexander is thinking about the common matter of the four 'simple bodies'.)

139 Alexander ap. Simpl. 211. 20–3 Diels. Moraux 2001: 137.

140 Drossaart-Lulofs 1965.

141 *φύσις δὲ ἢ ὅλως ἥ τε πρώτη ὕλη (καὶ αὕτη διχῶς, ἢ ἡ πρὸς αὐτὸ πρώτη ἢ ἡ ὅλως πρώτη, οἷον τῶν χαλκῶν ἔργων πρὸς αὐτὰ μὲν πρῶτος ὁ χαλκὸς, ὅλως δ' ἴσως ὕδωρ, εἰ πάντα τὰ τηκτὰ ὕδωρ) κτλ.*

established that, differently from its Stoic counterpart, the Aristotelian concept of prime matter is unambiguously 'not a body', even though it does not exist without a body.[142] Another way of putting this would be to say that 'qualityless', ἄποιος, is a stronger concept for Peripatetics than for the Stoics with their technical concept of corporeal 'quality'.[143]

4

In this section I would like to draw attention to several aspects of Alexander's discussion of the elements which may be based on earlier traditions of Peripatos and reflect some of the agenda of the Stoic doctrine of principles.

Alexander opens his treatise *De Anima* with a summary of Aristotelian doctrine of principles. He introduces the principles of matter and form of natural bodies and shows the derivation of the four elements from these principles. This latter procedure is not found in Aristotle, but has a parallel in the Antiochean 'concordance'.[144]

> Since there is a difference in natural bodies (namely, some of them are simple, others composite), the matter of the composite bodies and their substrate is itself a natural body composed of form and matter (for every natural body is composed of these). As for the simple bodies, they have no composite substrate, for otherwise [a body] would itself be composite. And if their substrate is not composite, it is not a body, given that every body is composed of form and matter. (3. 21–7 Bruns)

The 'formless and shapeless' matter, corresponding to the prime matter of the 'Antiochean' consensus, is defined here in a more technical way as the matter underlying the elements:

> Thus, the substrate of simple bodies and their matter is some simple nature lacking form, which by its own account lacks figure, form and shape. Because it is and is said to be formless that is called form which having come to be in it brings to an end its mentioned privation; and such nature one would call matter in a strict sense. (3. 27–4. 4 Bruns)

We may note that 'lacking qualities' for Alexander means also 'lacking dimensions': Alexander points out that prime matter does not exist separately from any elemental form.[145] The order of derivation of prime matter in *De Anima* is the reverse of the one we found in Antiochus' summary: Alexander calls the prime matter 'matter in a strict sense', but the derivation starts from the

[142] Alexander, *An.* 3. 26–4.20, cf. Aristotle's discussion of 'possible body' in *GC* 1. 5 (320a27–b17).

[143] Frede has argued that Stoic οὐσία has spatial dimensions (see Frede 2005: 222–3). Cf. Cooper at pp. 97–8 and Gourinat Ch. 3 above; on Peripatetic matter, cf. Moraux 2001: 229.

[144] Cf. *Acad.* 1. 25 and discussion at pp. 139–40 above.

[145] *An.* 4. 9–22 Bruns. Cf. the point made about matter as the passive principle in the 'Antiochean' synopsis in Cicero, *Acad.* 1. 24, p. 137 above.

elements. This may be in line with the way Aristotle treats the matter of the elements.[146]

Alexander does explain the role of form in the generation of elements. Just as the natural form generally is the substance of a thing, in the same way primary qualities (hot/cold, dry/moist) in proper combinations are the natural forms of the elements:

For of fire, since it is a simple natural body, the form is heat and dryness, as well as the lightness supervening upon these, and the matter is the substrate of these, which being by its own nature none of these, can receive equally these and their opposites (and owing to this nature, there are transformations of simple bodies into each other). (5. 4–9 Bruns)

The property of lightness which supervenes on the 'agent' qualities of heat and dryness[147] is the principle of motion and as such a 'power' of fire which accounts for its natural motion upwards. This model of 'supervening power' is used by Alexander in his exposition of Aristotle's definition of soul. Soul is defined as the power supervening on the multiple bodily structures, whose material constitution ultimately goes back to the combinations of four 'simple bodies':

So that if there is going to be, beside the simple bodies, some composite natural body, it must have more simple bodies as its underlying substrates; and this multitude is the multitude derived from the variety of forms that are in them, and therefore such bodies are composite. And the nature and form of a thing which has several different forms underlying it with matter must be more manifold and more perfect, since each particular nature in the bodies underlying it[148] contributes something to the form which is common to them all. For such a form is in a way a form of forms and as it were a perfection of perfections. Therefore those should not marvel at the variety of forms in the natural bodies who have the causes of their variation clearly in their underlying substrates. For it is plausible that both the multitude of forms in the bodies underlying them and their varied mixture contain the causes of so great a variation.[149]

The combinations are in accordance with a specific composite 'form', a 'formula' that corresponds to each particular species.[150] Using this approach, Alexander develops a scale of beings similar to one reported by some sources for Chrysippus.[151] But while the Stoic ladder is built on the principle of increasing pneumatic tension, Alexander's is that of increased complexity which can be expressed in terms of elemental combinations. This kind of analysis does not

146 Cf. Rashed 2005 on *GC* 2. 1, 329a27–3: 'Bref, la "matière première" est une modalité des "corps premiers", et non ceux-ci de celle-là'.

147 Cf. *GC* 2. 2, 329b18–32.

148 Alex. *An.* 8. 10–11: i.e. each underlying form contributes to the form that supervenes on the whole. The syntax of the Greek is ambiguous, but the meaning is not affected.

149 *An.* 8. 5–17 Bruns. Discussions in Moraux 1942: 30–43; Donini 1971; Accattino and Donini 1996; Caston 1997: 347–53; Moraux 2001: 354–9.

150 *An.* 9. 11–26 Bruns.

151 *SVF* 2. 458 (Philo, *Leg. Alleg.* 2. 22–3; *Quod Deus Sit Immut.* 35–6), cf. LS 47.

seem warranted by any of Aristotle's treatises.[152] On the other hand, it fits well with doxographical expositions of Aristotelian physics of the Imperial age, such as the physical fragments of Arius 'Didymus' discussed above.[153]

The fifth element is never mentioned in this derivation of the elements, save for a brief remark towards the end of the account, to the effect that the soul of gods, if indeed it were called soul, would be called so homonymously.[154] Extant fragments from Alexander's discussion of the souls of heavenly bodies suggest that the homonymy is based on the functions of the soul. In the case of heavenly bodies, Alexander emphasizes, against the Stoics, that there are no parallels to the sublunary life functions such as nutrition and sense perception.[155] It seems also that there is no homonymy in the case of 'nature' (as opposed to 'soul'): the souls of heavenly bodies are their natures, comparable to the lightness of fire and heaviness of the earth.[156]

In this account of the elements in *De Anima*, physics seems to be completely separated from theology. Rather, in the proem to the treatise Alexander explains at some length why physics in fact is the best theology.[157] But among the school works we do find a number of texts showing that properly theological concerns are still very much alive in Alexander's theory of elements, in connection with the problem of providence. I am going to look at the texts which discuss the relation between divine power and the cosmic elements.

Quaest. 2. 3 is entitled 'What the power is that comes to be from the movement of the divine body, in the body adjacent to it which is mortal and subject to coming to be'. Heavenly body is supposed to be the source of divine providence operating to some extent in the sublunary world: this thesis Alexander sets out to prove. The problem goes back to the Hellenistic Peripatetic thesis (discussed above) that providence extends only till the sphere of the moon.[158]

The author considers two solutions, based, respectively, on two different models of the aether's function within the cosmos. According to the first theory,

152 In fact, the inclusion of the elements in the theory of substance is sometimes perceived by Aristotle as problematic, cf. *Metaph.* Z 16, 1040b5–10.

153 See p. 156 above.

154 *An.* 28. 26–8 Bruns: ἡ γὰρ τῶν θεῶν ψυχή, εἰ καὶ ταύτην δεῖ ψυχὴν καλεῖν, ὁμωνύμως ἂν ταύτῃ ψυχὴ λέγοιτο.

155 Simplicius, *In Cael.* 54. 6–12; Moraux 2001: 192.

156 Simplicius, *In Cael.* 380. 29–381. 2. Discussion in Accattino 1991: 45; Moraux 2001: 176–80, 194–7, 214.

157 *An.* 1. 1–2. 25.

158 Alexander in the treatise *On Providence* preserved only in Arabic cites this as Aristotle's view: 'for it is clear with regard to the account of the Philosopher that he says that the providence exists until the sphere of the moon (*ilâ nahwa falaki al-qamar*)'. In Greek sources, we find an identical formula in Clement: Οὐδὲν δὲ οἶμαι χαλεπὸν ἐνταῦθα γενόμενος καὶ τῶν ἐκ τοῦ Περιπάτου μνησθῆναι· καὶ ὅ γε τῆς αἱρέσεως πατήρ, τῶν ὅλων οὐ νοήσας τὸν πατέρα, τὸν καλούμενον ὕπατον ψυχὴν εἶναι τοῦ παντὸς οἴεται· τουτέστι τοῦ κόσμου τὴν ψυχὴν θεὸν ὑπολαμβάνων αὐτὸς αὑτῷ περιπείρεται. Ὁ γάρ τοι μέχρι τῆς σελήνης αὐτῆς διορίζων τὴν πρόνοιαν, ἔπειτα τὸν κόσμον θεὸν ἡγούμενος περιτρέπεται, τὸν ἄμοιρον τοῦ θεοῦ θεὸν δογματίζων. *Protr.* 5. 66. 4–5.

divine power accrues to sublunary things after the four elements have been formed (48. 22–4). This terse formula should not be taken as any admission of the beginning of the cosmos: 'before' and 'after' should refer to a history of individual composites whose physical constitution ultimately rests on the four combinations of the hot and cold and the dry and the moist. The divine power, on this picture, is a fifth separate physical factor, quasi-elemental power *sui generis*, which enters the constitution of bodies composed of the four elements and accounts for special properties, such as mental properties:

> The simple bodies too contribute to the coming-to-be of the bodies that come to be from them, and so too does the divine power in which they have a share according to their proximity. It is on account of this power that these no longer possess in themselves a principle only of motion in accordance with overcoming tendency, but have acquired in addition also a certain soul-borne motion which possesses its origin and coming-to-be from the divine power in which they have a share. (48. 29–49. 4 Bruns, tr. Sharples, modified)

This seems to be a quasi-physicalist picture of the heavenly influence, perhaps parallel to the reports of Critolaus we have discussed.[159] But there is a further detail: the power acts differently on different bodies depending on their elemental constitution: those whose constitution is finer and purer get more, others less.

The criticism of this solution in our treatise is as follows. On this picture, only such natural substances will benefit from this power as are formed by blending and alteration of the simple bodies, because the scope of influence of this divine nature is restricted to certain kinds of elemental mixture. It does not work for simple bodies in their unmixed state—not even if they are combined by juxtaposition. For the effects to take place, 'chemistry' is needed.[160]

We also get some details about the way the divine body was supposed to produce psychic powers.[161] The divine power exercises some selection in acting upon the elements. The range of selection is defined by the elemental constitution of different natural substances, presumably because mixture is the main method by which divine power propagates itself through the region of nature. A remarkable claim here is that the divine nature is mixed with the bodily mixture and modified in this process.[162] This process has parallels with the Stoic 'total pervasion', the mechanism by which the active principle acts upon

[159] See p. 149 above.

[160] 'So the compound bodies, which come to be by the mixture of the simple bodies and by alteration, possess a power of this sort which is in a way mixed and combined with the mixture of the bodies that possess it, and in accordance with this they come to be more perfect bodies and share in a more divine nature and principle; but as many bodies as seem to come to be compound by the composition and juxtaposition of the simple bodies are not also able to share in a power and nature of this sort, each of the simple bodies preserving its own nature in actuality in this sort of composition and mixture. For the divine power does not contribute to the being of bodies that are compounded in this way, remaining in each of them as it was before their composition' (49. 18–28, tr. Sharples).

[161] *Quaest.* 2. 3. 49. 4–14 Bruns.

[162] *Quaest.* 2. 3. 49. 25–7 Bruns.

matter.[163] As a whole, this kind of providential design cannot be described as Stoic, at least not as what we know as mainstream Stoic, because it excludes the divine principle from the elemental generation.[164] This solution to the main problem is rejected by the subsequent discussion in the *Quaestio*, to which we shall turn shortly.

There is another text in Alexander's corpus which seems to invoke the same half-way physicalist Peripatetic theory, and which is rightly cited by scholars in parallel with this solution. In the last section of the treatise *De Intellectu* (*Mantissa* 2), the author tells us about a Peripatetic doctrine[165] according to which divine intellect pervades the whole cosmos and permanently operates 'in matter as one substance in another, in actuality'.[166]

When, from the body that was blended, there comes to be fire or something of this sort as the result of the mixture, which is able to provide an instrument for this intellect, which is in this mixture—for it is in every body, and this too is a body—then this instrument is said to be intellect potentially, supervening on this sort of blending of bodies as a suitable potentiality for receiving the intellect that is in actuality. (112. 11–16 Bruns, tr. Sharples)

Human thought is the activity of the divine intellect operating by means of the human capacity of thinking (which is the potential intellect).[167] Alexander says that this account has several characteristic Stoic 'faults': (*a*) god is said to be in the lowest beings (presumably because the intellect is said to pervade all matter);[168] (*b*) in sublunary things there is providence by direct divine intervention;[169] (*c*) thinking is not 'up to us'. We shall note, however, that despite conceptual and terminological affinities with the Stoic theory, the thesis proposed by our unknown Peripatetic is different from the Stoic one: with the Stoics, the active principle pervades the matter as a whole 'by the whole', making no exceptions. Peripatetic divine nature in our theory is mixed not with all bodies without exception but only with those whose elemental constitution makes them suitable for such a mixture. There is no complete pervasion of the cosmos by the divine principle. But Alexander does not want to allow even for a 'partial' mixture (he goes against this view also in the treatise *De Mixtione* which is written in part

[163] Here we may recall Alexander's complaint that some Peripatetics found themselves attracted by the Chrysippean doctrine of mixture (*Mixt.* 3. 216. 9–15). On the distinction between blending and juxtaposition, see *Mixt.* 2. Cf. Hahm 1977; Frede 2005.

[164] See Hahm 1977; Frede 2005.

[165] Developed in response to the objection (probably by Atticus, cf. fr. 7 des Places) that the Aristotelian god being immobile will not be able to deliver his power to the things within the cosmos. See Rashed 1997.

[166] *Mant.* 2. 112. 10 Bruns.

[167] *Mant.* 2. 112. 23–5 Bruns.

[168] *Mant.* 2. 113. 12–14 Bruns.

[169] The proponent of this doctrine seems to believe in the providence which comes about from heavenly motion, as is Alexander's own view. So the Stoic view which is criticized by Alexander also for being superfluous may be just an implication of this theory of 'partial mixture' rather than a professed belief.

for the benefit of the Aristotelians susceptible to the Stoic reasoning). His own view of divine influence seems to be concisely stated in the second solution to *Quaest.* 2. 3.

The second solution is based on a more consistently physicalist view of the cosmos, and it claims to make a stronger case for the divine providence than the first one, because according to it, the action of the divine power is much more pervasive and covers the whole of the cosmos. Here the author seems to be after the notion as powerful as the Stoic one, except that the divine influence he has in mind reaches the sublunary cosmos by being propagated in the succession of movements rather than via 'total pervasion'.[170] The author refers to the fact that divine body in its motion heats the tinder sphere (as Aristotle describes this, controversially, for the sphere of the sun in *Meteor.* 1. 3), thus producing the primary qualities: heat and dryness.[171] Together with their opposites, i.e. coldness and moisture, these form the qualitative basis for the whole of the physical cosmos. Furthermore, 'divine body' refers to the whole ensemble of the heavenly bodies, so that the elemental generations and changes display some degree of regularity for which they depend on the perfect regularity of the heavens. This order is propagated to the composites, and thus pervades the whole cosmos. In this way divine power is said to inform the matter with the first and simple forms, from which then the full variety of composite natural substances comes to be. Again we are told explicitly that 'all those bodies which have more finer and purer bodies in their blending have a more perfect form, and all those which have in themselves less of such a body, and more of the passive and denser, have a more imperfect form'.[172] In this second solution, mental properties are explained by a finer physical constitution, but we have to understand that this latter is itself a function of the divine cosmic order. Thus providence is exercised by the Aristotelian god in an 'oblique' way: it accounts for the coming to be of the elements and elemental processes, whose regularity gives rise to the regular compounds. The Aristotelian idea of regularity of sublunary processes dependent upon the heavenly revolutions is interpreted by Alexander in terms of providence.[173] Differently from the Stoic providence, this one ultimately operates on the level of species, not individuals.

[170] M. Rashed (2007: 285–93) speaks of the 'mechanism' as a key feature of Alexander's approach (since it emphasizes the transmision of movement).

[171] The fact that these qualities are produced, and not imparted by the heavenly sphere to the sublunary realm, is emphasized in Alexander's commentary on *Meteor.* 1. 3, where he argues against the view that the sun heats 'by quality', using some of the reasoning techniques which have affinities with those of Stoic physics. See Alex. *in Netsc.* 18.6–19.9 Hayduck.

[172] *Quaest.* 2. 3. 50. 23–7 Bruns.

[173] In his commentary on *Meteor.* 2. 3 (257b32–258a3), where Aristotle says that the processes of evaporation from the sea must be ordered, Alexander remarks: *σημειωτέον δὲ πάλιν ὅτι τάξιν τινὰ καὶ ἐν τοῖς ἐν γενέσει καὶ φθορᾷ οὖσιν εἶναι λέγει, ἥτις τάξις παρὰ τὰ ἄστρα αὐτοῖς τῆς τοιαύτης μεταβολῆς γίνεται, ταῦτα δέ ἐστι τὰ θεῖα. εἰ δὲ τοῦτο, οὐκ ἀπρονόητα κατ᾽ αὐτὸν τὰ τῇδε* (83. 6–9).

CONCLUDING REMARKS

Stoic and Peripatetic physics must have a number of common features: strong teleological commitment, rationalist ethics, and theology, together with denial of traditional Platonic doctrine of forms and shared belief in physical continuum, almost guarantee that there will be parallels and affinities between the two systems. Yet it seems, from this brief survey, that these common features do not easily translate into the details of two cosmological doctrines. Everything works differently: the doctrine of the principles, theories of matter and elements, and the parallel theories of providence. Even in the Hellenistic Peripatos, when physics in a narrow sense is not among the school priorities, the background of the key tenets and arguments reported by the sources seems to be their distinct school agenda rather than common principles highlighted in the *ad hoc* doxographies in the engaged philosophical discussions. The critical tendencies within the school (the case of Xenarchus and the fifth element) do not necessarily amount to the rejection of the system, despite the fact that the system revised on the basis of criticisms may be significantly (for some, perhaps irreconcilably) different from the criticized original.

Alexander's engagement with the Stoics is probably both the closest and the most critical compared to all his known predecessors. It is the closest because he is probably fully conscious of the shared values, and it is the most critical because he is particularly keen to articulate all the relevant differences, both in terms of systematic exposition and Aristotelian exegesis. Although he does not tell us much about his Stoic sources, he does know them well, and not only as the targets for his criticism, but also as a possible inventory of argumentative and expository techniques. The structural parallels between his and Stoic arguments and terminology can be quite striking: the nature's ladder starting from the four elements, and the full-scale action of the divine power in the *Quaest.* 2. 3 are just two examples. But they also underscore the difference of his approach: the ladder is constructed in a Peripatetic way and has a role of illustration, and the divine providence pervades nature by providing its material framework. An interesting document, the so-called Vitelli fragment 2,[174] could be used as a late anachronistic footnote to the Antiochean harmony among the three schools. In it, Alexander is replying to the criticisms of a Stoic named Heraclides (who held the Stoic philosophical chair in Athens around Alexander's time or a little earlier).[175] As the title makes clear, Heraclides attacked the Aristotelian doctrine of the fifth body:

By the same Alexander, from the treatise *Against Heraclides*, a review of Aristotle's arguments concerning the fifth substance; where he argues against a certain Stoic

[174] Named after G. Vitelli who first described and published it (Vitelli 1902; English tr. Sharples 1994: 89–94).

[175] For this Heraclides we also have epigraphic evidence: *IG* II^{2}3801 (cf. *IG* II^{2}3989); see Oliver 1977: 164–5; Sharples 1990: 93–4 and nn. 80, 82.

philosopher who criticizes Aristotle saying that Aristotle differs from Plato in his view of gods and soul's immortality. (fr. 2. 1–4 Vitelli)

Alexander says, after a rhetorical opening:

That in this respect the Stoic school is in a greater disagreement with Plato than Aristotle is, everyone can easily see. For Plato thinks that the first god is incorporeal, and says that he rests in self-contemplation and thought, and there are some secondary gods managing the coming to be and being of other things, and what Aristotle says is in agreement with this; [Plato] also says that the soul is some incorporeal and imperishable substance, and each of these [claims] about it is proven also by Aristotle. They [Stoics], on the other hand, make god a body, and postulate that this body passes through all things; further, they say that soul is body, and that it is perishable, and that some souls perish immediately along with the perishing [bodies] that have them, while others are preserved until the most absurd conflagration. (fr. 2. 8–18 Vitelli)

The Stoic charges against Aristotle and Alexander's grievances against the Stoics are exploiting familiar features of both respective systems. What is more remarkable here is that both Heraclides in his criticism and Alexander in his reply seem to be claiming some common Platonic heritage, and these claims apparently will not conflict with the improvements on the Platonic system which must be made in each school.*

REFERENCES

Accattino, P. (1991) 'Alessandro di Afrodisia e gli astri: l'anima e la luce', in *Atti della Accademia delle Scienze di Torino, Classe di Scienze Morali, Storiche e Filologiche*, 122: 79–94.

—— and Donini, P. L. (1996) *Alessandro di Afrodisia: L'anima* (Rome and Bari: Laterza).

Algra, K. (1995) *Concepts of Space in Greek Thought* (Leiden: Brill).

—— (2000) 'The Treatise of Cleomedes and its Critique of Epicurean Cosmology', in M. Erler and R. Bees (eds.), *Epikureismus in der späten Republik und der Kaiserzeit* (Stuttgart: F. Steiner), 164–89.

—— (2004) '*On Generation and Corruption* I.3: Substantial Change and the Problem of Not-Being', in F. de Haas and J. Mansfeld (eds.), *Aristotle; On Generation and Corruption: Symposium Aristotelicum* (Oxford: Clarendon), i. 91–121.

Arnaldez, R. (introd., notes) (1969) *Philo: De aeternitate mundi*, trans. J. Pouilloux (Paris: Editions du Cerf).

Barnes, J. (1997) 'Roman Aristotle', in J. Barnes and M. Griffin (eds.), *Philosophia Togata* (Oxford: Clarendon Press), ii. 1–69.

* I am grateful to all the participants of the conference at UNAM, particularly to Keimpe Algra, Thomas Bénatouïl, John Cooper, David Hahm, Ricardo Salles, and Brad Inwood; to my commentator Alejandro Tellkamp for a helpful discussion of my paper; to Ricardo Salles for his effort in organizing the conference and editorial help; and to Brad Inwood, Ricardo Salles, and the anonymous readers for their written comments which saved me from a number of errors and unclarities; the remaining ones are, of course, my own.

Berryman, S. (1996) 'Re-Thinking Aristotelian Teleology', Ph.D., University of Texas, Austin.

Bobzien, S. (1998) *Determinism and Freedom in Stoic Philosophy* (Oxford: OUP).

Bowen, A. C., and Todd, R. B. (eds.) (2004) *Cleomedes' Lectures on Astronomy* (Berkeley, Calif.: University of California Press).

Brittain, C. (2001) *Philo of Larissa* (Oxford: OUP).

—— (ed.) (2006) *Cicero on Academic Scepticism* (Indianapolis, Ind.: Hackett).

Broadie, S. (2004) '*On Generation and Corruption* I.4: Distinguishing Alteration', in F. de Haas and J. Mansfeld (eds.), *Aristotle; On Generation and Corruption: Symposium Aristotelicum* (Oxford: Clarendon), i. 123–50.

Brunschwig, J. (1988) 'La théorie stoïcienne du genre suprême', in J. Barnes and M. Mignicci (eds.), *Matter and Metaphysics: Fourth Symposium Hellenisticum* (Naples: Bibliopolis), 21–127.

Caston, V. (1997) 'Epiphenomenalisms Ancient and Modern', *Philosophical Review*, 106(3): 309–63.

—— (1999) 'Aristotle's Two Intellects: A Modest Proposal', *Phronesis,* 44(3): 199–227.

Chaniotis, A. (2004) 'New Inscriptions from Aphrodisias (1995–2001)', *American Journal of Archaeology,* 108: 377–214.

Charles, D. (2004) 'Simple Genesis and Prime Matter', in F. de Haas and J. Mansfeld (eds.), *Aristotle; On Generation and Corruption: Symposium Aristotelicum* (Oxford: Clarendon), i. 151–70.

Donini, P. L. (1971) 'L'anima e gli elementi nel Alessandro di Afrodisia', *Atti della Accademia delle Scienze di Torino*, ii. 105 (Turin: Accademia delle Scienze), 61–107.

—— (1982) *Le scuole, l'anima, l'impero: La filosofia antica da Antioco a Plotino* (Turin: Rosenberg & Sellier).

Dorandi, T. (1999) 'Chronology' in K. Agra, J. Barnes, J. Mansfeld, and M. Schofield (eds.), *The Cambridge History of Hellenistic Philosophy* (Cambridge: Cambridge University Press), 31–54.

Drossaart-Lulofs, H. J. (ed.) (1965) *Nicolaus Damascenus on the Philosophy of Aristotle* (Leiden: Brill).

Eijk, Ph. J. van der (2005) 'Divine Movement and Human Nature in Eudemian Ethics 8.2', in Ph. J. van der Eijk, *Medicine and Philosophy in Classical Antiquity: Doctors and Philosophers on Nature, Soul, Health and Disease* (Cambridge: CUP), 238–58.

Falcon, A. (2001) *Corpi e movimenti* (Naples: Bibliopolis).

Fazzo, S. (2002) *Aporia e sistema: La materia, la forma, il divino nelle* Quaestiones *di Alessandro di Afrodisia* (Pisa: ETS).

Frede, M. (2005) 'La Théologie stoïcienne', in G. Romeyer Dherbey and J.-B. Gourinat (eds.), *Les Stoïciens* (Paris: Vrin), 213–32.

Freudenthal, G. (1995) *Aristotle's Theory of Material Substance* (Oxford: Clarendon).

Furley, D. J. (1989*a*) 'Aristotelian Material in Cicero's *De natura deorum*', in W. W. Fortenbaugh and P. Steinmetz (eds.), *Cicero's Knowledge of the Peripatos* (New Brunswick, NJ: Transaction), 201–19.

—— (1989*b*) 'Lucretius and the Stoics', in D. J. Furley, *Cosmic Problems* (Cambridge: CUP), 183–205.

Görler, W. (1989) 'Cicero und die "Schule des Aristoteles"', in W. W. Fortenbaugh and P. Steinmetz (eds.), *Cicero's Knowledge of the Peripatos* (New Brunswick, NJ: Transaction), 246–63.

Görler, W. (1990) 'Antiochos von Askalon über die "Alten" und über die Stoa. Beobachtungen zu Cicero, *Academici posteriores* 1, 24–43', in P. Steinmetz (ed.), *Beiträge zur Hellenisctischen Literatur und ihrer Rezeption in Rom* (Stuttgart: Steiner), 123–39.

—— (1997) 'Cicero's Philosophical Stance in the Lucullus', in B. Inwood and J. Mansfeld (eds.), *Assent and Argument: Studies in Cicero's Academic Books* (Leiden: Brill), 36–57.

Gottschalk, H. (1987) 'Aristotelian Philosophy in the Roman World', in W. Haase (ed.), *Aufstieg und Niedergang der römischen Welt,* II.36.1 (Berlin: de Gruyter), 1079–174.

Griffin, M. (1997) 'The Composition of the *Academica*: Motives and Versions', in B. Inwood and J. Mansfeld (eds.) *Assent and Argument: Studies in Cicero's Academic Books* (Leiden: Brill), 1–35.

Hahm, D. E. (1977) *The Origins of Stoic Cosmology* (Columbus, Ohio: Ohio State University Press).

—— (1982) 'The Fifth Element in Aristotle's *De philosophia*: A Critical Re-examination', *JHS* 102: 60–74.

Hankinson, R. J. (2002/3) 'Xenarchus, Alexander, and Simplicius on Simple Motions, Bodies and Magnitudes', *BICS* 46: 19–42.

Kupreeva, I. (2003) 'Qualities and Bodies: Alexander against the Stoics', *OSAP* 25: 296–344.

—— (2004) 'Aristotelian Dynamics in the 2nd Century School Debates: Galen and Alexander of Aphrodisias on Organic Powers and Movements', *Bulletin of the Institute of Classical Studies*, suppl. 83, 1: 71–95.

Long, A. A. (1985) 'The Stoics on World-Conflagration and Everlasting Recurrence', *Southern Journal of Philosophy*, 23 suppl.: 13–37.

Mansfeld, J. (1979) 'Providence and the Destruction of the Universe in Early Stoic Thought', in M. J. Vermaseren (ed.), *Studies in Hellenistic Religions* (Leiden: Brill), 129–88.

—— (1988) 'Diaphonia: The Argument of Alexander in De Fato Chs 1–2', *Phronesis*, 33: 181–207.

—— (1990) 'Doxography and Dialectic: The *Sitz im Leben* of the *Placita*', in W. Haase (ed.), *Aufstieg und Niedergang der römischen Welt,* II.36.4 (Berlin: de Gruyter), 3056–229.

—— (1992) *Heresiography in Context: Hippolytus'* Elenchos *as a Source for Greek Philosophy* (Leiden: Brill).

Moraux, P. (1942) *Aléxandre d'Aphrodise Exégète de la Noétique d'Aristote* (Liège and Paris: Vrin).

—— (1963) 'Quinta Essentia', in *RE* 24: 1171–263.

—— (1973) *Der Aristotelismus bei den Griechen: Von Andronikos bis Alexander von Aphrodisias* (Berlin: de Gruyter), i.

—— (1984) *Der Aristotelismus bei den Griechen: Von Andronikos bis Alexander von Aphrodisias* (Berlin: de Gruyter), ii.

—— (2001) *Der Aristotelismus bei den Griechen: Von Andronikos bis Alexander von Aphrodisias* (Berlin: de Gruyter), iii.

Mueller, I. (1994) 'Hippolytus, Aristotle, Basilides', in L. P. Schrenk (ed.), *Aristotle in Late Antiquity* (Washington, DC: Catholic University of America Press).

Oliver, J. H. (1977) 'The Diadoche at Athens under the Humanistic Emperors', *AJP* 98/2: 160–78.

Olivier, F. (1895) *De Critolao Peripatetico*, diss. (Berlin: Schade).
Polito, R., 2006. 'Matter, Medicine, and the Mind: Asclepiades vs. Epicurus', *OSAP* 31: 285–335.
Rashed, M. (1997) 'A "New" Text by Alexander on the Soul's Motion', in R. Sorabji (ed.), *Aristotle and After* (London: *Bulletin of the Institute of Classical Studies*, suppl. 68), 181–5.
—— (ed., trans., comm.) (2005) *Aristote: De la génération et la corruption* (Paris: Les Belles Lettres).
—— (2007) *Essentialisme: Aléxandre d'Aphrodise entre logique, physique et cosmologie* (Berlin: de Gruyter).
Rescigno, A. (ed.) (2004) *Alessandro di Afrodisia: Commentario al De caelo di Aristotele: Frammenti del Primo Libro* (Amsterdam: Hakkert).
Reydams-Schils, G. (1999) *Demiurge and Providence: Stoic and Platonist Readings of Plato's Timaeus* (Turnhout: Brepols).
Runia, D. T. (1981) 'Philo's *De Aeternitate Mundi*: The Problem of its Interpretation', *Vigiliae Christianae*, 35: 105–51.
Sandbach, F. H. (1985) *Aristotle and the Stoics* (Cambridge: CUP).
Sedley, D. N. (1989) 'Philosophical Allegiance in the Greco-Roman World', in J. Barnes and M. Griffin (eds.), *Philosophia Togata* (Oxford: Clarendon), i. 97–119.
—— (1998) *Lucretius and the Transformation of Greek Wisdom* (Cambridge: CUP).
—— (2002) 'The Origins of Stoic God', in D. Frede and A. Laks (eds.), *Traditions of Theology* (Leiden: Brill), 41–83.
Sharples, R. W. (1983) *Alexander of Aphrodisias On Fate* (London: Duckworth).
—— (1990) 'The School of Aristotle?', in R. Sorabji (ed.), *Aristotle Transformed* (London: Duckworth), 83–111.
—— (ed.) (1992) *Alexander of Aphrodisias: Quaestiones 1.1–2.15* (London: Duckworth).
—— (ed.) (1994) *Alexander of Aphrodisias: Quaestiones 2.16–3.15* (London: Duckworth).
—— (1995) 'Counting Plato's Principles', in L. Ayers (ed.), *The Passionate Intellect: Essays on the Transformation of Classical Literature* (New Brunswick, NJ: Transaction).
—— (1997) 'The Peripatetic School', in D. Furley (ed.), *From Aristotle to Augustine* (London: Routledge), 147–88.
—— (1998) *Theophrastus of Eresus: Sources for his Life, Writings, Thought and Influence.* Commentary, iii/1 (Leiden: Brill).
—— (2001) 'Dicaearchus on the Soul and on Divination', in W. W. Fortenbaugh and E. Schütrumpf (eds.), *Dicaearchus of Messana: Text, Translation and Introduction* (New Brunswick, NJ: Transaction), 143–73.
—— (2002) 'Aristotelian Theology after Aristotle', in D. Frede and A. Laks (eds.), *Traditions of Theology* (Leiden: Brill).
Sorabji, R. R. K. (1983) *Time, Necessity and the Continuum* (London: Duckworth).
—— (1999) 'Aspasius on Emotion', in A. Alberti and R. W. Sharples (eds.), *Aspasius: The Earlierst Extant Commentary on Aristotle's Ethics* (Berlin: de Gruyter), 96–106.
—— (2007) 'Time, Place and Extracosmic Space: Peripatetics in the First Century B. C. and a Stoic Opponent', in R. W. Sharples and R. Sorabj (eds.), *Greek and Roman Philosophy 100 BC–200 AD*, vol. 2 (London: Institute of Classical Studies), 563–74.
Tieleman, T. (2007) 'Methodology', in R. J. Hankinson (ed.), *The Cambridge Companion to Galen* (Cambridge: CUP), 49–65.

Todd, R. B. (1976) *Alexander of Aphrodisias on Stoic Physics* (Leiden: Brill).
—— (1978) 'Monism and Immanence: Foundations of Stoic Principles', in J. Rist (ed.), *The Stoics* (Berkeley, Calif.: University of California Press), 137–60.
Untersteiner, M. (ed.) (1963) *Aristotele Della filosofia* (Rome: Storia e Letteratura).
Vitelli, G. (1895) 'Frammenti di Alessandro di Afrodisia nel cod. Ricard. 63', *Studi Italiani di Filologia Classica,* 3: 379–81.
—— (1902) 'Due frammenti di Alessandro di Afrodisia', in *Festschrift Theodor Gomperz* (Vienna), 90–3.
Waszink, J. H. (ed.) (1947) *Tertulliani de anima* (Amsterdam: J. M. Meulenhoff).
Wehrli, F. (ed.) (1969) *Die Schule des Aristoteles: Texte und Kommentare,* Heft 10 (Kritolaos und seine Schüter) (Basel: Schesabe).
Wolff, M. (1988) 'Hipparchus and the Stoic Theory of Motion', in J. Barnes and M. Mignucci (eds.), *Matter and Metaphysics* (Naples: Bibliopolis), 471–545.

PART III

THE ETHICS AND RELIGION OF STOIC COSMO-THEOLOGY

7

Does Cosmic Nature Matter?

Some Remarks on the Cosmological Aspects of Stoic Ethics

Marcelo D. Boeri

1. INTRODUCTION: MAKING ROOM FOR COSMOLOGICAL ETHICS

Cosmological ethics is a fascinating idea. It presupposes that universal nature and our own natures have something in common, that cosmic nature has a rational structure (or is the same as universal reason) and our natures—which, according to Chrysippus, are microcosmic instances of universal nature insofar as our natures are *parts* of it—have such a structure as well. This idea also assumes that our practical life has something to do with universal nature as long as universal nature can have a normative character that works as a criterion for moral action, and that human reason, in being an instance of universal reason and thereby akin to it, is able to know universal reason and to consider events from a universal reason viewpoint. Cosmological ethics is also an approach that strongly relies on the ability of human reason to know the whole system of the world, and that assumes that human beings have a special place in such a system. Nowadays many people are inclined to sympathize with ecologist movements; the Stoic recommendation of living in agreement with nature in order to attain a smooth flow of life, therefore, should sound appealing and reasonable to those who are willing to support the ecologist agenda. However, to live in accordance with nature in Stoic terms is more than just being a sympathizer with the ecological

The first draft of this chapter was written as a contribution to the III Workshop on Ancient Philosophy, organized by Ricardo Salles and held at the Universidad Nacional Autónoma de México (June 2006). A different draft was presented at Brown University (Philosophy and Classics), in Apr. 2007. For their comments and remarks I am grateful to Deborah Boedeker, Mary L. Gill, Pura Nieto, Kurt Raflaubb, Glenn Rawson, and, esp., David Konstan. For their critical remarks I am also indebted to the anonymous referees of this volume, Ricardo Salles, and Thomas Bénatouïl. The final version of this piece was written with the financial support of Fondecyt (project 1085103, Chile).

movement. Being a Stoic follower of nature involves a serious effort to know the way the cosmos works and to enquire into the probable place that one, as a privileged part of cosmic nature, should occupy in it. It also involves being able to consider oneself as part of a whole and to understand what happens to oneself as a particular event in the life of the whole, no matter that in the short term a particular fact or situation of one's life appears to be inconvenient or painful.

In recent years there has been a debate concerning whether or not Stoic ethics does depend on claims about cosmic nature. Some scholars have defended the thesis that Stoic ethics should be considered free from cosmic nature (and, therefore, as not dependent on providential cosmology; I shall call this interpretation 'the heterodox view').[1] Others have argued heavily for the respectability and reasonableness of the cosmic viewpoint in Stoic ethics (I shall call this position 'the orthodox view').[2] For the reasons I will be giving in what follows I tend to believe that the latter view is the correct one, not only due to the overwhelming number of texts where it is explicitly stated that cosmic nature should be taken into account, but also because of Stoic philosophical 'holism'. Such a 'holistic' conception of philosophy takes for granted that all the parts of philosophy (ethics, physics, and logic) are so interrelated with each other that, to some extent, no part can be considered independently of the others. This approach of philosophy contributes to making the idea of a cosmological ethics understandable, and accounts for the fact that the Stoics took the role of universal nature in the specification of the human end to be reasonable (see DL 7. 87–8). In fact, if all the parts of philosophy are interrelated and no part can be considered independently of the others, it should not be odd (at least for a Stoic) to find ingredients belonging to cosmology within the domain of ethics. The oddity of this kind of assertion, though, is what has led some distinguished interpreters of Stoicism (like Julia Annas) to assume that the idea itself of a cosmological ethics is hard to believe, since it 'does lead us . . . to accept counterintuitive conclusions such as that nothing but vice is bad, and that emotions like regret are all mistaken'.[3] However, this sort of apparently paradoxical thesis is what the Stoics explicitly argued for; if this is so, one should try to avoid one's Humean assumptions in interpreting Stoic ethics. Otherwise, one could not assess properly the scope of views that sound counterintuitive to a modern mind, but not necessarily to an ancient Stoic mind.[4] Nevertheless, it should be recalled that a number of Stoic theses were weird even to some

[1] Notably Engberg-Pedersen 1986: 145–83, at 149–50; 1990: 40–2; Annas 1993: 160–5; more recently Irwin 2003: 346, and again Annas 2007.

[2] Representative scholars of this view are Long 1996*a*: 134–55, at 135–47; 1996*b*: 85–106, at 101–3. Inwood 1985: 212–15; Cooper 1999: 427–48, at 439–44; Betegh 2003: 273–302, at 274, 276, 290–293, 298–9.

[3] Annas 2007: 69–70.

[4] For a different approach see, in addition to Annas's work, Brad Inwood's remarks in Ch. 8 below. Inwood's view, though, is very nuanced since, although he maintains that few people—both ancient and modern—'think it sound to hold that learning cosmology will actually help us deal

ancient philosophers; but one might reasonably suspect that this is due to the fact that the Stoic starting points and general assumptions were different from those of the Platonic and Aristotelian philosophers who found weird the Stoic approaches to particular issues. No matter how counterintuitive or outrageous a Stoic thesis may sound: outrageousness or to posit tenets that are at odds with the common conceptions (to use Plutarch's expression) were never reasonable criteria for assessing a philosophical theory.

Now despite the fact that the Stoics appear to endorse the assertion that a happy life and a real human life requires understanding the way the cosmos works and the place that humans, as privileged parts of cosmic nature, should occupy, it is far from being clear, in my view, in what manner an understanding of the cosmos would contribute to performing rational actions and living a rational life, which is the same as living according to nature. In this chapter I wish to investigate the aforementioned relevance of the claims about cosmic nature for ethics in connection with the general conception of reality the Stoics had. In addition to arguing in favour of the relevance that the cosmic viewpoint has for early Stoic ethics, I would like to suggest that the cosmic perspective is relevant for Stoic ethics for systematic reasons related to the conception the Stoics have of philosophy. As far as I know, the role of cosmic nature in late Stoicism (of Epictetus and, particularly, of Marcus Aurelius) has never been questioned. To put it roughly, in Epictetus and Marcus Aurelius philosophy begins by pointing out the basic problem of the relationship between human being and the universe, and reaches an ethical affirmation where the reference of the human life to the cosmos is crucial.[5] This kind of Stoicism traces the course of cosmic events back to a 'cosmic will' with ethical aims, and claims that human beings must be subordinated to the cosmic will, of which they are a part. What I want to argue here is that both Epictetus and Marcus Aurelius, when speaking of the role of cosmic nature for ethics, are following a doctrine already present in and advanced by the older Stoics.[6]

with personal grief', he also argues for the seriousness of the respectable *intellectual* necessity of physics in Stoicism (see esp. p. 203; his emphasis).

[5] I am thinking of passages like Marcus Aurelius 9. 1, where injustice is explained as an impious act on account of the fact that universal nature (ἡ τῶν ὅλων φύσις) has constituted rational animals for the sake of one another, to help one another 'according to their deserts' (κατ' ἀξίαν). So, the one who transgresses the will of nature (τὸ βούλημα ταύτης), Marcus argues, is guilty of impiety against the highest divinity. Moreover, in dealing with lies Marcus points out that both the one who lies intentionally and the one who lies unintentionally are guilty of impiety: the former because he acts unjustly by deceiving, the latter inasmuch as he is not in harmony (διαφωνεῖ) with universal nature, and inasmuch as he yields disorder (ἀκοσμεῖ) in being in conflict (μαχόμενος) with the nature of the cosmos. In all these cases a moral failure is seen as a sort of discordance with universal nature.

[6] To be sure, the theme of a cosmological ethics can sound a little bizarre to our Humean minds. However, as shown by Betegh 2003, and more recently and extensively by Carone (2005: esp. 53–78), it is an issue that was taken seriously by Plato. It seems very clear to me that the Stoics developed Plato's suggestion and took for granted some of his assumptions, such as that reason is

What I have called 'the orthodox view' can have distinct emphases; for instance, although I agree with the general approach of the orthodox view, I shall not endorse the rather extreme claim of A. A. Long—one of the champions of the view—that Stoic ethics *is grounded* in or *parasitical on* Stoic physics.[7] Annas has given several reasons to reject this sort of foundationalism when understanding Stoic ethics; but the main one is that, if ethics depends upon physics, there would be an important asymmetry between these two parts of philosophy (the former being the *explanandum* and the latter the *explanans*), violating so the 'integrated picture' that all parts of philosophy are mixed and that none of them is preferred (*προκεκρίσθαι*) to any another (DL 7. 40).[8] But for this very reason one should admit that the thesis of a cosmological ethics is reasonable within the Stoic assumptions of reality, or so I shall argue.[9] My chapter is divided into two main parts: the first one (sections 2–3) deals with some textual arguments intending to show that some assumptions of what I have called 'the heterodox view' are wrong. The second part (sections 4–5) focuses on what probably is the most problematic point of the cosmological ethics of the Stoics: how the particular tenets of Stoic ethics are related to cosmic nature or, in other words, how some ethical tenets can be somehow connected with our knowledge of cosmic nature.

2. THE PARTS AND THE WHOLE: TURNING TO THE VIEWPOINT OF THE WHOLE

Let me start by quoting in full three well-known passages where the relevance of cosmic nature for ethics in early Stoicism is emphasized:

(A) But since animals have the additional faculty of impulse, through the use of which they go in search of what is appropriate (*τὰ οἰκεῖα*) to them, what is natural for them is to be administered in accordance to their impulse. And since reason, by way of a more perfect management, has been bestowed on rational beings, to live correctly[10] in

regarded as being a sort of divinity in us (see Plato, *Timaeus* 90a2–4, where it is maintained that 'the most important' part of our soul ([*τὸ δὲ δὴ περὶ τοῦ κυριωτάτου παρ' ἡμῖν ψυχῆς εἴδους*), which is a god's gift, is our *δαίμων*; see also 90c5-6: the person who keeps 'well-ordered the *δαίμων* dwelling in himself is entirely happy'). In the first part of Ch. 8, Inwood concludes that the idea that 'there is a single, general Stoic view on the question of the role played by physics should probably be shelved' (see p. 206). Although I agree with him that it is hard to determine a *single* or *general* Stoic view on the role played by physics in Stoic ethics, my chapter can be taken as a modest attempt to account for the *probable* role of physics in Stoic ethics.

[7] Long 1996*c*: 179–201; 1968: 341, quoted by Annas 2007: 66.

[8] Annas 2007: 66–7, 74–5.

[9] To some extent, this is taken into account by Annas (2007: 67–8), but in the nuanced version of the 'orthodox view' I defend in this chapter I take the issue of the mixed Stoic presentation of philosophy differently from Long's approach.

[10] Cooper (1999: 439 n. 30) notes the ambiguity with regard to the placement of the adverb *ὀρθῶς*, that might go with *τὸ κατὰ λόγον ζῆν* (this interpretation) or with *γίνεσθαι*.

accordance with reason comes to be natural for them. For reason supervenes as the craftsman of impulse.[11] Therefore Zeno in his book *On the Nature of Man* was the first to say that living in agreement with nature[12] is the end, which is living in accordance with virtue. For nature leads us towards virtue. Similarly Cleanthes . . . , Posidonius and Hecaton. Further, living in accordance with virtue is equivalent to living in accordance with experience of what happens by nature, as Chrysippus says in *On Ends* book 1: for our natures are parts of the nature of the whole. Therefore, living in agreement with nature (*τὸ ἀκολούθως τῇ φύσει ζῆν*) comes to be the end, which is in accordance with the nature of oneself and that of the whole (*κατά τε τὴν αὑτοῦ καὶ κατὰ τὴν τῶν ὅλων*), engaging in no activity wont to be forbidden by the universal law (*ὁ νόμος ὁ κοινός*), which is the right reason pervading everything and identical to Zeus, who is this director of the administration of existing things. And the virtue of the happy man and his good flow of life are just this: always doing everything on the basis of the concordance (*συμφωνία*) of each man's guardian spirit (*δαίμων*) with the will (*βούλησις*)[13] of the administrator of the whole. . . . The nature consequential upon which one ought to live is taken by Chrysippus to be both the common and, particularly, the human. But Cleanthes admits only the common nature, as that which one ought to follow, and no longer also the particular. (DL 7. 86–9; tr. Long and Sedley, LS 57A and 63C)

(B) They also say that the whole virtue, which is related to the human being, and happiness is a consistent (*ἀκόλουθος*) and concordant (*ὁμολογουμένη*) life by nature. And Zeno characterized the end thus: living in concordance (*ὁμολογουμένως*), that is, living according to a single and harmonic (*σύμφωνον*) reason, since those who live in conflict [with reason] are miserable. (Stob. *Ecl.* 2. 75. 8–76. 1, ed. Wachsmuth; my tr.)

(C) It is not possible to discover any other beginning of justice or any source for it other than that from Zeus and from the universal nature (*κοινὴ φύσις*), for thence everything of the kind must have its beginning if we are going to have anything to say about good and evil . . . For there is no other or more suitable way (*οἰκειότερον*) of approaching the theory (*λόγος*) of good and evil or the virtues or happiness [than] from the universal nature and the dispensation of the universe (*κόσμος*). . . . For the theory of good and evil must be connected with these, since good and evil have no better beginning or point of reference and physical speculation (*φυσικὴ θεωρία*) is to be undertaken for no

[11] This is the final section of the classic passage for the study of *οἰκείωσις* in Stoicism (DL 7. 85–6). Note that the word *οἰκείωσις* does not appear in the passage at stake, although distinct verbal forms of *οἰκειῶ* (*οἰκειούσης*, *οἰκειῶσαι*) do occur.

[12] According to Stobaeus' testimony, Zeno would have said that the end is simply 'living in agreement', and Cleanthes, Zeno's successor as head of the school, would have extended this formulation into 'living in agreement with nature' (see Stob. *Ecl.* 2. 75. 11–76. 8, ed. Wachsmuth; included in LS 63B).

[13] It might be possible to render *βούλησις* by 'wish'. In fact, for the Stoics *βούλησις* is a peculiar type of desire, i.e. 'reasonable desire' (*εὔλογος ὄρεξις*; Stob. *Ecl.* 2. 87. 21–2; at this point the Stoics seem to be following Aristotle; cf. *De Anima*, 414b2, 432b5–6, 433a26–7). In the case of Aristotle it is pretty clear that desire, when related to the rational part of the soul, is *βούλησις*, while when related to the irrational part it is *ἐπιθυμία* or *θυμός*. For the Stoics *βούλησις* is the kind of desire that is experienced by the sage person (actually, it is one of the three *εὐπάθειαι*, along with *χαρά* and *εὐλάβεια*), the one whose soul is rationally disposed as a whole. For the evidence see DL 7. 116 (*SVF* 3. 431; LS 65F) and Plutarch, *De Stoic. Repug.* 1037f–1038a (*SVF* 3. 175; cf. LS 53R), who advances the idea that the sages' reason is not different from the law (presumably, the universal law).

other purpose than for the discrimination (διάστασις) of good and evil. According to Chrysippus, then, physical theory turns out to be 'at once before and behind' ethics, or rather the whirligig of the arrangement is utterly bewildering if the former must be placed after the latter, no part of which can be grasped without it. (Plutarch, *De Stoicorum Repugnantiis* 1035c–d, tr. Cherniss)

One might arguably say that interpretations where the claims about cosmic nature play a significant role for ethics depend upon passage A. It is the letter of presentation of the Stoic naturalism as well as of Stoic cosmological ethics insofar as it suggests why human practical life requires knowledge of both physics and theology. Physics is important in order to know the manner in which the cosmos (or nature) works; but theology is also relevant as long as the Stoics maintained that, in a sense, the cosmos is god himself peculiarly qualified (ἰδίως ποιόν) consisting of the whole substance.[14] Apart from this, theology, a branch of physics according to the Stoics, is important because the agent must make an effort to understand the will of Zeus, in accordance with which his own daemon should be in order to have a 'smooth flow of life'. As indicated by passage B, virtue is a consistent and harmonic life by nature, this suggesting the Stoic worry about emphasizing both the internal concordance (i.e. concordance or harmony with oneself) and the external concordance (i.e. concordance or harmony with the whole), a point on which Cleanthes especially insists in his *Hymn to Zeus*.[15] To be sure, the thesis that cosmic nature plays a significant role for ethics seems to have been assumed by Posidonius as well. In fact, he appears to have held that only the φυσικός can grasp the structure of the universe and be an αἰτιολογικός in the strict sense, inasmuch as he is the only one who knows the way in which all the parts of the universe are interrelated.[16] 'Just as light, Posidonius says when interpreting Plato's *Timaeus*, is apprehended by sight which is luminous, and voice by hearing which is airy, so too the nature of the whole (ἡ τῶν ὅλων φύσις) should be apprehended by the reason which is akin to it' (ὑπὸ συγγενοῦς τοῦ λόγου; my translation).[17] As is plain from another Posidonius passage, the reason which is akin to the nature of the whole is clearly 'the daemon which is akin (συγγενὲς δαίμων) and has a nature which is similar to [the daemon] that rules the whole world'.[18] Within the context, what

[14] DL 7. 137; Stob. *Ecl.* 1. 184. 8 ss. ed. Wachsmuth = *SVF* 2. 527.

[15] See *Hymn to Zeus*, vv. 20–1: 'For thus you have fitted together (συνήρμοκας) all good things with the bad, so that there is one eternal rational principle (λόγος) for them all' (tr. Inwood-Gerson). The most detailed and recent work I know about Cleanthes' *Hymn to Zeus* is the paper by Thom 2001: 477–99.

[16] Frag. 254 (= DL 7. 132–3), ed. Theiler; see also Frag 18 (= Simplicius, *In Aristotelis Phy.* 291. 21–292. 31), ed. EK.

[17] Frag. 85, ed. EK.

[18] Frag. 187. 7–8, ed. EK. Alesse argues that homology in Posidonius is just the coherence with the best part of the soul (i.e. the rational part), as far as it stands for the divine element in us. Hence Posidonius' way of understanding the moral end would keep the Stoic notion of coherence in a different manner from the one in which such a coherence was understood by early Stoicism (see

Posidonius endeavours to show is that the cause of emotional states—which are understood as being 'discord', 'lack of homology' (ἀνομολογία)—lies in the fact that human beings do not follow their daemon, i.e. the reason present in the human soul which is akin to universal reason. It doesn't matter that the Stoic Aristo of Chios was intent on proving that physical matters were beyond us insofar us they cannot be known and contribute no benefit.[19] It does not matter either that Panaetius, when declaring divination to be unreal (ἀνυπόστατος) and implying his doubt on the causal power of providence,[20] was willing to put into doubt the cosmic sympathy and, thereby, the role of cosmic nature for ethics. If both Aristo and Panaetius question the relevance of physics or of providential cosmic nature (and hence the relevance of physics and cosmic nature for ethics), it is because those theses were maintained and defended by some members in the school. This being so, we have both evidence for early (Zeno, Cleanthes, and Chrysippus) and middle Stoicism (Posidonius) that the Stoics were interested in emphasizing the relevance of cosmic nature for ethics.

There are several topics in passage A that must be noted.

(i) If for irrational animals what is according to nature is what is according to impulse (and so they 'have to' live according to impulse),[21] and for humans what is according to nature is what is rational, it follows that human beings, in order to have a real human life, *ought to* live rationally, since this is a way to fulfil their inner nature.

(ii) Given that our human nature coincides with reason, and living is a practical activity, it seems plausible that for humans living according to nature is living according to virtue. In fact, virtue is the best thing we can attain at the practical level.

(iii) We, like the rest of living beings, have a sort of tendency to accomplish our own nature; that is why nature leads us to virtue, because being a virtuous

Alesse 1994: 256–7). Even accepting Alesse's view that Posidonius' focus is on the rational part of the soul, the bulk of Posidonius' homology continues to be in the divine, which is the pattern of the correct behaviour. I am not convinced by Alesse's contention, since her interpretation neglects the emphasis upon the contrast 'nature of the whole–reason which is akin to us' (in Frag. 85, ed. EK).

[19] Eusebius, *Praep. Ev.* 15. 62. 7–11 (*SVF* 1. 353).

[20] DL 7. 149 (test. 139, ed. Alesse); Epiphanius, *De Fide* 9. 45 (test. 134, ed. Alesse = Frag. 68 ed. Van Straaten). On the meaning of the adjective ἀνυπόστατος applied to divination see Alesse, 1997: 270, who argues that the term, common in sceptical refutations, suggests that Panaetius applied ἐποχή to divination. This probably means that, for Panaetius, divination is not an existent (ὄν), but a subsistent item (ὑφιστάμενον).

[21] This expression is potentially misleading insofar as it is a natural fact (not a requirement or a duty) that animals live according to their impulse. There is no doubt that nature in animals has not the same normative character as in humans (in the sense of fulfilling a duty), since living in accordance with impulse is what animals do, and this does not imply any kind of effort to fit into what they *have* to do. But in the case of animals one might understand the 'to have to' in a wide normative sense, i.e. without assuming that such 'to have' means fitting into an actual norm, so to speak, since 'living in accordance with impulse' is something that nature has disposed and established as well. I am grateful to Thomas Bénatouïl for urging me to clarify this point.

agent is to realize one's own nature (this point is explained by Musonius Rufus, Frag. 17 Hense, briefly discussed below in section 3). At this point, in connection with the assertion that nature drives us to virtue, I would like to mention briefly one issue that turns out to be relevant to the heterodox view, namely, that neither Sextus Empiricus nor Stobaeus, two important sources for Stoic ethics, give importance to cosmic nature.[22] It should be noted, though, that at least in Stobaeus' extract of Stoic ethics two important references to nature could be found (probably meaning 'cosmic nature') in ethical context. Stobaeus, probably citing the first century BC Stoic philosopher Arius Didymus, argues that the end of all the virtues consists in living in agreement with nature (*τὸ ἀκολούθως τῇ φύσει ζῆν*, the same expression we find at passage A to make reference to universal nature), for the human being from nature possesses inclinations (*ἀφορμαὶ παρὰ τῆς φύσεως*) for discovering what is appropriate. And each virtue, in concordantly performing (*τὸ σύμφωνον πράττουσα*) what is appropriate, makes the agent live in consistency with nature.[23] As far as I can see, the most natural way to understand this passage is as a slightly different version of the formula of the end ('living in agreement with nature'). It is arguable that the human being from nature possesses inclinations for discovering what is appropriate because nature is taken to be a criterion of what is appropriate, this criterion being both our own nature—which can mirror universal nature—and universal nature, of which our natures are parts and on which they depend. This view is not new in Greek philosophy at the moment the Stoics take it as theirs; Plato had already employed the account of the whole and the parts while arguing that the powers and properties of the parts are explained by reference to the whole (not vice versa). Even though Plato does not explicitly maintain that the whole universe is a *criterion* for morality, he does emphasize that the universe is the point of reference with respect of which the parts acquire their value and reality (see *Philebus* 29b–33a; cf. also *Timaeus* 90b1–d7). Finally, Plato himself might have inspired the Stoic thesis that the human being from nature possesses inclinations for discovering what is appropriate; in fact, in *Philebus* 21b6–8 he contends that if someone should choose what is not really good, he would do so unwillingly (from ignorance or necessity) and against what is by nature truly choiceworthy, implying this way that there is a tendency towards what is really good, which agrees with one's reason.

(iv) If living according to nature means living according to both one's own nature and the nature of the whole (to the extent that our natures, as

[22] Cf. Annas 1993: 160 n. 4.

[23] Stob. *Ecl.* 2. 65. 11–17 (ed. Wachsmuth). It sounds a little odd to say 'the end of all these virtues' (*Πασῶν δὲ τούτων τῶν ἀρετῶν τὸ τέλος*; 65. 11–12; as pointed out by Long 1983: 63 n. 12). The more promising meaning of this stance is that exercising virtues brings about living according to nature for humans.

Chrysippus says, are parts of the nature of the whole), and if our actions must be performed following what universal law indicates that it should be done, then, universal law (which is universal nature), by saying what should be done and what should not be done, becomes a practical criterion of our actions. The heterodox view attempts to establish that the idea that we are simple parts of a larger whole properly belongs to the late Stoicism of Marcus Aurelius and Epictetus.[24] But, as becomes clear from passage A,[25] the emphasis on the fact that we are parts of a larger whole is already present in Chrysippus and, if Galen is to be believed (in Posidonius Frag. 187, EK), it was also present in Posidonius, a long time before the later Stoics. It sounds more natural to consider Marcus Aurelius' and Epictetus' continuous remarks with regard to us as being 'parts of a whole' like developments inspired by Chrysippus' doctrine.

(v) Finally, if virtue and happiness consist in performing everything in concordance with each one's daemon and with the will of the administrator of the universe, and if we effectively can reach that concordance or harmony, it seems that the Stoics are suggesting that human beings have the possibility of sharing the perfect rationality and happiness of god.[26] If that is the case, it would also be natural to attain the understanding of all events from a cosmic viewpoint, a thesis manifestly present in later Stoicism[27] but also significantly advanced by some early Stoics. If Plutarch is to be trusted, Chrysippus was willing to claim that some apparent evils, if more closely examined, can be seen as goods. So, when cities are too populous, people are moved to the colonies or a war against someone is initiated; so Chrysippus seems to have maintained that 'god gives occasions for destruction to begin' (Plut. *De Stoic. Rep.* 1049b; tr. Cherniss); such a destruction, however, is contemplated in the 'economy' of the whole and, in this sense, it involves a beneficial result. What we should do, then, is to move from the perspective of the part to the perspective of the whole; this is exactly what understanding the administration of the universe means and, as passage A shows, bringing one's own daemon (i.e. each one's own reason) into agreement with the

[24] Annas 1993: 162.

[25] Quoted and commented on by Annas 1993: 160. She does not make, though, a detailed commentary on Chrysippus' assertion that our natures are parts of the nature of the whole; she just says that this thesis is common to later Stoicism, and that in the early versions of Stoic ethics the strategies that are derived directly from the idea that we are parts of a larger whole 'are conspicuously absent' (162). Passage A, though, part of which content can reasonably be attributed to Chrysippus, seems to prove the opposite.

[26] The Stoics say neither that human beings can share the rationality of god nor that they can be happy like gods. They rather say that the *sage* people's happiness does not differ from the divine happiness of Zeus; Chrysippus appears to have suggested that 'momentary' (*ἀμεριαία*) happiness does not differ from the happiness of Zeus, and that the happiness of Zeus is not more choiceworthy (*αἱρετωτέρα*) or more venerable (*σεμνοτέρα*) than that of wise people (see Stob. *Ecl.* 2. 98. 19–99. 2; ed. Wachsmuth).

[27] Marcus Aurelius 8. 46, 5. 8.

will of the manager of the universe. In Epictetus' words this is 'the system in which men and gods are associated' (*Diss.* 1. 9. 4), a rational structure where all the oppositions (including virtue and vice, abundance and dearth) are for the sake of the harmony of the universe (1. 12. 16).[28] But the issue of turning to the perspective of the whole also goes back to Cleanthes (*Hymn to Zeus*, vv. 11–22), this proving that the recommendation of turning to the perspective of the whole was a tenet originally conceived of in early Stoicism, and developed and deepened by late Stoicism.[29]

Passage A also shows why in the Stoic view, or rather in the Chrysippean view, there is not a significant difference between our natures and the universal nature; this is also the reason why I think that the Stoics took the universal law to be *not* something alien or external to us[30] but, on the contrary, something that peculiarly belongs to us as rational beings. In the *scala naturae* that text A presents (see especially DL 7. 85–6), starting out from the plants and ending up with humans, reason appears as the peculiar component of humans which enables them to control their impulses, to the extent that reason is added to us 'as a craftsman of impulse' (DL 7. 86). I consider that this can be understood as the possibility humans have to moderate their impulses and to retard the satisfaction of some appetite. Indeed this is to act rationally, because reason in us is, like the cosmos, a mark of order, and for a human being to retard the satisfaction of a desire or even to remove some irrational desire are specific marks of rationality. Reason, as a craftsman of our impulses, models such impulses and helps the agent to avoid developing 'excessive impulses'.

3. EARLY AND LATE STOICISM ON THE PERSPECTIVE OF THE WHOLE

As we have seen, Chrysippus claimed that we are microcosmic parts of universal nature, and if we are portions of universal nature, our natures must be akin to universal nature as well.[31] That we are akin to universal nature can be inferred, I submit, from passage A quoted above; but the thesis that there is a certain affinity between human beings and the rational order of cosmos is also explicitly suggested

[28] In a similar vein, see also Marcus 5. 8, 2. 17.

[29] Cleanthes' *Hymn* is not taken into account by Annas when she argues that the thesis that we are parts of a larger whole (and thereby the recommendation that we should turn to the perspective of the whole) belongs to late Stoicism.

[30] As Annas thinks (1993: 160–2). In this sense, I would subscribe to J. M. Cooper's claim that 'Zeus's or nature's law is our law, too, the law of human beings as such'; 'Once we recognize and take seriously our own rationality . . . we should see ourselves as bound by this law' (2004: 212–13).

[31] For the Platonic antecedent of this view see Plato's *Timaeus*, where it is argued that we are wholes (33a), that the cosmos is a whole (32d), and that both we and the animals are parts of the cosmos (30c).

by one important source reporting on the Stoic notions of god and cosmos. When distinguishing different meanings of the word *κόσμος* according to the Stoics, Eusebius (probably quoting the first-century BC Stoic Arius Didymus) says that the cosmos can be understood as the dwelling (*οἰκητήριον*) of gods and human beings, and as the structure (*σύστημα*) composed of gods and human beings, and of those things that come into being for the sake of them. Just as a city is said to be twofold, the argument runs, the dwelling as well as the structure composed of those who dwell there along with the citizens, so too the cosmos is like a city constituted out of gods and human beings, the former being those who hold the leadership and the latter those who are subordinated to that leadership. But between them 'there is a reciprocal community (or even an 'affinity'; *κοινωνία*) on account of the fact that they take part in reason, which is natural law' (*κοινωνίαν δ' ὑπάρχειν πρὸς ἀλλήλους διὰ τὸ λόγου μετέχειν, ὅς ἐστι φύσει νόμος*).[32]

In my view, it is pretty clear that both in passage A and in the Eusebius passage just quoted some tenets that became common in Marcus Aurelius' and in Epictetus' late Stoicism are plainly established. There are particularly three connected assertions that were especially significant in the cosmic nature's viewpoint for ethics in Marcus Aurelius and Epictetus, and that are already present in the sources reporting early Stoicism: (i) the thesis that our natures are parts of the universal nature, (ii) that human nature and universal nature share certain affinity, and (iii) that there must be a certain concordance or harmony (*συμφωνία*) between each individual's daemon and the will of Zeus. Marcus Aurelius repeats many times the dependence of our natures (as parts) upon the universal nature, but an emblematic passage for this statement in Marcus is the following:

> This must you always remember: what is the nature of the whole, what is my own nature, and how my own nature is related to that, *and what kind of a part it is of what kind of a whole* (*ὁποῖόν τι μέρος ὁποίου τοῦ ὅλου οὖσα*), and remember that there is no one who is able to prevent you from always acting and saying those things that are consistent with nature, of which you are a part (*τὰ ἀκόλουθα τῇ φύσει, ἧς μέρος εἶ*). (2. 9; my tr.)

The topic of harmony or concordance between the personal daemon and the will of Zeus is also clear in Marcus when he argues that the particular nature must be in tune with the universal nature provided that they are the same.[33]

[32] Arius Didymus, *apud Eusebium, Praep. Ev.* 15. 15. 3–5 (*SVF* 2. 528; LS 67L), cited and commented on by M. Schofield 1999: 66–7. If Schofield is right (and I think he is), there are reasons for assuming that the formulations in Eusebius' passage derive from Chrysippus (67). The idea of the affinity between god and human being is, once more, Platonic in character (see Plato, *Gorgias* 508a1: *θεοὺς καὶ ἀνθρώπους τὴν κοινωνίαν συνέχειν*).

[33] Marcus Aurelius 5. 3: 'Keep a straight course, being consistent with your own nature and universal nature (*ἀκολουθ ῶν τῇ ἰδίᾳ καὰ τῇ κοινῇ*); *the path of these two is one*'. See also 4.23: 'O world, I am in tune with every note (*Πᾶν μοι συναρμόζει*) of thy great harmony' (tr. M. Staniforth).

The topic of god within us—an issue which is Platonic in character[34]—is also present in Epictetus, when he states that Zeus has assigned to each person a tutor (ἐπίτροπος, i.e. his own personal daemon, who is his personal guardian). Now what I would like to emphasize once more is that, taking into account the passages just cited, we have good reason to suspect that these developments are not original to Epictetus or Marcus Aurelius; a more reasonable approach to this topic would be, it does seem to me, to assume that they took into consideration passages like text A and that they draw on the early Stoics' theses. Not only is the terminology quite similar, but also the bulk of the argument: (1) universal nature and particular natures are common and the same in substance; (2) if universal nature is rational and depicts order, we ought to be rational and depict order; (3) our rationality is a 'gift' given by Zeus, and constitutes a sort of 'inward divinity' (δαίμων) dwelling in us that must be our ruler and guidance.[35]

Let me turn now to passage C, where unequivocally it is stressed the relevance that cosmic nature had for Chrysippus as a starting point when investigating moral matters; this passage also emphasizes the importance of engaging in physics when one intends to deal with matters regarding good and evil. So physics looks like an important discipline when doing ethics. Annas quotes the passage and, following Brunschwig, argues that 'the quotations come from *physical*, not *ethical* treatises', from which she infers that cosmic nature has a role, although not as part of ethical theory.[36] But this remark does not account for the strong emphasis put by Chrysippus both on universal nature and on physical theory as starting points for knowing good and bad, the virtues, and happiness. In her new paper on Stoic ethics Annas continues to maintain that this passage does not provide evidence for the claim that Stoic physics is foundational for Stoic ethics (a view that, as indicated at the outset of this chapter, I find reasonable), but now she emphasizes the fact that Plutarch can be pulling passages out of context and that what one should think is that Chrysippus is writing works that mix physics and ethics for pedagogical purposes. This being so, then, we may be certain that 'there was no such established thing as *the* way to present Stoic ethics'.[37] It does seem to me that Annas is entirely successful in showing that we should rely on the 'integrated picture' (her expression) of Stoic philosophy rather than on a picture where ethics heavily depends on physics. At any rate, I think that the Plutarch passage is suggesting much more that a mere *order of presentation* of philosophical topics. According to Annas, the emphasis on physics should be understood just as a matter of presentation of Stoic ethics; in other words, this is just a way of presenting Stoic ethics, so one should not infer more than that while considering passages such as A, B, and C. Let us look at my passage C and let us take that passage to be one in which Chrysippus

[34] Plato, *Phaedo* 107d; *Republic* 617e; *Timaeus* 90b–d; see also n. 6 above.

[35] Cf. Marcus Aurelius 5. 27.

[36] Annas 1993: 164 (her emphasis).

[37] Annas 2007: 79, 82–3, 85.

is just putting emphasis upon one particular order of presentation of ethics. In this case, Chrysippus would be just stressing the physical side as a necessary condition for understanding virtue (like justice in Chrysippus' text), insofar as physical theory is to be undertaken for the sake of discriminating goods and evils. That is, if ethics is presented from the physical perspective, the emphasis on cosmic nature should not be regarded as being the only perspective from which virtue or other ethical items can be grasped. It would be just a matter of emphasis, i.e. the emphasis proper of the part of philosophy from which ethics is being presented. Annas puts emphasis upon the fact that there is a sharp contrast between the presentation of Stoic ethics that makes use of physical considerations and providence, on the one hand, and modern ethical theories, which abjure any such use, on the other hand.[38] However, even following her own argument, one should admit that the presentation of Stoic ethics which strongly relies on the cosmic viewpoint was entirely reasonable for the Stoics; that is to say, even admitting with Annas that the presentation of ethics based on physics *is just one way* of presenting ethics, one should take for granted that for the Stoics it is a very reasonable way of presenting it, and it is as reasonable as any other way. Now if this is so, one should probably draw the conclusion that, at least within the physical approach as a means to present Stoic ethics, the use of physical considerations and providence were certainly admissible. Indeed, this point cannot be accepted by any modern approach that rejects even the possibility of considering a providential view of the universe when dealing with ethical matters. A major problem I see in Annas's interpretation is that she (not the Stoics) takes this latter approach to be the *only reasonable* one when dealing with Stoic ethics, so that, in her opinion, the view taking into account the cosmological approach must be dismissed. One point that she stresses in her study is that, although the orthodox view starts by underlining that Stoic ethics must be understood in terms of Stoic physics, and by giving a significant role to a providential view of cosmic nature, such an account ends up by discussing impulse, virtue, emotion, and so on, introducing all the particular topics of Stoic ethics in their own right without appealing to *pneuma* or the cosmos.[39] This is an important point in Annas's argument, but it fails to notice the fact that if one takes into consideration the relevance of the cosmological approach to Stoic ethics some apparently counterintuitive Stoic tenets (such as 'nothing but vice is bad' or 'life and health, and their opposites death and disease, are neither good nor bad') become understandable. This is, in fact, the explanation we sometimes find in different Stoic philosophers when they have to account for the apparently counterintuitive thesis that pain, death, and so forth are not evils.[40] In addition, the approach to Stoic ethics that abjures any physical or providential consideration does not do justice to a Stoic thesis which is closely related to the

[38] Ibid. 86. [39] Ibid. 67.

[40] See Plutarch, *De Stoic. Rep.* 1049b (= *SVF* 2. 1177), reporting a Chrysippean view.

orthodox view: the stance that universal nature extends to all things, that is, that whatever that comes to be (whether in the whole universe or in any of its parts) necessarily has come to be according to that nature and its reason.[41] On the other hand, I fail to see that Cicero does not present Stoic ethics via cosmic nature in *De Finibus* 3, as Annas insists.[42] In Cicero's presentation of Stoic ethics there is at least one passage (*De Finibus* 3. 73; I discuss it briefly below) where cosmic nature plays an important role, so that the Cicero passage can be paralleled to my text C.

Now if one follows Annas's suggestion and privileges the integrated picture of the Stoic system and, accordingly, one accepts that some ingredients of physics might be relevant to ethics,[43] one still should wonder how this is possible. At least three Stoic sources report that the Stoics think there are three 'most generic' virtues: logic, physics, and ethics, these virtues being the three parts of philosophy.[44] To our common sense it sounds a little weird that physics could count as a virtue. Physics is a certain kind of knowledge; but this does not suffice to make physics a virtue, since physics is not knowledge of the good.[45] The Stoics declared virtue to be a certain form of knowledge (i.e. knowledge that can be identified with the knowledge of the good), and in its most technical sense it is a 'consistent disposition' (διάθεσις ὁμολογουμένη; DL 7. 89), or a 'consistent disposition of the soul which is coherent with itself with regard to the whole life' (text B). This self-coherence that a virtuous soul must have is explained on account of the fact that happiness, the Stoics claim, resides in virtue, since virtue is a soul that has been made for the sake of 'homology' (ὁμολογία) of one's whole life. In other words, without coherence or agreement, there is no happiness, which is the same as saying that without coherence there is not a real rational life. Now if the insistence on having a consistent life is so strong, it is reasonable to assume that human beings have, for the most part, a life that is not in agreement with nature. As suggested by Cleanthes, the human being is the only one who is capable of disregarding the law and, in doing so, he leads his own life against his own nature. Everlasting reason—or simply god, from whom we humans 'have obtained by lot an imitation of god'[46]—Cleanthes claims, is what 'the wicked flee from and neglect, ill-fated, since they always long for the possession of good things and do not see the common law of god, nor do they hear it; and if they

[41] Plut. *De Stoic. Rep.* 1050c, quoting Chrysippus (= *SVF* 2. 937). See also Plutarch 1050a: 'For no particular thing, not even the slightest, can have come about otherwise than in conformity with the universal nature and its reason' (tr. Cherniss). I find this sort of passage not without insight, as it provides support for Long's reading. For the emphasis upon the unified view (or 'sympathy') the Stoics are interested in displaying of the cosmos as a real living being see Ch. 3 above by Meyer (esp. pp. 80–5 and the passages quoted and discussed there).

[42] Annas 2007: 85–6.

[43] Of course, she doesn't make this connection.

[44] Cicero, *De Finibus* 3. 72–3; DL 7. 92; Ps.-Plutarch, *Placita* 1. 2.

[45] As underlined by Menn 1995: 1.

[46] *Hymn to Zeus*, v. 4, reading θεοῦ μίμημα with LS 54I. For the textual difficulties this well-known crux presents see Thom 2001: 487 n. 42.

obeyed it sensibly (σὺν νῷ) they would have a good life (βίος ἐσθλός)'.[47] In a similar vein, Seneca points out that good can be present only where there is reason, or rather, where there is a perfectly developed reason. Man's good is not present in him unless he has a perfect reason (*ratio perfecta; Ep.* 124. 11); the agent will reach the good when he or she has reached reason (*Ep.* 124. 8–9) and, while divine good is perfect by nature, human good can become perfect by practice or by the agent's effort, since god's good is given by nature and man's good is acquired by one's care (*unius bonum natura perficit, dei scilicet, alterius cura, hominis; Ep.* 124. 14). This last remark in Seneca's argument suggests that the human being has the possibility of doing something different from what he or she should do. Thus while irrational animals must just follow the 'conduct' of the species, the human being, because of being rational and having the same nature as god (even though human nature differs from god's nature on account of being mortal), can 'build up' his own nature by permanently adjusting his own behaviour to what is most appropriate to his nature.

The universal homology is tackled by the Stoics from the following three approaches: physics, ethics, and logic. In fact, λόγος manifests itself in the following three domains (which coincide with the three parts of philosophy): in nature (understood in terms of the rational order of the world), in language (as the privileged place of its manifestation insofar as the core function of language seems to be both to articulate and to give meaning to the reality through the

[47] Cleanthes, *Hymn to Zeus*, vv. 21–25 (tr. Inwood-Gerson). It is noteworthy that human beings, the privileged parts of cosmic nature, are the only ones able to act contrary to the law of nature. However, the disagreement produced by this privileged part is somehow necessary to allow Zeus' ability to restore order, this order being understood as the balanced tension of opposites (see *Hymn to Zeus* vv. 18–21). In the interpretation of these lines, I side with the scholars who claim that the background for these verses should be Heraclitus' notion of the *coiunctio oppositorum* (Long 1996*d*: 35–57). In addition to the fact that Cleanthes seems to be quoting verbatim some Heraclitus' fragments in his *Hymn* (cf. vv. 17–25 and Heraclitus B 1, 10, and 50, DK), I fail to see why if the λόγος (understood in an Heraclitean manner) encompasses good *and* evil, it would not be possible for the κακοί to act 'without' it' (Thom 2001: 490). Cleanthes can be incorporating Heraclitus' *logos* into his own Stoic view, according to which *logos* in humans involves the capacity of acting against what universal *logos* prescribes, i.e. by acting against their own nature. On the other hand, Thom's remark fails to explain how Cleanthes argues that no deed is performed independently of the rational order of god (see *Hymn to Zeus*, v. 15: Οὐδέ τι γίγνεται ἔργον ἐπὶ χθονὶ σοῦ δίχα), implying that the fact that the κακοί act ignoring what universal law prescribes is contemplated in the providential plan of Zeus. In other words, what the wicked do is part of universal order. It is true that no deed takes place without god, *except what the wicked do in their folly* (*Hymn to Zeus*, vv. 15–17); but at the same time god is described as being able to put in order what is in disorder. In my view, this is stated as if the possibility human beings have to act against the rational order and the ability god has to put in order what is in disorder were recurrent processes that constitute such an order. Besides, one should take into account the fact that it is god himself who produces a creature (human being) able to be in conflict both with the rational order of universe and with him. Thus a much more natural way of approaching the issue would be by considering that, to some extent, it is necessary that some parts of the whole are at odds with the rational order. What Cleanthes probably has in mind is Plato's idea that evil cannot be eradicated, for there must always be something opposed to the good (*Theaetetus* 176a5–6); this is, again, a Heraclitean view (both in Plato and in the Stoics).

discourses in which existent things are expressed and accounted for), and in action (when λόγος embodies itself in the figure of the wise person, the entirely rational agent). The Stoics spoke of 'reason' and 'rational' in two complementary manners: on the one hand, in a cosmological sense, according to which nothing occurs independently of the rational organization of the cosmos, to the extent that everything happens in accordance with the divine λόγος ruling and pervading all the things.[48] Thus, it might be said that for the Stoics all that is natural is rational and all that is rational is natural.[49] On the other hand, there is an anthropological sense of reason and rational, according to which human beings—from a certain specific age but virtually as from birth[50]—have the faculty of becoming aware of their participation in the cosmic order. For humans such taking part in cosmic order is possible by acknowledging their own condition, which, at a certain moment, can involve as a consequence a morally worthy behaviour since the agent will be able to adjust his or her actions according to what he or she recognizes as being suitable to living in agreement. There is an interesting argument attributed to Musonius Rufus that is helpful to understand this point. When an old man was asked what the best provision of old age would be, he responded: 'the same as the one of youth age as well: to live methodically and according to nature' (τὸ ζῆν ὁδῷ καὶ κατὰ φύσιν; my translation). In order to understand this, Musonius argues, what one should do is comprehend that the human being's nature is not directed at pleasure (as is not the nature of irrational animals either).[51] A living being is living according to nature when it depicts its own perfection or excellence (ἀρετή) by means of which the living being acts according to its own nature. Now since each one's nature leads each one towards his or her virtue or excellence, it is reasonable to maintain that a human being is not living according to nature when she is living in pleasure, but when living in her own perfection or excellence (Frag. 17. 89. 12–90. 1, ed. Hense). The argument applies both to rational and irrational beings, and shows that each one's excellence is the criterion ruling each one's life. It also exemplifies the way in which universal nature can be a criterion for moral action in humans: given that human nature is part of universal nature and somehow *is* universal nature, and since both universal and human nature display what is perfect or excellent,

[48] Cleanthes, *Hymn to Zeus*, vv. 7–8: 'All this cosmos . . . obeys you'; 'With it you direct the universal reason which runs to all through all things' v. 12; 'No deed is done on earth . . . , god, without your offices', v. 15 (tr. Long and Sedley).

[49] This assertion, as indicated above (n. 21), is potentially misleading insofar as a plant e.g. is something natural but one would fail to see in what sense it is rational. However, a plant or an animal is 'rational' not in the sense of having a rational faculty, but due to the fact that they depict part of the rational order of the cosmos.

[50] Cf. *Scholium to Plato's Alcibiades* 1. 121e; Iamblichus, *De Anima*, in Stob. *Ecl.* 1. 48. 8, p. 317. 21 ed. Wachsmuth (= *SVF* 1. 149); DL 7. 55–6.

[51] Not even a horse, Musonius contends, would consider that it attains its own end simply because of the fact that it can eat, drink, or copulate, but when performing what is proper to a horse (Frag. 17. 5–8, ed. Hense). This appears to be an argument addressed against the Epicurean tenet that pleasure is the end (see also DL 7. 85–6).

it follows that to live according to virtue (i.e. the perfection or excellence proper of the rational living being) is the same as living according to nature. Therefore, in order to conduct a rational life a human must live according to virtue, insofar as the exercise of virtue shows what is best in her. And what is best in a human being is the exercise of her rationality.

The Stoics maintain that we do physics 'whenever we investigate the cosmos and what is in the cosmos'.[52] Chrysippus seems to have been even more emphatic with regard to the relevance of physics for ethics when asserting that physical speculation (*φυσικὴ θεωρία*) is to be undertaken for the sake of nothing else than the distinction concerning the good and the evil (as passage C stresses). If this is so, physics turns out to be a very important field of knowledge for ethics and, more specifically, for the person who is interested in developing a rational life. Certainly, Annas would object that, even though one starts by considering Stoic ethics as depending on Stoic physics and by giving a significant role to a providential view of cosmic nature, such a view ends by discussing the specific topics of ethics, i.e. virtue, vice, emotion, and so on, without appealing to *pneuma* or the cosmos. However, in the holistic perspective of all the parts of philosophy I am interested in emphasizing here, a Stoic might argue that there are some topics belonging to physics that might be relevant to our understanding of ethics. For instance, virtues are states of the soul (a body, according to the Stoics); but as bodies, they are causes of incorporeal effects: it is because of prudence that being prudent occurs (*γίνεται*; Stobaeus, *Ecl.* 1. 138. 23–5, LS 55A). Moreover, Cleanthes seems to have maintained that tension (of *πνεῦμα*) is the impact of fire, and that if it becomes convenient in the soul to achieve what is fitting it is called 'power' (*δύναμις*). Such a power, Cleanthes goes on to argue, when it arises in situations that manifestly require persistence, is continence; when it arises in situations requiring endurance, it is courage (Plut. *De Stoic. Repug.* 1034D–E; LS 61C). In other words, the virtuous dispositions of a rational agent are understood in terms of *pneuma*'s good tension (*εὐτονία*).[53] Indeed, these textual examples do not endorse the thesis that ethics is *grounded* in physics, but they are useful to note the relevance that the theory of causality and of *pneuma* had within the practical domain. Both causality and pneumatic theory are properly discussed within Stoic physics; nevertheless, they were not limited to the purely physical account and they did extend to the ethical sphere as well. These passages also make it clear why the Stoics did not feel uncomfortable when introducing some physical issues within their ethics.

This however does not shed light on how physics is a virtue, although it implicitly suggests that, in order to investigate the cosmos and what is in it,

[52] Ps.-Plutarch, *Placita* 1. 2; *SVF* 2. 35; LS 26A.

[53] See also Galen, *PHP* 270. 10–24; 272. 9–274. 26, ed. De Lacy. The same thing can be seen in the standard definition of two basic emotional states, such as pain ('a contraction of the soul') and pleasure ('an expansion of the soul'). Both contraction and expansion are *pneuma*'s movements.

i.e. in order to do physics, we should have a cognitive disposition as well as the knowledge of the principles governing the cosmos as an ordered totality. A passage in Cicero' *De Finibus* can be helpful in clarifying this point. Cicero explains, through his Stoic spokesman Cato, that the Stoics reasonably (*non sine causa*) have called physics 'virtue', and this is so because the one who is about to live according to nature (*convenienter naturae*) must start from the whole world and from its management (*procuratio*). Cato underlines that no one can judge truly (*vere iudicare*) on good and evil unless he has known the whole plan or purpose (*ratio*)[54] of nature, and also the life of the gods, as well as whether human nature is or is not in agreement with that of the universe (*De Finibus* 3. 73). This important passage from Cicero shows at least three points: first, it makes it clear that physics is a virtue because it is a form of knowledge centred on the κόσμος, a knowledge which in turn is the starting point for those whose concern is to live virtuously. Second, this knowledge that physics consists in is relevant for the practical life of the agent, since it is at least a necessary condition for the understanding of good and evil. So, although physics is not knowledge of the good, it is anyway a knowledge that contributes to the knowledge of the good; besides, it doesn't matter that physics is not defined as 'knowledge of the good'. The Stoics were also willing to assert that dialectic is a virtue (DL 7. 46) and, of course, 'virtue' applied to dialectic does not mean 'knowledge of the good', either. Third, while this passage confirms, once more, the importance of cosmic nature for ethics (this time in Cicero's *De Finibus*) it does not imply, *pace* the heterodox view, that physics only concerns the understanding we can have of ethics. Annas argues that 'if cosmic nature were a first principle for ethics', we should find 'direct derivations of particular ethical theses purely from cosmic nature', but this is exactly what we do not find.[55] But to my judgement, this passage from Cicero shows that physics for the Stoics is concerned with more than the understanding we can have of ethics; it rather presents physics as a necessary field of knowledge to deal with ethics and, as I hope to show below, it gives some basic patterns for practical ethics. Thus, even though we cannot find some particular tenets directly derived from cosmic nature, cosmic nature plays an important role in setting criteria for decision-making on concrete situations of action (see below, section 4).

It is likely that a promising starting point for clarifying the role of cosmic nature for ethics is to examine the Stoic sage's life, and this is so because the

[54] For this interpretation of *ratio* see Glare 1996: *s.v. ratio*, 10: 'A plan of action, purpose'.

[55] Annas 1993: 165. In her recent paper on Stoic ethics she points out that Cicero's *De Finibus* 3, unlike DL 7. 85–9, is a key text where Stoic ethics is not *presented* via cosmic nature (Annas 2007: 85). But, as I have already pointed out above, this remark fails to consider the relevance that cosmic nature has in *De Finibus* 3. 73. Like in the Plutarch passage commented on above, I would say that, even though one should admit her remark with regard to the order of *presentation* of Stoic ethics, that does not mean that cosmic nature does not have relevance for ethics. At least within the physical presentation of ethics considerations related to cosmic nature and providence are relevant.

Stoic sage can be taken to be a standard of a perfect rational life or, in other words, as perfectly mirroring universal reason. To be sure, the many astonishing characteristics of the sage depict him as a perfectly rational being and as a good exemplar of the universal reason he exemplifies.[56] The amazing qualities of the sage show that, at least in theory, he is the only one who truly lives in accordance with nature. Of course, the universal nature does not perform any of these activities, but the idea is that all that happens in nature as a whole is balanced and perfectly done, and this is 'rationality' or rather a macrocosmic version of what rationality is. The Stoic goal of living according to nature should presuppose 'the ability to make correct judgments about facts and values',[57] and this is something that only the sage can perform, because he is the only one whose judgement is always correct rather than just sometimes, or even at random.

But one might wonder what this has to do with physics. The Stoic sage, whose psychological disposition is 'right reason', understands the causal nexus of fate, since he understands the principle of universal causality, according to which all the phenomena are accounted for and all the events and particular facts of the world can be predicted as well.[58] Everything is determined not only physically—for it is part of a continuous causal series in a universe which is continuous—but also providentially. This is the reason why, at least theoretically, if one knew the laws of the causal interaction and the manner god acts, i.e. if one were a Stoic sage, one would be able to predict all future events. This being so, chance events are just phenomena whose causes are unknown to us.[59] Chrysippus, cited by Epictetus, argues:

so long as the consequences are unclear to me, I always hold to the things best adapted to secure what is in accordance with nature, for god himself created me with the faculty for selecting (*ἐκλεκτικόν*) such things. If I really knew that it was fated (*καθείμαρται*) for

[56] All the astonishing characteristics of the Stoic sage are conveniently listed in Stob. *Ecl.* 2. 111–12.

[57] Long 1996*b*: 94. In addition to these extraordinary qualities we should also recall three other characteristics the wise person has: lack of precipitancy (*ἀπροπτωσία*), unhastiness (*ἀνεικαιότης*), and irrefutability (*ἀνελεγξία*; DL 7. 47). Irrefutability, defined as strength in argument to avoid being carried away by argument into the contradictory of one's own thesis, reminds of Plato's claim that 'the truth is never refuted' (*Gorgias* 473b10–11).

[58] This argument can be regarded as being a sort of 'phenomenical argument' in order to explain the existence of fate. Indeed, it is a fact empirically evident that the world has an organic unity (cf. Alexander of Aphrodisias, *On Fate* 192. 8–13, ed. Bruns, and Ch. 3 above). From the Stoic point of view, the success of divination also appears to be a fact empirically obvious (cf. Diogenianus, *apud Eusebium, Praep. Ev.* 4.3.1 = *SVF* 2. 939). As Diogenianus notes, Chrysippus' argument sounds circular, since he tries to prove the existence of fate out of divination, but his belief in divination as an effective method to predict future events presupposes the doctrine of fate. See also Cicero, *On Divination* 1. 34.

[59] Cf. Alexander of Aphrodisias, *On Fate* 174. 1–11; ed. Bruns, and especially Cicero, *Academica* 1. 29. In other passages the theoretical possibility of predicting future events from antecedent causes is put forward (Cicero, *On Divination* 1. 127–8), something that only the wise person might perform. Divination and its species are indicated in Stobaeus as a characteristic of the Stoic sage person (see *Ecl.* 2. 67. 13–19; 114. 16–21, ed. Wachsmuth). In both passages divination is defined

me to be ill at this moment, I would have an impulse to be so: for the foot too, if it had mind, would have an impulse to be muddied.[60]

The analogy is clearly aimed at showing that a rational being is to the cosmos, as a foot is to the man. It also underlines the relation between one's reason and cosmic reason. If one is aware of what is fated for oneself, one knows the providential plan of god, and as a consequence of an active exercise of his rational capacity one can understand his painful situation *sub specie rationis dei*. And this knowledge is knowledge of the cosmos, i.e. it is 'physics'. So this passage again shows (against the heterodox view) that cosmic nature plays a relevant role for ethics in Chrysippus' Stoicism.

One of the problems one has to face at this point is that being a Stoic sage is something extremely hard, as recognized by the Stoics themselves. As a matter of fact, Zeno, Cleanthes, and Chrysippus explicitly acknowledged this. In the same line of thought, the innovative Panaetius recognized this difficulty at the same time that he attempted to humanize the picture of the Stoic sage.[61] The description of the Stoic sage as the one who makes use of virtues during *all his life* gave the early Stoics the bad press of presenting their ethics (and politics) as an absolutely impracticable theory: in fact, the Stoic sage—with his characteristics of infallibility, rational perfection, absolute coherence (insofar as in him the perfect ὁμολογία with nature is embodied)—is someone extremely rare.[62] If Chrysippus, Plutarch mocks, does not show as a virtuous person either himself or any of his own acquaintances or teachers,[63] what about the rest of human beings? I think that the Stoics were willing to suggest that even ordinary people might attain this ideal, as long as they also are rational beings and thereby particular instances of universal nature. This is what at least the Stoic Epictetus suggests when arguing that engaging in argument is relevant insofar as it has a connection with how we should behave in our lives (*Diss.* 1. 7. 1),

as '*knowledge* considering the signs coming from the gods'. The Stoic thesis of universal causality is well attested in several sources; cf. Diogenianus, *apud Eusebium, Praep. Ev.* 6. 7. 8. 1–39 (*SVF* 2. 925 and 998); Plut. *De Stoic. Repug.* 1049f–1050d (cf. *SVF* 2. 937), 1056c. The detailed discussion of this and other aspects presupposes important issues concerning the connection among ethics, logic, and physics that cannot be developed here. For a more detailed treatment of these topics, allow me to refer to my 2001: 728–50 and to Boeri and Vigo 2002: 32–51.

[60] *Diss.* 2. 6. 9–10, tr. R. Hard slightly modified. For a useful discussion of this text I refer to Menn 1995: 23–4.

[61] 'Since we do not live with men who are perfect and clearly wise, but with those who are doing splendidly if they have in them mere imagines of virtue, I think that we must understand this too: no one should be wholly neglected if any indication of virtue appears in him,' Panaetius, cited by Cicero, *On Duties* 1. 46 (tr. E. M. Atkins).

[62] See Stob. *Ecl.* 2. 100. 2, 112. 1–5 (ed. Wachsmuth).

[63] *De Stoic. Rep.* 1048e (*SVF* 3. 662 and 668); Seneca, *Ep.* 42. 1; Alexander, *On Fate* 199. 16–22, ed. Bruns; Sextus Empiricus, *M.* 9. 133 and esp. 7. 433, where he ironically says that the Stoics Zeno, Cleanthes, and Chrysippus counted themselves among the base people, not among the wise (see also Sextus, *PH* 3. 250, DL 3. 32, and Cic. *Acad.* 2. 145).

and this engagement does not seem to be reserved to a certain kind of people, but it is arguably applicable to all human beings. To the question 'is it possible to be a perfectly rational person?' we should give at least two responses. On the one hand, an affirmative one: if every human shares the same rational nature, then all of us, due to the mere fact that we are humans, are able to develop our rationality towards the correct direction and therefore everyone is capable of becoming a sage. On the other hand, the response should be negative, since although people usually develop their rationality, not everyone develops it in the Stoic sense of 'right reason'. Hence not everyone becomes a Stoic sage.

My remarks on the connection between ethics and physics can be objected to in several ways. For example, one might object that it is not a good argument to say that, given that for the Stoics all parts of philosophy are not separate from one another (DL 7. 40), then, there must be a close connection between ethics and physics. For the same close connection could be emphasized with regard to ethics and logic. And I would admit that point, which, however, does not constitute a serious objection to what I am suggesting. First of all, because Stoic ethics or, more precisely, Stoic psychology of action (which rarely can be separated from ethics in Stoicism) in some important details presupposes some notions originally belonging to logic (like the theory of *λεκτά*). Second, as indicated above, dialectic (or logic), like physics, is said to be a virtue,[64] and a virtue 'which contains other virtues as species'.[65] These virtues are *ἀπροπτωσία* ('non-precipitancy'), *ἀνεικαιότης* ('unhastiness'), *ἀνελεγξία* ('irrefutability'). I want to focus briefly on the first two specific virtues just listed: 'non-precipitancy' and 'unhastiness'. The virtue of 'non-precipitancy' is defined as 'knowledge of when one ought to assent and when not';[66] 'unhastiness' is defined as 'a strong minded rationality (*ἰσχυρὸς λόγος*) with respect to what is likely (*εἰκός*), so that one does not give in to it'.[67] Non-precipitancy and unhastiness describe excellences proper to the sage, and they underlie the Stoic claim that the sage never errs. On the contrary, those who are not well trained in handling their impressions have precipitancy (*προπέτεια*) and veer into disorder (*ἀκοσμία*) and hastiness (*εἰκαιότης*). Now, although these two virtues are 'dialectical virtues', they play a significant role for the Stoic psychology of action (and thus for Stoic ethics) in examining the psychological state the virtuous person should have. Given the Stoic insistence on the rationality of nature, with the divine *logos* immanent everywhere, the possession of these particular dialectical virtues on behalf of the sage is another example of the presence of the universal reason in the agent who has understood the structure of the cosmos and, accordingly, has translated such a structure into his cognitive dispositions.

[64] Cic. *De Finibus* 3.72; DL 7. 46. [65] DL 7. 46 (tr. Inwood-Gerson).
[66] Ibid. [67] Ibid.

4. COSMIC NATURE AND THE PROPER OBJECT OF ETHICS

The other complicated issue when trying to show the role that cosmic nature plays for ethics in Stoicism is focused on how the particular tenets of Stoic ethics are related to cosmic nature. For example, how cosmic nature is connected with the thesis that virtue alone is good, vice alone bad, and that all else is indifferent; or that while virtue is chosen, the indifferents are just selected; or that while some indifferents contain a selective value (*ἀξία ἐκλεκτική*), other indifferents contain a disvalue which is non-selective (*ἀπαξία ἀπεκλεκτική*). We usually pursue things like life, health, pleasure, wealth, reputation (the so-called 'preferred indifferents'; *προηγμένα*), and avoid things like death, illness, pain, poverty (the 'dispreferred indifferents'; *ἀποπροηγμένα*). And we do so because, although life, health, pleasure, and wealth are not goods by themselves, it is 'natural' that we pursue them, insofar as the preferred indifferents (or rather 'the preferred indifferents for themselves'; *δι' αὑτά*) are 'according to nature' (DL 7. 107). Moreover, the preferred indifferents are those containing a value, this value being a 'contribution to the life in agreement' (*σύμβλησιν πρὸς τὸν ὁμολογούμενον βίον*; DL 7. 105). This important detail shows that preferred indifferents were relevant for the happy life understood as a life in agreement for they contribute to attaining such an agreement.[68] One might suspect that in the theory of indifferents there is a trace of the rationality that nature confers upon the agent. In fact, it is natural, and hence rational, to choose to have a natural ability (*εὐφυΐα*) rather than not to have such a natural ability, wealth rather than poverty, soundness rather than lack of soundness. The natural disposition of human beings to select the preferred indifferents also points out that, even though the only real good is virtue, virtuous life depends, to some extent, upon the assumption of things that are neither good nor bad, and of things that contain an intrinsic value or disvalue as well as a value and disvalue added by the agent. In fact, as observed by the heretic Aristo, depending on certain circumstances, the agent can choose a dispreferred indifferent rather than a preferred indifferent. Aristo, as reported by Sextus, would have argued against the Zenonian and Chrysippean theory of indifferents by focusing on two related points. (i) To call some items—such as health—'preferred' is equivalent to calling it 'good'; hence the difference between good and preferred indifferents is solely in name. (ii) The indifferents between virtue and vice are not some of them preferred *by nature* (*φύσει*) and some dispreferred; rather, they depend on the circumstances (*παρὰ τὰς περιστάσεις*)—which vary with the times—such that the things which are

[68] This raises the problem of how something which is not truly good can contribute to attaining the agreement with nature. I shall not develop this issue here.

said to be preferred are not *absolutely* (πάντως) preferred, and the items which are said to be dispreferred are not necessarily (κατ᾽ ἀνάγκην) dispreferred.[69] The purpose of Aristo clearly was to deny that the value of intermediate items between virtue and vice could be determined on account of their conformity to nature. If the sage's judgement can attribute value to what is dispreferred (e.g. sickness), it follows, in Aristo's view, that the preferred indifferents are not according to nature. Aristo's criticism does not affect the bulk of the theory, as long as it maintains that indifferents neither benefit nor harm, which is the same as arguing that no indifferent is absolutely beneficial (cf. DL 7. 102). Aristo's criticism, centred on the thesis that there is no indifferent that is preferred by nature, due to the fact that depending on the circumstances the sage will choose sickness (a dispreferred indifferent) rather than health (a preferred indifferent), also does not affect the orthodox view. In fact, the orthodox view does not say that preferred indifferents are preferred *by nature*, in the sense that they come to be *absolutely* (πάντως) preferred in any circumstance.[70] For instance, the Stoic Musonius is willing to argue that if one in old age were able to admit death without fear and with confidence (τὸ προσδέχεσθαι τὸν θάνατον ἀφόβως καὶ θαρραλέως), one would provide something significant for living without pain and according to nature.[71] In other words, even though one is naturally inclined to prefer life instead of death, in old age, when death is approaching, one should accept death as a natural stage in life.[72] In addition to this, the sage evaluates the variability indicated by Aristo with regard to the fact that, depending on the circumstance, the person will choose, for instance, being sick instead of being healthy.[73] And the sage is the one who is aware of the rational order, which he has incorporated into his own character and, accordingly, he is the only one well qualified to determine what is good or beneficial at a given circumstance.

It does seem to me that we have good reasons for contending that the issue of indifferents was particularly important both in Stoic ethics in general and in practical ethics in particular. If Aristo's remark is correct and the sage will choose a dispreferred indifferent rather than a preferred one in a concrete situation, then, Stoic ethics appears to be more practicable than was thought by critics of Stoicism in antiquity. The consideration of the indifferents became so significant for the Stoics that it was incorporated into two reformulations of the end. In

[69] Sextus Empiricus, *M*.11.64–7.

[70] In the sense that they are absolutely and, thereby, invariably preferred, but in the sense that we are naturally inclined to choose preferred indifferents instead of dispreferred indifferents.

[71] Musonius Rufus, Frag. 17 (p. 92. 14–17, ed. Hense).

[72] This is reminiscent of Socrates' claim at the beginning of Plato's *Crito* 43b10–11: 'It would not be fitting at my age to resent the fact that I must die now' (tr. Grube).

[73] Sextus, *M*.11.66: ἕλοιτ᾽ ἂν μᾶλλον ὁ σοφὸς τὸ νοσεῖν κατὰ τοῦτον τὸν καιρὸν ἢ τὸ ὑγιαίνειν. The example is that of those who, even being in sound health, serve under a tyrant and, because of this, are destroyed in their rationality. In this case health is neither absolutely preferred nor is sickness something absolutely rejected.

fact, Diogenes Laertius reports that Diogenes of Babylon maintained that 'the end consists in reasoning well (εὐλογιστεῖν) in the selection of things according to nature' (DL 7. 88). Panaetius, for his part, stated that 'the end is to live according to the inclinations (ἀφορμαί) given to us by nature'.[74] In both cases, I submit, things according to nature and the inclinations given to us by nature must be the preferred indifferents, of which we are aware once we have developed our reason, since our reason somehow *is* our nature. And this is so because, as Antipater says, we select the preferred indifferents 'according to a preferential reason' (κατὰ προηγούμενον λόγον),[75] this 'preferential' or principal reason being our developed reason indicating what is convenient in a real rational life. What guides this process in a mature human being, who has changed his initial instinctive impulses of self-preservation for rationalized impulses coinciding with the preferred indifferents, is reason and its preference for the selective value, 'according to which, when the given circumstances allow it (διδόντων τῶν πραγμάτων), he selects these things rather than those; such as health instead of sickness, life instead of death, and wealth instead of poverty'.[76] I regard this reference to 'circumstances' as being important because it compels us to take into account that the preferred indifferents are not always and in every case preferable. It will be the task of the prudential reason of the wise person to decide when such preferred indifferents are effectively preferable.

Now, I think, we are in a better position to start understanding how cosmic nature is related to some of the basic tenets of Stoic ethics. First, if we concentrate again on the distinction between what is really good (virtue) and what is good in the derivative sense of being preferred 'according to a preferential reason', we can notice that virtue alone is really good because (i) virtue is the sage's state of the soul (i.e. the soul of the agent who has been able to unify her nature with the nature of the whole),[77] and (ii) because good is benefit and virtue as well as virtuous action are benefits (cf. Sextus, *M*. 11. 22–3; LS 60G). Second, if the person actually has unified her nature with that of the whole, she will be

[74] Clement of Alexandria, *Stromateis* 2. 21, p. 183 (Stählin) (Testimony 53, ed. Alesse = Frag. 96 ed. Van Straaten). Alesse argues in favour of rendering ἀφορμαί by 'attitudini', and refers to Liddel *et al.* 1949: *s.v.* ἀφορμή (see Alesse 1997: 183). She explains her choice by arguing that such ἀφορμαί 'are present just in man, since they are proper of rational nature', 'it is not completely correct to interpret them (i.e. the ἀφορμαί) as instinctive inclinations, natural and spontaneous impulses'. Of course, I'm not understanding ἀφορμαί as instinctive items: if one takes them to be the natural inclinations of a rational being, one has no need to interpret them as '*instinctive* inclinations'. They simply are the inclinations belonging to a rational being.

[75] Or 'on the basis of a *predominating* reason' or 'in accordance with the *principal* reason' as Inwood translates (1985: 202, and Inwood's tr. of Stobaeus' extract of Stoic ethics, in Inwood and Gerson 1997: *ad loc.*, respectively).

[76] Antipater, quoted by Stob. *Ecl.* 2. 83. 13 ss, ed. Wachsmuth.

[77] If my account regarding how 'nature' should be understood in those passages where nature and reason are closely related (and also the fact that our natures are parts of the universal nature, and that this universal nature is not something alien or external to us) is sound (see the end of sect. 3 above), it appears to me that the concordance or coherence demanded in DL 7. 89 might be interpreted both in terms of a self-concordance and concordance with the whole.

able to accept that, under certain circumstances, she should choose 'to come out from life' or, as Musonius reminds us, she should accept death without fear and with confidence. Both to come out from life and to accept death are reasonable actions insofar as they contribute to the administration of the whole and, at the same time, to keep rationality in universe (recall once more the example of a person who serves under a tyrant, and to avoid being destroyed in his rationality by the tyrant, he can reasonably choose to commit suicide). This is the same as arguing that the agent is now capable of considering events from a universal reason viewpoint, because the agent somehow *is* universal reason. But, of course, the person does not need to commit suicide in order to start regarding events from a universal reason standpoint. In a more trivial sense—but not less important—one can do that when one understands that the right thing to do is to act with moderation or with justice; this understanding—which is both intellectual and practical[78]—guarantees the agent will live a real rational life and display in his behaviour the rational order of the whole.

5. CONCLUDING REMARKS

The theory underlying the idea of rational unity—a tenet that in the domain of political theory gives place to the Stoic cosmopolitanism—is explained out of certain systematic assumptions related to a philosophical approach based on a new conception of rationality: the thesis, alien to the thinkers of the classic period, that all the universe is pervaded by reason to such extent that there is nothing that is not an instance of universal reason. Within this rationality inherent to the cosmic system, humans are parts especially privileged, insofar as they may generate reflective processes and, in this way, they are able to understand the structure of the world. The common principle that all humans take part in teaches us, according to the Stoics, that the only thing that places us at a superior level with regard to our fellow human beings is having a better disposition of character, that is to say, a virtuous character. Both our common origin and rational structure also prove that nobody is nobler than anyone else, except with regard to a more correct and more suitable disposition for good arts (*rectius ingenium et artibus bonis aptius*). This is so, argues Seneca, because the world (*mundus*) is 'the unique father for everyone, and the primary origin (*prima origo*) of each one is reduced to it, whether it is a distinguished or an obscure position'.[79] In its context this passage accounts for and develops the Stoic thesis that distinctions—based on nobility titles or cradle privileges—among

[78] For this approach in early Stoicism see Stob. *Ecl.* 2. 63. 11–12, ed. Wachsmuth, and DL 7. 126, 130. For the same view in late Stoicism cf. Musonius Rufus, Frag. 6 (p. 22. 7–9, ed. Hense). A brief discussion of this topic can be found in Boeri 2005: 402–5.

[79] Seneca, *On Benefices* 3. 28. 1.

human beings are totally arbitrary and conventional. For the Stoics the just is by nature (DL 7. 128), and given that all human beings were born endowed with rationality, all of them are equally capable of exercising their rationality in the right way. As Seneca puts it, emphasizing once more our condition of 'parts of the whole', our shared rationality is something divine in us, who are parts of god (*Ep.* 92. 30; *SVF* 2. 637). Appropriation or familiarization (οἰκείωσις), understood in its 'social' dimension—i.e. the appropriation or familiarization which in the first stage of life is concerned with the 'egoist' interests of the living being with himself, but later can be developed towards the familiarization with the other members of the species in a sort of 'altruistic move'—is a proper picture of how god or the universal reason expects us to love our fellow human beings.

The Stoic theses, no matter how counterintuitive they look, can be the result of a legitimate confidence in 'right reason', a reason that, as other philosophers thought in the history of philosophy, would be able to guarantee an increasing and permanent moral progress which, ultimately, would end by abolishing the permanent armies and war, as well as secure the possibility of a 'world citizenship', where there would not be National States, but a World State. This is the translation that the Enlightenment made of the cosmopolitanism and rational naturalism of the Stoics.[80] Nothing of what in fact happened in the last century or of what is happening in the present one probably would make us think that the Stoic ideal of cosmopolitanism and government by reason is possible. This state of affairs, however, not only does not deactivate the Stoic ideal of a world community of rational beings who are not willing to inflict pain or harm on the other members of their species; maybe it even strengthens that ideal. A Stoic philosopher would be willing to argue that what in fact is going on does not prevent him from thinking of what *should* happen. A Stoic would also argue that human beings continue to behave like irrationals with respect of their other fellow men precisely because they refuse to accept that, because of our being members of the same community of rational beings, we are able to train our characters properly and that, once our characters have been trained, we are also able to recognize in each of ourselves the features that characteristically describe our humanity and that of others. From a less optimistic point of view one

[80] I am thinking above all of Kant. In *Toward Perpetual Peace* he quotes the example of the dog tied to a cart (in Seneca's version, *Ep.* 107. 11) within a context of political discussion: *fata volentem ducunt, nolentem trahunt*: 'Fates lead the willing, drive the unwilling' (8. 365, ed. *Kant's Gesammelte Schriften*, Berlin: Walter de Gruyter & Co. 1900). In Kant's view, nature is the guarantee of perpetual peace since in its mechanistic course it displays a *finality* (*Zweckmässigkeit*) that introduces concord (*Eintracht*) in discord (*Zwietracht*), no matter what man wishes. When Kant asserts that nature wills that this or that happen, he suggests very stoically, that he does not mean that nature can impose a *duty* to do something—since practical reason *without coercion* can do that—but that nature does what it does, no matter what we want. The thesis that nature follows a mechanistic course, but within a teleological order, sounds very Stoic, too. Even though the Stoics argue for a rigid causal order, they do not describe such an order as a purely mechanistic system. The teleological factor appears clearly outlined in the notion of providence, which can be identified with fate.

might reply that the desire to destroy our fellow human beings and, eventually, the world is actually part of our nature. If this is the case, the view that some acts of aggression or certain terrible physical or psychological pains (like torture) inflicted on other human beings are 'inhuman' is self-refutable, since only human beings perform this kind of acts. However, a Stoic can always respond that to maintain that aggression is what characteristically describes human nature is a false belief, and that this belief is false can be proved from the fact that if everybody thought and acted in that way, we would destroy the cosmos, i.e. the 'order', which means that we would end up by destroying ourselves since each of us is a microcosmic instance of the whole world. This is the ideal that, in Marcus Aurelius' words, says:

All things are reciprocally interwoven . . . , everything is coordinated and confers order to the same cosmos (συγκοσμεῖ τὸν αὐτὸν κόσμον). For cosmos is a unity made up of all things, and god is one, pervading all things. And there is only one reality (οὐσία) and only one law, i.e. universal reason (λόγος κοινός) of all the living beings that are intelligent, and there is only one truth, since perfection (τελειότης) of the living beings that are alike in kind and that participate in the same reason is one, too. (7. 9; my tr.)

REFERENCES

Alesse, F. (1994) *Panezio di Rodi e la tradizione stoica* (Naples: Bibliopolis).

——(1997) *Panezio di Rodi: Testimonianze,* ed., tr., and commentary by Francesca Alesse (Naples: Bibliopolis).

Annas, J. (1993) *The Morality of Happiness* (Oxford: OUP).

——(2007) 'Ethics in Stoic Philosophy', *Phronesis,* 52: 58–87.

Betegh, G. (2003) 'Cosmological Ethics in the *Timaeus* and Early Stoicism', *Oxford Studies in Ancient Philosophy,* 24: 273–302.

Boeri, M. D. (2001) 'The Stoics on Bodies and Incorporeals', *Review of Metaphysics,* 54(4): 723–52.

——(2005) 'Socrates and Aristotle in the Stoic Account of *Akrasia*', in R. Salles (ed.), *Metaphysics, Soul, and Ethics in Ancient Thought:. Themes from the Work of Richard Sorabji* (Oxford: OUP), 383–412.

——and Vigo, A. G. (2002) 'Die Affektenlehre der Stoa', in A. Engstler and R. Schnepf (eds.), *Affekte und Ethik: Spinozas Lehre im Kontext* (Hildesheim: Georg Olms Verlag), 32–59.

Carone, G. R. (2005) *Plato's Cosmology and its Ethical Dimensions* (Cambridge: CUP).

Cooper, J. M. (1999) 'Eudaimonism, the Appeal to Nature, and "Moral Duty" in Stoicism', in J. M. Cooper, *Reason and Emotion: Essays on Ancient Moral Psychology and Ethical Theory* (Princeton: Princeton UP), 427–48.

——(2004) 'Stoic Autonomy', in J. M. Cooper, *Knowledge, Nature, and the Good: Essays on Ancient Philosophy* (Princeton: Princeton UP), 212–13.

Engberg-Pedersen, T. (1986) 'Discovering the Good: *Oikeiosis* and *Kathekonta*', in M. Schofield and G. Striker (eds.), *The Norms of Nature: Studies in Hellenistic Ethics*

(Cambridge and Paris: CUP and Éditions de la Maison des sciences de l'homme), 145–83.

Engberg-Pedersen, T. *The Stoic Theory of Oikeiosis* (Aarhus: Aarhus UP).

Glare, P. G. W. (ed.) (1996) *Oxford Latin Dictionary* (Oxford: OUP).

Inwood, B. (1985) *Ethics and Human Action in Early Stoicism* (Oxford: OUP).

—— and Gerson, L. P. (1997) *Hellenistic Philosophy: Introductory Readings* (Indianopolis and Cambridge: Hackett; 1st edn. 1985).

Irwin, T. (2003) 'Stoic Naturalism and its Critics', in B. Inwood (ed.), *The Cambridge Companion to the Stoics* (Cambridge: CUP), 345–64.

Liddel, H. G., Scott, R., and Jones, H. S. (1949) *A Greek English Lexicon* (Oxford: Clarendon Press).

Long, A. A. (1968) 'The Stoic Conception of Evil', *Philosophical Quarterly,* 18: 329–43.

—— (1983) 'Arius Diymus and the Exposition of Stoic Ethics', in W. W. Fortenbaugh (ed.), *On Stoic and Peripatetic Ethics: The Works of Arius Didymus* (New Brunswick, NJ, and London: Transaction), 41–65.

—— (1996) *Stoic Studies* (Cambridge: CUP).

—— (1996*a*) 'The Logical Basis of Stoic Ethics', in Long, *Stoic Studies* (Cambridge: CUP), 134–55.

—— (1996*b*) 'Dialectic and the Stoic Sage', in Long, *Stoic Studies* (Cambridge: CUP), 85–106.

—— (1996*c*) 'Stoic Eudaimonism', in Long, *Stoic Studies* (Cambridge: CUP), 179–201.

—— (1996*d*) 'Heraclitus and Stoicism', in Long, *Stoic Studies* (Cambridge: CUP), 35–57.

Menn, S. (1995) 'Physics as a Virtue', *Boston Area Colloquium in Ancient Philosophy,* 11: 1–34.

Schofield, M. (1999) *The Stoic Idea of the City* (Chicago and London: University of Chicago Press).

Thom, J. C. (2001) 'Cleanthes' *Hymn to Zeus* and Early Christian Literature', in A. Y. Collins and M. M. Mitchell (eds.), *Antiquity and Humanity: Essays on Ancient Religion and Philosophy (presented to Hans Dieter Betz on his 70th Birthday)* (Tübingen: Mohr Siebeck), 477–99.

8

Why Physics?

Brad Inwood

My problem begins with Socrates, and a widespread view about his philosophical activity. Aristotle makes the point twice in famous texts (revised Oxford translation, slight changes). 'Socrates, however, was busying himself about ethical matters and neglecting the world of nature as a whole . . .' (Aristotle, *Metaphysics* *A* 987b1-2). 'In the time of Socrates . . . people gave up inquiring into nature and philosophers diverted their attention to politics and to the virtues which benefit mankind' (*Parts of Animals* 642a28–31). Cicero shared this view of Socrates. Three famous texts stand witness. *Tusculan Disputations* 5. 10 gives us the catchphrase: 'But Socrates was the first to call philosophy down from the heavens, to locate it in cities, even to bring it into homes, and to compel philosophy to enquire about life and ethics, good and bad things.' In *Academica* 1. 15 Varro says:

> I believe (and everyone agrees on this) that Socrates was the first to call philosophy away from hidden matters which nature herself had concealed—the preoccupation of all philosophers before him—and to introduce it to ordinary life so that it could enquire about virtues and vices and generally about good and bad things. He held [and this is the important part] that the heavens are far beyond our knowledge or (if they can indeed be known) that they have no bearing on living well.

Readers of the *Apology* and the *Phaedo* know how hard it is to come to a firm view on the real historical situation, but the dominant ancient view of Socrates is clear enough.[1] When we think of the Stoics as Socratic philosophers, then, one

I am grateful for comments and critical reaction from several audiences before which versions of this paper have been read: Cornell University (Philosophy and Classics), Concordia University in Montréal (Philosophy), Brown University (Philosophy and Classics), the University of Edinburgh (the 12th A. E. Taylor Lecture in Ancient Philosophy), and at the Seminar on Moral and Political Thought at Johns Hopkins University. I am especially grateful for the helpful comments I received from participants at the conference on 'Dios y el Cosmos en la Filosofía Estoica' organized by Ricardo Salles at the Universidad Nacional Autónoma de México, 3–5 July 2006, Salles first among them, but also Keimpe Algra, Thomas Bénatouïl, Gábor Betegh, John Cooper, David Hahm, and Inna Kupreeva.

[1] Even allowing for the 'Socratic' providential teleology in Xenophon's *Memorabilia* 1. 4 and 4. 3.

way we can approach them is by asking what could possibly have moved them to abandon this Socratic legacy and throw themselves into the study of the natural world as though it could be known and really did matter to human happiness. How Socratic were the starting points of Stoicism? This is really a question about Zeno of Citium. Diogenes Laertius tells us that when he came to Athens from Cyprus in his early twenties (7. 31) he had already been fired up by reading Socratic books (7. 28) and (7. 2–3) and that after arriving in Athens he was drawn to a bookshop where book 2 of Xenophon's *Memorabilia* was being read out. Attracted by it, he asked where he could find men like the ones described there. The bookseller pointed to Crates of Thebes, a well-known Cynic philosopher, and said 'Follow him.' Now, Cynics certainly took their cue from nature rather than convention. But they didn't study the natural world to do so.

Without this context, my question might otherwise seem puzzling. Why should[2] a Stoic study physics? It might seem odd to ask because the answer is apparently so obvious. Stoics, we read in many sources, hold that happiness is a matter of following nature. If physics is the study of nature, then unless we want to go wandering about aimlessly we need to know what it is we are following. So physics would, on this naïve view, be of great value *at least* in an instrumental sort of way.[3] We study it because otherwise we just won't know what we are doing when we pursue the happy life—which is what we are really after. Compare the Epicureans. Epicurus held (*KD* 11–13), that we need the study of nature (*phusiologia*) *only* to the extent that is necessary to dispel the fear of death and

[2] In this chapter I will from time to time move back and forth between a motivational formulation and a justificatory formulation of the question 'Why physics?' and this duality is meant to be reflected in the ambiguous word 'should' here. Clearly the connection between the two versions of the question is complicated and should not be taken for granted. I do, however, think that there is a clear sense in which these two versions of the question amount to the same thing. For (at least in Seneca and also quite commonly in the ancient world, if not also in the minds of many today) the question 'Why should I do X?' is asked against the background of an assumption that expending scarce and valuable time on something which yields no benefit is irrational and contrary to our interests; the concern with the scarcity of time is virtually ubiquitous in Seneca's *Letters* and made explicit at 117. 32 (see below). The demonstrated absence of a justification of some sort for X ought, then, to eliminate the motivation to do X in a rational agent. And throughout Seneca and I both assume that an appeal, even an implicit appeal, to human rationality is, if convincing, a decisive argumentative move.

[3] One might well ask whether 'instrumental' is the right way to characterize this kind of motivation, but I think it is a reasonable way to capture the issue I have in mind. Consider e.g. the Stoic distinction between goods which are 'goal-like' (τελικά) and those which are 'instrumental' (ποιητικά) as set out in Diogenes Laertius 7. 96, at Stobaeus, *Ecl.* 2. 71–2, and in a few other key texts (*SVF* 3. 106–9). The distinction drawn in *Ecl.* 2. 71 between goods which produce (ἀπογεννᾶν) happiness and those which complete it (συμπληρο̂υν), being *parts of* happiness itself, is particularly rich and interesting. The distinctions among kinds of goods made in these doxographical passages are not the same as the distinctions among kinds of motivation which I am exploring here; but the recognition (surprising to some, perhaps, who may have overestimated the significance of Stoic eudaimonism) that instrumental values are not foreign to Stoic moral theory should render my use of the terminology here permissible. Whether it is ultimately the best way to express the distinction I develop over the course of this discussion is another matter, but for the time being it seems to be the best that I can come up with.

suspicion about the heavenly bodies (are they potentially vengeful gods rather than merely physical phenomena?). If a Stoic were to study physics *merely* to find out what he or she should follow in order to be happy, then this would bear a superficial resemblance to Epicurean motivations. Similar concerns arise if we think of the motivation for studying physics expressed over twenty years ago by Nicholas White in 'The Role of Physics in Stoic Ethics'.[4] Although in the end White's position is refreshingly aporetic, his main line of enquiry deals with an attempt to specify which beliefs about nature one would need to hold in order to be immune from passions. Being seriously committed to the goodness of the whole of which one is a part is supposed to make one free, not from pain of various sorts occasioned by events in the world, but free from 'minding it'. But not only is it odd and rebarbative to see the heart of Stoic philosophy as a set of intellectual props for such an attitude;[5] but as White concedes, there is, as well, virtually no evidence that the Stoics actually held this view about physics and (what is worse) one can only *imagine* why they might want to defend their theory in this way. Few people, ancient or modern, think it sound or plausible to hold that learning cosmology will actually help us deal with personal grief—even if that were held to be the point of doing philosophy. So if that's the reason for a Stoic to commit to the study of physics it will be a pretty poor show all around. As White concludes (p. 72) 'we can see at least some room for the later Stoics [and I would say *any* Stoics] to deny, or to omit to believe, that physical theory was required by the ethical views that they retained from the early Stoa without feeling forced to admit that they had thereby abandoned the central part of the Stoic position'. I won't work through White's argument nor do I accept all of it. But the challenge to our assumptions about the respectable *intellectual* necessity of physics in Stoicism is serious.

It gets worse. According to some sources for Stoicism, physics is not just a part of philosophy for the Stoics, as it is also for the Epicureans. It is also said to be a virtue in its own right. Cicero, who provides the only usefully coherent account of this idea, reports what he takes to be a standard Stoic view (*De Finibus* 3. 73):[6]

Physics too not unreasonably has earned the same distinction [being named a virtue], precisely because anyone who is going to live consistently with nature must start out from

[4] White 1985. [5] Compare e.g. the remarks in Brennan 2005: ch. 1.

[6] This general issue has been discussed at length by Stephen Menn (1995). This claim about virtue is also reported as a variant view (*alloi de*) at DL 7. 92 but with no explanation or rationale. According to Ps.-Plutarch in the proem to book 1 of his *Placita* (*Doxographi Graeci*, p. 273) logic, ethics, and physics are virtues and this grounds the tripartition of philosophy. But this is somewhat incoherent, since it is *sophia* not philosophy which is a form of *epistēmē*. The rationale given by Ps.-Plutarch conflicts with the general treatment by Menn, which assumes reasonably enough that virtues are forms of *epistēmē*. If the status of logic, physics, and ethics as virtues is intended to support the tripartition of philosophy (here described as an *askēsis* in contrast to an *epistēmē*), then a great deal more must be said about how such a practice relates to the knowledge which constitutes wisdom or the virtues. Ps.-Plutarch's account, it appears, is confused. Menn's discussion also slides over the issues of practical vs. theoretical knowledge and assumes too readily that bodies of knowledge, including craft knowledge, can motivate agents suitably and that there are such things as hormetic cataleptic presentations.

the cosmos as a whole and the providential care taken for it. Nor can anyone make true judgements about what is good and bad except on the basis of a knowledge of the nature and life of the gods too and of whether or not man's nature is congruent with that of the universe. No one can see the significance (and it is very great) of the ancient precepts of the sages . . . without physics. . . . Nor can one understand piety to the gods nor how grateful we should be to them without a detailed account of nature.

This suggests a similarly limited kind of motivation for studying physics. Physics is a virtue because without knowledge of at least some parts of physics one will simply fail at one's ethical enterprise, which according to Cicero includes a proper appreciation of god and his providential care for human beings, along with a deep understanding of the wisdom of ancient sages. If this is the full justification for regarding physics as a virtue, then it certainly looks as though the point put into Cato's mouth by Cicero[7] is quite compatible with an instrumentalist account of the motivation to study physics. And this would be puzzling if we expect that three parts of philosophy (all regarded as virtues by Cicero here) to be on a level with each other. On the other hand, *knowledge* of the natural world is an intellectual excellence (the ability to get anything right is an excellent characteristic to have, after all); looked at this way, physics would be a virtue in a quite different sense.[8] But in fact, it is presented by Cicero as being subservient to the goals to be achieved by ethics. Is this the Stoic position on the question?

Not straightforwardly. For Cicero himself, earlier in the same book, gives us a quite different sense of the reason Stoics had for studying physics. In sections 17 and 18 of book 3 Cato claims that, in addition to humans' pre-rational attachment to self-preservation and self-enhancement, there is also in us a basic drive for learning and truth, a drive which seems to be on a par with the *oikeiōsis* which is the starting point for the development of our ethical nature. It is natural to compare this fundamental drive with the conception of human nature evidently held by Aristotle and others, especially Academics like Cicero:[9] they thought that human nature is just built for knowing things. The *locus classicus* for this sort of view is probably the opening of *Metaphysics* A. All men by nature yearn to know. Since for Aristotle the complete actualization of our natural capacities is our *telos* and since knowledge includes a grasp of principles and causes (*archai, aitiai*) of nature and what is beyond it, Aristotle's assumptions about human nature entail that knowing physics and metaphysics is at least a formally necessary condition

[7] I emphasize this in order to avoid giving the impression that this is Cicero's own final and settled view of the matter. His views seem to me to be much more complicated and in *De Officiis*, where he speaks in his own voice, he sometimes subordinates theoretical to practical wisdom in a way that suggests that he held a different view, at least for the purposes of that work. See e.g. 1. 13–19 and 1. 153–8. 1. 19 expresses particular caution about the perils of excessive devotion to theoretical wisdom.

[8] Menn 1995 does not seem to see this as a distinct way in which physics could be a virtue, though Nussbaum in her commentary (1995: 36) makes this point with elegant understatement.

[9] But see n. 7 above: Cicero does not consistently express this view in his own right. This is not surprising for an Academic sceptic.

for complete happiness.[10] Reflection upon such causes and explanations is the sort of activity which also brings us the greatest and most consistent attendant pleasure, since the greatest pleasure accompanies the best exercise of our highest capacities.

Cicero's sympathy for this view of human nature is confirmed elsewhere in his discussions of Stoicism. In *De Officiis* (1. 11–13), when outlining (presumably on the basis of Panaetius' version of Stoicism) the foundations in human nature of the four cardinal virtues, Cicero not only emphasizes the instrumental merits of learning about the natural world, but also makes dramatic claims about an intrinsic human drive for knowledge just for its own sake (1. 13):

> The most important characteristic of human beings is the searching pursuit of truth. And so whenever we are free from unavoidable business concerns we are eager to see, to hear or to learn something, and we regard the understanding of facts—either hidden facts or awe-inspiring facts—as essential to living happily. From this we can grasp that what is true, straightforward and transparent is most suited to human nature.

This innate human drive for understanding natural phenomena and their hidden causes is termed a 'lust for seeing the truth' (*veri videndi cupiditas*). Cicero is apparently quite sympathetic to the idea that learning about the world has intrinsic rather than merely instrumental value. We might wonder whether this is a straightforward report of 'early' Stoic views or whether it reveals the influence of Panaetius (who is often thought to represent an opening up of Stoicism to the influence of Aristotle and Plato) and Posidonius, the familiar of Cicero who was described by the Stoicizing geographer Strabo (2. 3. 8) as deeply engaged with a pursuit of causal explanations and as being full of Aristotelian inclinations.

What, then, *is* the correct Stoic view about the reason we have for studying physics? *Is* there a single Stoic view about the relationship of physics to the rest of philosophy? Julia Annas, who has often challenged conventional wisdom about the relationship of physics to ethics, has recently reminded us of the internal variety of Stoicism and suggested that a narrow insistence on recovering *the* Stoic position on the topic might be a mistake.[11] With this I heartily agree. The variety of opinion among Stoics even on some quite central issues is something we do well to recall. For instance, Aristo of Chius, like Cleanthes a student of Zeno's, rejected the systematic study of physics and logic in terms reminiscent of the Socrates of Cicero's *Academica*, 'saying that the one was beyond our powers and that the other was nothing to us and that only ethics mattered to us' (DL 7. 160). His free-thinking has earned him the label of

[10] Aristotle's position on the necessary ingredients for happiness is notoriously difficult. Perhaps it is only the greatest and most complete happiness which requires contemplation of eternal truths and happiness (rather than the greatest happiness) accompanies activity according to other virtues, perhaps only non-intellectual virtues. I take no position on that vexed issue here.

[11] See Annas 2007. Further impetus for scepticism about the existence of a Stoic orthodoxy consistent over time can be found in Ludlam 2003; Ludlam's deflationary argument may be excessive at times, but at the very least it serves to underline the intellectual variety of the school at all periods.

'renegade' in modern studies—although that term should surely be reserved for someone like the Dionysius who embraced hedonism, rather than being applied to Aristo, who arguably stuck closer to the Socratic core of Stoic thinking (with its focus on ethics *rather than* physics) than many others in his school.

Aristo, however, failed to win recognition as Zeno's successor. This seems to have led him to lecture independently for a while (as Chrysippus did during Cleanthes' years as head of the school, DL 7. 179) and to seeing his keener students labelled 'Aristonians' (DL 7. 161)—just as early followers of Zeno were labelled Zenonians (DL 7. 5).[12] Chrysippus, although he challenged Cleanthes' views, seems to have made a point of snubbing Cleanthes' rival Aristo from an early point in his own career (DL 7. 182); ancient biographical sources note Chrysippus' loyalty to Cleanthes as well as his success in reshaping school doctrine (see especially DL 7. 179). This is how Aristo's version of Stoicism came to be stigmatized as deviant. This seems to me to mean little more than that Chrysippus was at least temporarily successful in his attempt to seize the intellectual patrimony of Zeno.[13] We should not think that it tells us much about what Stoicism without qualification *really* is.

If this is the case, and if even the founder of the school began as a Socratically inclined follower of the Cynics, then the idea that there is a single, general Stoic view on the question of the role played by physics should probably be shelved. Somehow and to some degree Zeno eventually turned to the study of nature to complement the Cynic's narrower vision of philosophy,[14] a competing vision which Aristo defended into the next generation. Cleanthes, who may have emphasized theology as much if not more than the rest of physics, and Chrysippus have heavily shaped our modern view of what the school stands for. But by the time of Panaetius and Posidonius, this stranglehold on orthodoxy was loosened. Later Stoics (Musonius, Epictetus, and Seneca) continued to take some inspiration from the Cynic and Aristonian strands in the school's history,

[12] The tendency to have groups of students named after an innovative master, even within a single school, was persistent. Cicero, at *Lucullus* 70, has his speaker Lucullus allege that one of Antiochus' motivations for developing innovative doctrines was to have his followers dubbed 'Antiocheans'. For the Aristonians cf. Ioppolo 1980: 23. For the wider phenomenon of philosophers aligning themselves decisively with a master to whom they felt allegiance, see Sedley 1989.

[13] Chrysippus seems to have brought not only two nephews from Soli to the school in Athens but also Hyllus of Soli (see *Der Neue Pauly,* s.v. Chrysipp). His success at taking control of the school's intellectual agenda is reflected by the fact that his students Zeno of Tarsus and Diogenes of Babylon both became heads of the school, and that Diogenes' student Antipater of Tarsus succeeded him. It is not surprising, then, that Chrysipppus' version of Stoicism became canonical; the sheer volume of his works and his intellectual talent complemented his organizational strengths.

[14] J. M. Rist argued that the motivation for this change came from Zeno's study with Polemo in the Academy (Rist 1969: ch. 4); Rist was following C. O. Brink. In recent years David Sedley has also emphasized the influence of early Academics on the foundations of Stoicism and this is certainly compatible with the manifest importance of the *Timaeus* to Stoic cosmology (Sedley 2003: esp. pp. 10–13).

so we must remain sensitive to the complexity of legitimate Stoic positions even on what seem to be central issues of doctrine.

So I am going to give up for now on analysing an overall Stoic view on the role of physics, which seems not to have existed.[15] Instead, I will turn to Seneca, whose views on the issue are sufficiently complicated and interesting to warrant our attention. Seneca approaches this issue in at least three different contexts. (1) Sometimes he addresses explicitly and theoretically the question of the parts of philosophy and how they are related to each other. The clearest case of this is in *Letters* 89 and 90. (2) At other times he sets out the motivations for studying physics in a positive and hortatory voice; there are many such passages, but the preface to book 1 of the *Natural Questions* is one of the clearest and most dramatic; it is in this context that theological issues play more of a role in his account, but considerations of god's role in nature do not seem to offer independent normative considerations beyond what follows from the Stoic claim that human nature and divine nature are essentially connected. (3) Perhaps most often Seneca portrays himself as being on the defensive: after or while expatiating on some theme from physics or 'metaphysics' he stops himself, challenges the usefulness of his enquiry, and then defends himself against the charge. Passages of this type dominate our usual assessment of Seneca's fully considered position on the role of physics in philosophy and John Cooper has recently restated the view quite forcefully.[16] I want to put these views into a fuller and more well rounded context before attempting a general characterization of Seneca's or any general Stoic view on the matter.

At the opening of letter 89 we learn that Lucilius has asked about the parts of philosophy.[17] Seneca reassures Lucilius that this is a genuinely useful question to ask: anyone hustling after wisdom must learn this (*ad sapientiam properanti necessariam*). The study of the parts makes a grasp of the whole easier, he says. 'I wish that philosophy could present itself to us as a whole, just as the entire appearance of the cosmos comes into view.' If philosophy could be seen as a unity we would come spontaneously to love it and put less significant concerns in perspective. But that kind of global perspective on philosophy is not attainable by *us*, though the sage can manage it, so we must settle for a part-wise view, which Seneca compares to the way we come to know the less apparent features of the natural world, the *secreta mundi* (89. 1–2).[18] Seneca then assures us that his division of philosophy will not be excessively fine-grained—for that would make it as hard to grasp as it would be if no division at all were undertaken (89. 2–3) and goes on to distinguish wisdom from

[15] The best recent discussion of the general Stoic position is Betegh (2003).

[16] Cooper 2004.

[17] We are reasonably well informed about this topic from other sources. See e.g. the texts collected at LS 26. The statement in DL 7. 39 that the partition is of *ho kata philosophian logos* is also found in the division of the Academic Eudorus (Stob. *Ecl.* 2. 42. 8–13).

[18] On which see my discussion of the *Natural Questions* in Inwood 2005: ch. 6.

philosophy, the pursuit of wisdom (89. 4–7); Seneca acknowledges that there is a legitimate variety of accounts on this point but reports with greater detachment the view of those Stoics who (relying on the definition of philosophy as 'the study of virtue' and no doubt on the idea that wisdom is the perfected state of the human mind) maintain that there can be no separation of wisdom and philosophy. If virtue is the goal we pursue, we apparently cannot do so without having virtue, so that wisdom and philosophy turn out to be extensionally though not intensionally equivalent (89. 8). But the main point here is the relationship of the three parts of philosophy: ethics, physics, and logic. Attempts by the Peripatetics to add other parts turn out to be redundant—politics and economics fall under ethics. Epicureans are chastized for not accepting logic as a proper part (89. 9–11). Of greater interest to us are the reports of the Cyrenaics (who try to deny physics but wind up reintroducing it, Seneca claims, as a part of ethics) and Aristo of Chios, whose claim that physics and ethics and even the parainetic branch of ethics[19] are not necessary is roundly criticized (89. 12–13).

This leaves the familiar tripartition embraced by Stoics and Platonists (though the latter are not mentioned here). Ethics is divided into three subdivisions: axiology assigns the proper value to things, the second part deals with the regulation of desires (*impetus*), and the third part with actions—what to do when and under what circumstances (89. 14–16). Physics is dispatched quite quickly, by comparison with this account of ethics, and its compass is very limited:

> The physical part of philosophy is divisible in two, corporeals and incorporeals. And each of these parts is divided into its own 'levels', if I may put it so. The topic dealing with bodies has two levels, dealing with things which act and things which are generated from them—the elements are what is generated. The topic dealing with the elements is, as some think, simple, but others think that it is divided into matter and the cause of all changes on the one hand and the elements on the other. (Seneca, *Letter* 89. 16)

That is all he has to say about physics here—not a word here about cosmology, providence, theology, and so forth. More is said, in fact, about the subdivisions of logic in 89. 17, but Seneca abruptly breaks away from the details of his reply to Lucilius (he says he does not want this to turn into a *quaestionum liber*) and in the rest of the letter (89. 18–23) urges Lucilius to devote himself primarily to the more practical side of ethics. Although Seneca had begun the letter by conceding the value of Lucilius' query about a somewhat technical question (the parts of philosophy) he concludes with the advice to 'study, not in order to know anything *more* but in order to know it *better*'. The overall impression is that physics itself is not that much direct help when abstracted from philosophy as a whole. Philosophy itself, not any one division of it, is

[19] Cf. *Letters* 94–5.

the knowledge of things human and divine and their causes (89. 5, cf. *Letter* 90. 3). *Philosophy as a whole* turns out to be the really valuable study. Ethics can be abstracted with profit from philosophy as a whole, but physics, at least as Seneca here describes it, seems to have no particular value except as an integral part of the whole. The kind of physics which might really help with life as a whole is 'big picture' physics (the aspects of physics *not* itemized by Seneca in 89. 16), philosophy itself taken as a whole (like the cosmos as a whole) and thought of as the study of gods, humans, and how the whole system goes together.

This general impression is strongly supported in the next letter, clearly intended as a companion piece. In *Letter* 90. 34–5 we learn that wisdom's job is understanding the big picture, the truth about nature, in a way that other animals cannot. What seems to correspond to physics in this unified account of wisdom is theological and cosmological insight; what correspond to logic are basic epistemological procedures; and ethics is summarized as recognition of the importance of controlling pleasures and of personal autonomy if one is to attain happiness.

It is evident in this letter too that studying physics, looked at in relative isolation in this way, is not a particularly attractive or important part of a philosophical life. We can understand philosophy as a whole—as Annas might say, an integrated whole—and when we do we will have the kind of understanding of the natural world which really is rewarding and enables us to live the happy life of which we are capable. But the maimed and limited study of physics which Seneca describes in *Letter* 89 when he pulls it out of its context, this kind of highly technical physics does not seem to have anything like the payoff which we can get from the integrated approach to philosophy and the grasp of physical theory in that larger context. Thinking just of *Letters* 89 and 90, I confess that I cannot readily understand why Seneca told Lucilius that his question about the parts of philosophy was a good one. Nothing in the sequel seems to me to have justified that polite answer.

Another perspective on physics emerges from the preface to *On Providence*, also addressed to Lucilius. Seneca distinguishes the narrow question he has asked (why do bad things happen to good men if the world is governed by providence?) from the larger set of issues of which it is a part (1. 1). Because Lucilius' question is narrowly focused, it is not necessary in the immediate context to deal with these larger issues and the rest of physics, much of it quite detailed (1. 2–4)—though Seneca does not deny that sooner or later *proof* is required and even says that the specific question raised would be handled more conveniently in the setting of a general work (*in contextu operis*, 1. 1). There is even a note of criticism when Seneca says that *Lucilius* is prepared to settle for 'a small part being ripped out of the whole' (*a toto particulam revelli*). Seneca outlines all the bits of physics that he does not intend to deal with, so that the work as a whole deals with man's relationship to god (a theme in ethics, perhaps, but certainly a larger physical and theological theme as well).[20] The list of physical topics passed over as not necessary for the immediate and narrow discussion includes: the *proof* that the cosmos relies on a protector, that the heavenly bodies do not move by chance but in accordance with an eternal law, that the atomistic theory about chance aggregation of matter is false as applied to explain the orderly stability of the world as we see it; he also passes over the standard themes of meteorology (rainfall, thunder and lightning, earthquakes, tides and various other marine phenomena—the list reads like a partial table of contents for the *Natural Questions*). All of these themes, he says, can be reserved for another occasion, all the more so since Lucilius did not express doubts about whether the world is providentially and teleologically ordered, but rather was just complaining about the apparently harsh treatment of good men under that order.[21]

Consider what Seneca says about physics here: not that one does not need proofs of all these doctrines in order to live a perfected human life, merely that they are not necessary for the narrow question Lucilius asks here. And Seneca does complain about the artificial narrowness of the question asked and makes clear his preference for a more thorough treatment. He holds, then, that a really convincing response to the kind of question that troubles and destabilizes any thinking person ideally demands an extensive treatment of physical doctrines; it demands the whole of physics and not just a part torn out of context. If we ask 'Why physics?' one answer is clearly: in order to achieve peace of mind about our place in the world, so that we may live a tranquil and happy life. In order to work for us this must be a comprehensive account of physics and of course also a *true* account,[22] but even so this reply is cast in what seem to be instrumental terms. Like Epicurus, Seneca justifies the study of physics as a merely necessary condition for human happiness. This kind of similarity to Epicurean motivations is also right on the surface in *NQ* 6, where the detailed investigation of earthquakes and all of their causes is justified primarily by concern to address the fear of death.[23]

[20] Later on (5. 6–7) several aspects of the doctrine of fate are specifically indicated as being necessary to holding the true and appropriate views about the relationship of man to god (*eo quidem magis quod scio omnia certa et in aeternum dicta lege decurrere*). Compare the discussion of fate in *NQ* 2.

[21] Some passages also claim that low-level physical explanation can have direct moral application. At *NQ* 4b13 Lucilius rudely demands a justification for having to spend so much time on trivialities (*ineptiae*) like the explanation of snow when there is 'real' moral work to be done. Seneca argues that the enquiry into the nature of snow and frost constitutes a reproach to those who abuse nature for the sake of their own luxury.

[22] See *NQ* 2. 35 for the dismissal of doctrines which are *only solacia aegrae mentis* and not also true.

[23] *NQ* 6. 1–3 have as their overall theme *scire ne timeas*. In 6. 1 the motivation for explanation is said to be *solacium* (see 2. 35). There is a strongly Epicurean colour to the whole book and its argument. Note also the interdependence of *fortitudo et doctrina* at 6. 32 and the book's concluding attack on the fear of death.

This is not Seneca's only story about the reason for studying physics (not even in this book—see below on *NQ* 6. 4. 2), as we shall see. But before we move on, let us consider a drawback of this motivation. If we study physics in order to get the payoff in ethics, this invites an argument that we might be able to get the same result with less effort, an argument to the effect that the moral outcomes could be just as good without studying physics or without studying it in quite so much detail. Aristo's Socratic approach is like this—in *Letter* 89. 13 he is reported as focusing on the redundancy of physics (and logic)—they are *supervacua*. This is the same criticism he made of the parainetic branch of philosophy, according to Seneca's report (*Letter* 94. 4–5, 11)—Aristo apparently had a general strategy for justifying his narrow focus on ethics. Letter 94 and 95 make clear how little sympathy Seneca has for this minimalist position on philosophy—he even finds a way to bring Posidonius' devotion to *aitiologia* into the discussion of the role of moral *decreta* at *Letter* 95. 65. And yet, in a somewhat different context (the preface to the final book of the *De Beneficiis*, probably written at about the same time as these letters), Seneca seems unusually sympathetic to Demetrius the Cynic's sweeping challenge to all physical theory.[24]

It is worth looking closely at Demetrius' position on the scope of philosophy. There is, Demetrius claimed, more *benefit* (*Benef.* 7. 1. 3) if you have only a few bits of philosophical learning but can deploy them quickly and easily than if you have more but cannot use them readily. The kinds of *praecepta philosophiae* he has in mind are indicated in 7. 1. 5: the tides, the seven-year cycles in human life, phenomena of optics, reproductive biology, the fates, and astrology. Truth, he echoes Democritus, is in the depths (7. 1. 6) and we are blessed by nature in that the essential bits of scientific knowledge are so easy to acquire. 'It isn't hard to discover anything except things whose only benefit lies in the simple discovery of them—everything which will make us better people and happy is either in plain view or very close to the surface.' The short list of such truths includes the claims that true wealth is internal to our own characters, that gods and men need not be feared, that death is not an evil, that everything is pretty straightforward for someone who dedicates his life to virtue, that our social nature and connection to the gods and divine order are the key to human success—this, he said, is the sum and substance of 'useful and necessary knowledge: the rest of our learning is a source of entertainment for our leisure time. Once our mind has withdrawn to a safe location it may then take a detour into topics that confer on our intellect sophistication, not strength' (7. 1. 7).[25]

[24] In *Letter* 71. 7 Seneca recalls that Socrates 'brought all of philosophy back to ethics and said that the highest wisdom is to distinguish good from bad'; this is not an extreme view of Socrates' indifference to ethics nor does Seneca go as far as Aristo. The balance of letter 71 illustrates the importance of 'big-picture' physics for reflection on and analysis of the vagaries of human life. Although the letter does not deal explicitly with the relationship of physics to ethics it concretely illustrates what is perhaps Seneca's commonest view on its relevance.

[25] Compare *Letter* 117. 19, quoted above.

On this view, for happiness we would need only a few quite general truths of physics to anchor our moral progress—again, our natural sociability and sameness of nature with the gods are truths discovered by the study of nature. But physics as a whole, in all its difficult scientific detail, *that* we do not need. The moral utility of physics is genuine and indispensable, but limited in scope. The hard and obscure bits of physics are enjoyable;[26] they do matter to us. But they do not contribute to our strength of character or to our happiness. Seneca's sympathy for Demetrius' position is perhaps overstated here, but only slightly. The instrumental motivation for the study of physics has limited scope. Although there can, in a eudaimonist system, be no stronger motivation than the belief that something is necessary to happiness, it is always vulnerable to being pared back to a minimum. Demetrius, Aristo, and even Epicurus argue that our needs in physics are limited. And for someone who thinks that physics is a virtue or that it is a part of philosophy of equal importance with ethics, this is a dangerous or unstable position to be left in.

But Seneca is not content with this account. When we ask of him 'Why physics?' he has another kind of response. Physics, Seneca sometimes holds, is worth studying in its own right because of the intrinsic value it has for us as rational human beings. Over and above the utility of physics in guiding our moral lives, physics *just is* worth studying because it suits our nature. This intrinsic value of physics to *us* is evident in a striking extended section of the *Consolatio ad Marciam*. In sections 17–18 Seneca tackles the question of human beings' relationship to the natural world. For life is an uncertain, often unpleasant prospect and philosophers had often questioned the value of living at all. Plato has Socrates recall Philolaus' special exhortation that we should not commit suicide in the face of the world's negative side.[27] Not surprisingly, consolatory literature also addresses this theme from time to time. Here Seneca explores the question of whether we would choose to live at all if we knew in advance what the world is like, what combination of wonderful and dreadful things it contains.[28] He does so by extolling the wonders of the natural world which it is possible for us to know and experience: geography, botany, astronomy, natural wonders of all kinds as revealed by rational enquiry are a huge natural attraction for us (18. 2–7), more than enough to offset the 'thousand plagues for body and soul: wars, pillaging, poison, shipwrecks, bad weather, and the bitter longing for one's loved ones—and their deaths, perhaps easy deaths and perhaps bound up

[26] See e.g. *Letter* 106. 3 for the kind of issue which Seneca thinks can only be justified by the pleasure it provides rather than by its moral utility: 'And so I shall pluck this too out of the established sequence of connected issues and if there are any others of the same sort I shall send them along to you on my own, even if you don't ask. What are these issues, you ask? Things which it is more pleasant than beneficial to know, like the one you are asking about: is the good a body?' Chrysippus, one should recall, cautioned against excessive reliance on pleasure as a justification for purely intellectual enquiry: see *SVF* 3. 702 = *De Stoic. Rep.* 1033d = LS 67X.

[27] Contrast the views of Hegesias 'the death-persuader' (DL 2. 86, 93–5).

[28] I have discussed this at some length in Inwood 2005: ch. 8.

with pain and suffering'. When we are faced with the prospect of all this misfortune, the wonders of nature as revealed by our observation of the natural world will induce us to say that we do want to live, even on these terms (18. 8). Here then is a morally urgent outcome from the study of physics: a knowledge of the natural world has a direct impact on our choices about the relative values of life and death. But this is not the direct instrumental application of physics to ethics that we have considered so far; unlike that simpler view, this one relies on our ability to perceive the *intrinsic* value of contemplating the natural world. Ethics, or some of the decisions which the moral agent must face sooner or later, is partly shaped by our appreciation for and enjoyment of knowledge that can only come from the study of nature. Even in *NQ* 6, a book dominated by the instrumental view of physics, we can find a version of this more robust motivation. In 6. 4. 2, for example, Seneca asks and answers the key question:

> What, you ask, will the reward for our effort be? The greatest possible: to come to know nature. And the investigation of this subject, despite its great potential utility, has in it nothing more beautiful than this: it captivates human attention by its splendour and we cherish it not because of the payoff it brings but because of a sense of wonder.

Seneca expresses a similar view about the benefits of studying physics in a famous section of his *De Otio*. In sections 4 and 5 Seneca describes the two 'republics' and the kind of allegiance we owe to each of them. The 'large and genuinely public republic' is the one 'in which gods and humans are contained, in which we do not look to this or that corner but rather we measure the bounds of our citizenship with the sun'; the other republic is our actual political state (4. 1). Some people serve both republics at once, others only one of the two.

> We can devote ourselves to this larger republic even when we are at leisure, or rather, we can do so even better when we are at leisure, so we wind up investigating what virtue is (one or many), whether nature or art makes men good, whether there is one thing which embraces the seas, the lands and their contents or whether god scattered around many bodies of this kind, whether there all the matter from which all things are generated is continuous and full or whether matter is separated with void interspersed with the solids, where god is located, whether he contemplates his work or manages it, whether he is located around it on the outside or is infused throughout the whole, whether the cosmos is immortal or should be reckoned among perishable and temporary things. What does the person who considers these questions do for god? He makes sure that his work does not go unwitnessed. (Seneca, *De Otio* 4. 2)

Just as Cicero's Balbus claims in *De Natura Deorum* 2. 37–8 that humans are born to contemplate and to imitate the cosmos, so Seneca here goes on to claim that living according to nature (our *telos*) comes down to living for what we are born for, for the contemplation of things, and for action (*De Otio* 5. 1: *natura nos ad utrumque genuit, et contemplationi rerum et actioni*). This is also similar to the view reported for earlier Stoics in DL 7. 130: that there are three ways of life, the theoretical, the practical, and the rational, and that the third of these is the one to choose, since 'the rational animal was created by nature fit (*epitēdes*) for action and contemplation'. Epictetus too acknowledges this double function of human beings (at *Diss.* 1. 6. 19–22).

Just as Cicero began his Stoic account in *De Finibus* 3 by asserting the naturalness of our drive for knowledge alongside that for self-preservation (3. 7), so in the *De Otio* Seneca says that our devotion to the study of nature can be confirmed by reflection on the *cupido ignota noscendi* in each of us.[29] Learning something remote and difficult is a reward (*merces*) and our eagerness to pursue investigations is a kind of compulsion (*cogit* 5. 2). A genuine drive for *theōria* is part of our natural endowment (5. 3)—our bodies are built for it in a way reminiscent of the account in the *Timaeus* (5. 4). The nature we are to study includes abstruse cosmological and astronomical topics (5. 4–6). Seneca repeats the claim that we are born for this life (5. 7) and that nature *wants* us to live this life (5. 8). 'And so,' he concludes, 'I live according to nature if I devote myself wholly to nature, if I contemplate and worship her. But nature did want me to do both, to act and to be at leisure for contemplation; I do both since there is an element of action even in contemplation.'

Here, then, is the other answer to 'Why physics?' Because the study of nature is just built into our nature by nature (which is of course equivalent to god). We cannot be fulfilled without studying the cosmos, not just because it gives us moral guidance and not just because the actual cities in which we might live are corrupt (*De Otio* 8), but because this is part of what we are *for*. This strand in Stoic thinking may seem to be somehow a foreign graft, since we see it so clearly in Plato and Aristotle. But not only is this attitude also reflected in other Stoic sources, it is in itself no less Stoic than Aristo's complete rejection of physics and logic, a view which Seneca resisted in *Letter* 95. 10–11, where Seneca again claims that philosophy is both contemplative and active and insists on the unlimited scope of human enquiry.[30]

Seneca recognizes two quite different reasons for studying physics and for attending to its doctrines, reasons which might well seem to be opposed to each other. Studying physics provides direct instrumental support to what we might call the enterprise of ethics, but it also fulfils something very important and

[29] Compare *Off.* 1. 13 *veri videndi cupiditas* above.

[30] Seneca's willingness to invoke Lucretius here is more an indication of his keenness to find Latin literary authorities for his views than any real indication that he is here restricting himself to Epicurean motivations for doing physics. Seneca does go on in *Letter* 95.12 to claim as well that general and otherwise disinterested pursuit of theory serves to support and stabilize our moral endeavour. The argument seems to go even further, if we are to take it literally. Philosophy is contemplative and active (95. 10); since it is contemplative it has its own *decreta* (95. 11), which clearly include doctrines about nature; these *decreta* stabilize ethics (95. 12) and even protect it. Even more: 'The difference between philosophy's doctrines and its precepts is the same as that between the elements and our limbs—the latter depend on the former; the former are the causes of the latter as they are of all things.' Although the *decreta* discussed in *Letters* 94 and 95 are wide-ranging and include some purely moral doctrines, in this context Seneca has been talking about only one kind of *decretum*, the kind that you can also find in astronomy and geometry and deal with cosmology as well as the more mundane aspects of physics (95. 10).

fundamental in our natures, the built-in drive for contemplation of nature. We see these motivations at greater length and even more explicitly in the final two texts I wish to consider, the *Natural Questions* and the later books of the *Letters to Lucilius*. These are long texts, so I will concentrate on the less familiar aspects of this double motivation.

In the preface to book 1 of the *NQ* Seneca makes grand claims for the branch of philosophy which deals with the gods—and theology is, of course, a branch of physics. Theology is as far above the philosophical study of human beings as philosophy itself is above other arts.[31] The *NQ* as a whole celebrates knowledge of nature, but this preface urges over and over again that it is a divine activity to understand the hidden secrets of nature, in particular subjects like astronomy, the nature of matter, theology, and other obscure topics like the details of meteorology. The prospect of this knowledge is far more important than the merely mundane business of getting one's morality straightened out. Seneca thanks nature for making this kind of discovery possible and even says that without being admitted to these secrets it would not be worthwhile to have been born (1 pref. 2–4).[32] Seeing the humble place of man in nature gives us perspective: our moral victories are modest when compared with the world of nature and its perfections (1 pref. 5–6) and the value of moral perfection lies in the fact that moral virtue prepares the mind for making astronomical discoveries and so taking our places beside the gods (1 pref. 6–8). The preface concludes with the claim that the ultimate value of physics lies in its ability to help us to transcend our mortality (1 pref. 16–17).

These are heady claims, hyperbolic perhaps, but not un-Stoic. Though strongly phrased, the basic picture Seneca paints is one he has given elsewhere. The emphasis on the need for leisure in order to indulge this pursuit of transcendence returns at 2. 53. 3 (*ad haec vacui revertemur*) and when Lucilius demands a more practical reason for studying physical subtleties at 2. 59 Seneca is able to give him the kind of answer he wants. It is the realization that our mortal life is small in comparison with the larger world to which we belong that enables us to deal with the miseries of this life. This is the practical benefit which flows from attending to our links with the divine. In the *Natural Questions*, then, two distinct views are taken about the value of physics, both that it is an intrinsic value and that *because of* the intrinsic value we can get practical value out of it when we consider the vicissitudes of our lives. The complexity of the case for studying physics is summed up at 3 pref. 18:[33]

[31] For this theme cf. also *Letter* 90, in which Seneca breaks with Posidonius in order to insist that merely useful arts and technologies cannot be attributed to philosophy.

[32] Cf. *Ad Marciam* 18 discussed above.

[33] The notion that otherwise obscure topics benefit us due to their role in simple mental exercise is one that Seneca takes seriously. In *Letter* 113. 1, though, Seneca makes it very clear that 'giving our wits a workout' is not a justification for boundless indulgence in overly specialized topics. See also *Letter* 117. 19.

> Here is how the study of the nature of things will benefit us. First, we will retreat from sordid matters; then we will withdraw the mind itself. . . from the body; next, when our intellect is trained on hidden obscurities it will be all the more able at dealing with obvious matters; but nothing is more out in the open than those healthy lessons which we learn as a defence against our wickedness and madness—conditions which we condemn but retain.

We begin from the simple appeal of avoiding what is lowly and dirty (*discedemus a sordidis*) and move on to the more familiar justification of physics in terms of its moral payoff. This dual motivation has already appeared in this preface. When asking rhetorically what is really the most important thing in life (*quid praecipuum*) Seneca says first 'to have seen everything in one's mind and . . . to have conquered one's vices' (3 pref. 10). Physics serves both ends.[34]

In the *Letters to Lucilius* we see a similar range of attitudes. Letter 65 (on the nature of causes) is perhaps the most interesting. In 65. 15 Seneca allows Lucilius to challenge him on whether there is any point in probing these questions. Is there, he asks, any moral payoff in these more abstruse physical questions? 'You say, "What pleasure to do you take in wasting time on those issues, ones that do not strip you of any passion or ward off any desire?" ' This gives Seneca the opportunity to articulate his thoughts on the issue. Sections 16 to 24 are occupied with his reply and are worth closer study than I can give them here. In summary, Seneca claims that he is dealing with morally important questions. He begins (like Socrates) from human affairs and *only then* moves on to cosmological and physical science (65. 15). And even this more advanced study is not a waste of time, for the main danger there is *overspecialization*. Providing one avoids that risk[35] cosmology is intellectually liberating—it affiliates our activities with the mind rather than the body and since the body is such a burden *any* theoretical abstraction is a recreational relief. Our nature is rewarded and fulfilled by taking refuge in theoretical enquiries. In 65. 18 Seneca echoes Socrates' acceptance of the Pythagorean idea that life is like a tour of military service—not our genuine calling as free intellects but still, for all that, our duty. 'Like a soldier under oath he thinks of this life as a tour of duty; and he has been trained to neither love nor hate life, and he puts up with mortal matters though he knows that higher things await him.'

[34] As 3 pref. 11 shows, it is the awareness of where we stand relative to the gods that enables us to rise above our failings and defects, which are, then, dealt with extensively until the end of 17. In books 4a and 4b Seneca gives Lucilius more personally targeted arguments. At 4a pref. 20–2 Seneca presents physical enquiry as a welcome and uplifting respite from his worries, but by 4b. 13 Lucilius is imagined as demanding a justification for the apparently needless detail (*ineptiae*) of meteorology—this is a move which clearly anticipates several of the letters—and Seneca's response to the challenge is straightforward moral utilitarianism. And yet at the beginning of book 7 quite explicitly (7. 1) and by implication at its end (7. 32) Seneca again puts far greater emphasis on the intrinsic value of enquiry for its own sake.

[35] Compare this to Chrysippus' caution about a hedonistic approach to theoretical studies: *SVF* 3. 702 = *De Stoic. Rep.* 1033d = LS 67X and n. 25 above.

Hence, Seneca argues, the investigation of nature, even the really abstruse aspects of cosmology, is a kind of reunion with one's genuine nature. A narrow focus on ethics would stunt our nature.

Do you ban me from an investigation of nature, drag me away from the whole and confine me to a part. Shall I not investigate the principles of all things? Who gave them form? Who made distinctions among things which were melded into one and enmeshed in passive matter? Shall I not enquire who is the artisan of this cosmos? How so great a mass was reduced to lawlike structure? Who gathered the scattered bits, who separated what was combined and brought shape to things lying in unsightly neglect? Where did this great light come from? Is it fire or something brighter than fire? Shall I not ask these questions? Shall I remain ignorant of my origins? Am I to see these things just once or am I to be born many times? Where am I to go from here? What residence awaits the soul when it is freed from the laws of human servitude? You forbid me to meddle with the heavens, i.e. you order me to live with bowed head. (Seneca, *Letter* 65. 19–20)

Seneca claims not only that doing cosmology fulfils our nature by integrating us with the divine, but that devoting effort to it liberates us *from* the body by changing our attitudes *to* the body:

I am greater than that and born for greater things than to be a slave to my body, which I think of as no different than a chain fastened about my freedom. So I position it as a defence against fortune, so that she will stop right there; I permit no wound to get through the body to *me*. This is the only part of me which can suffer wrongs. A free mind lives in this vulnerable dwelling. . . . to despise one's body is a reliable freedom. (65. 21–2)

The defence of serious cosmological activity concludes with an assertion of the close connection of god and our mind (in contrast to the body) and the way that an awareness of our place in the cosmos enables us to put misfortune and death into perspective (65. 23–4). Here we can see with particular clarity the close relationship between the intrinsic value claimed for the study of physics and its alleged moral payoff—as we have seen, these motiviations do not exclude each other.[36]

We see a similar combination of motivations in *Letter* 58. At section 25 Lucilius is imagined as asking 'What good will this technicality do for me?' and Seneca replies 'None, if you ask me. But just as the engraver relaxes, refreshes

[36] But, as Ricardo Salles asked in private correspondence, do these motivations go beyond mere compatibility? In a way, yes. For as suggested here the intrinsic pursuit of knowledge contributes to an understanding of our proper place in nature and this helps to put the miseries of human life into perspective and improves our understanding of the body–soul relation in a way that matters to our decisions about life. Seneca is certainly aware of this connection between intrinsic motivation and the advantage which derives from a 'proper' understanding of physics and cosmology. Yet Seneca is not as explicit as one would want about this connection and it would be a mild exaggeration to claim that this is part of his 'theory' on the question. Such suggestiveness when we might prefer overt theorization is something which readers of Seneca eventually come to recognize as a feature of his philosophical method.

and, as they say, "nourishes" his eyes, tired from lengthy concentration, so too we should sometimes relax our mind and refresh it with certain amusements. But let the amusements themselves be work and from them too, if you pay attention, you will gain something which could turn out to be good for you.'

Here technical philosophical enquiry that is not justified by any moral payoff is clearly distinguished from the anticipated practical benefit. As Seneca goes on to say in 58. 26, his normal practice is to ferret out possible payoffs from even the least likely sources:

> This, Lucilius, is what I normally do: from every notion, even if it is quite remote from philosophy, I try to dig out something and make it useful. What is more remote from the improvement of our habits than the discourse I just gave? How can the Platonic ideas make me better? What could I derive from them that might control my desires?

Seneca admits (agreeing, perhaps with Aristotle's reflections about the form of the good in *NE* 1) that Platonic metaphysics is a pretty unlikely source of direct moral benefit. But even that can, with sufficient ingenuity and hard work, yield a genuine payoff and in the sections which follow (58. 26–36) he gives his readers a tour-de-force of indirect moral benefits. First, Platonic metaphysics can show us that the most tempting of stimuli are all the more resistible because they are in fact not *really* real. The transience of material things makes them unsuitable objects of our ultimate desire; the transience of our own lives is a similar reminder that we should not be overly attached to our bodies. Once we see all of this we should be able to evaluate them more thoughtfully. There is a similar lesson to be learnt about the way our minds can control and benefit our bodies if we only have the sense to put the mind in charge. Even if we cared about nothing but personal longevity, then even on that narrow assumption a Platonic control of the body by the mind could deliver the goods.

Letter 124 contains a similar set of ideas, though it would take too long to develop it in detail. At the beginning (124. 1) Seneca asserts the importance of avoiding the extremes of technicality and welcomes the constraint imposed by the expectation of some moral payoff from the investigation of technical questions (this one happens to be metaphysical rather than cosmological, but that does not make a significant difference). Then at the end of the letter (124. 21) Lucilius is again envisaged as challenging the relevance of Seneca's discussion. One benefit of the activity is simple mental exercise: 'It exercises and sharpens the mind and, at the least, since it is bound to be doing something in any case, keeps it busy with an honourable employment. And it is also beneficial in that it slows down people who are rushing into moral error.' Fair enough, but he also goes on to make a more specific doctrinal argument, showing how an awareness of our similarity of nature with god as opposed to animals changes things for us, transforms the priorities that we adopt for our own lives; here, perhaps more than anywhere else, the divinity of our nature and its kinship to god seem to have a distinct motivating role. This very direct and doctrinal moral payoff accompanies the indirect benefits of mental exercise and distraction.

In Letter 117, also devoted to a topic which Seneca admits is potentially overspecialized (whether the good is a body), shows the same complicated and balanced approach to technical questions in physics. After airing the issue and disagreeing with his fellow Stoics at some length, Seneca hauls himself up short in 117. 18: 'For some time now I have been condemning myself and behaving like those whom I criticize, wasting words on an obvious issue. Who could be in any doubt that if heat is bad then being hot is bad? If cold is bad then being cold is bad? If life is good then living is good? All of that concerns wisdom but is not *in* wisdom. But we must spend our time *in* wisdom.' But even once he introduces the need to practise wisdom rather than just talk about it, Seneca does not immediately insist on direct moral utility (117. 19):

> Even if we want to digress a bit, wisdom has lots of room for quiet retreats. Let us investigate the nature of the gods, the nourishment of the heavenly bodies, the various paths of the stars, whether our affairs are moved in accordance with their motions, whether they are the source of movement for the bodies and souls of all things, whether even the things which are called fortuitous are actually bound by a definite law and nothing in this cosmos unfolds without warning or without order. These issues are already somewhat removed from the education of our characters, but they do uplift the mind and draw it towards the grandeur of the very things which it is considering. But the issues which I was discussing just a moment ago reduce the mind and degrade it. They do not, as you people think, sharpen the mind; they just make it thinner.

Here Seneca acknowledges the importance of the recreation of the mind by the doing of physics (as in letter 65) and marks the boundary between that motivation for physics and other issues for which even the argument from mental training does not suffice. In 117. 20 he identifies some topics which have intrinsic merit and claims that we *owe* those topics a certain amount of attention, so that diverting our energies to less valuable themes is a serious waste of resources. This theme continues at the end of the letter, after an extended (117. 20–31) discussion of how important the direct moral payoff is. At 117. 32–3, though, the theme of mental leisure and our responsibilities to use our time well returns, and the letter closes with a final reiteration of the importance of using our theoretical enquiry for the overall purpose of achieving moral improvement.

> Nature did not give us such a generous supply of free time that we have the luxury of letting any of it go to waste. And consider how much is lost even to those who are most careful; some is taken from each of us by our own health, some by the health of our friends and family; some is taken up by unavoidable business, some by public affairs; sleep takes its share of our lives. With such a limited and fast-moving supply of time, time which sweeps us away, what good does it do to squander pointlessly the majority of it? And add to this the fact that the mind is in the habit of amusing itself rather than healing itself and turning philosophy into a leisure activity when it is really a cure. I do not know what the difference is between wisdom and being wise. But I do know that it

makes no difference to me whether I know or not. Tell me, when I have learnt what the difference is between wisdom and being wise, will I be wise? Why, then, do you tie me down with the words of wisdom instead of with its deeds? Make me braver, make me more confident, make me equal to fortune, make me superior to it. But I *can* be superior if I direct all of my learning to that end.

The final letter to consider is perhaps one of the best known. In *Letter* 121 Seneca gives us one of our best and most detailed accounts of the theory of *oikeiōsis*. The doctrine figures importantly in discussions of Stoic ethics; on some accounts the doctrine appears as one of the more important arguments for parts of Stoic moral theory—Hierocles, for example, gave his account of *oikeiōsis* in men and animals at the beginning of a work entitled *Foundations of Ethics* and it features prominently in the opening sections of both Cicero's and Diogenes Laertius' accounts of Stoic ethics. Let's look at how Seneca introduces the theme (121. 1–3):

I can see that you will haul me into court when I set out for you today's little question, one that has engaged us for quite a while now. Once again you will shout, 'What does this have to do with ethics?' Shout away, then, while I, first of all, give you other opponents to prosecute, Posidonius and Archedemus (they'll accept the court's jurisdiction), and then say to you, 'it is not the case that everything which is ethical makes our character ethically good'. Some things bear on human nutrition, some on exercise, some on clothing, some on teaching, some on pleasure. But they all bear on human beings even if not all of them make humans better. Different things have different impacts on our character. Some things improve our character and make it orderly, while others investigate the nature and origin of our characters. When <I ask> why nature made humans, why she made us superior to the rest of the animals, do you think I have left character far behind? Not so. For how will you know what character you should have unless you find out what is best for a human being, unless you look into its nature. You won't really understand what you should do and what you should avoid until you have learnt what you owe to your own nature.

This topic seems to require some justification—he later describes it as 'a little more removed from our concerns'—but the justification seems quite easy to formulate. Here as elsewhere Seneca acknowledges a kind of instrumental motivation for studying physics. Knowing one's own nature is necessary for knowing what to do, and knowing human and animal nature is necessary for knowing one's own personal nature. To know oneself had always been taken as a necessary condition for moral success.[37] Seneca and other Stoics take an expansive view of what is required to learn that. Hence some topics are 'ethical' even though they don't directly improve us. When we investigate the nature and origin of human characters, we are contributing to ethics even though we do not thereby make our characters better. A knowledge of human nature is

[37] As Socrates pointedly says at *Phaedrus* 230a while justifying his lack of interest in investigating topics that are alien to his interests (*ta allotria*) until he has at least learnt about his own nature.

essential for determining what is best for a human being and that contributes to our knowledge of what our character should be. Knowing what we owe to our nature is an at least necessary condition for seeing what to pursue and what to avoid and how to live one's life.

John Cooper has taken a strong line about philosophical respectability with regard to logic and physics and finds Seneca seriously wanting.[38] He paints a picture of Seneca relying too much on rhetoric and not enough on reason-giving and conventional theorizing. Through many letters, he says, 'there runs a common thread: the uselessness of attending to or knowing about certain matters of philosophical theory that have no *palpable and immediate* ethical consequences'. I think that Cooper has given us too simple a characterization of Seneca's position and that he underestimates both Seneca *and* earlier Stoicism.

Too simple, because Cooper takes insufficient notice of Seneca's position in the history of his own tradition and pays too little attention to the complexity of Seneca's authorial strategy. As an amateur in a world of professional philosophizing he is often reflecting from the outside on the issue of whether there are or should be limits on the degree of technical detail in philosophical activity aimed at relevance to life. As an author, he often does what Plato also often does; he develops and exploits a character's point of view specifically in order to raise an issue and explore it, without necessarily having an unambiguous commitment to that position himself. In the letters and 'dialogues' his most compelling character is Seneca himself, but that should not license us to conflate the author and the character in his case any more than we would with Plato. Cooper also assumes a more or less uniform picture of early and orthodox Stoicism. Yet, on the very issue which engages us now, we have seen that many views coexisted within the school, even the minimalist Cynicizing view of Ariston of Chios who clearly saw himself (as others saw him) as being Socratic rather than un-Stoic. Failure of nerve and apostasy from the school can hardly be laid at the door of someone who felt drawn intellectually to Socrates, to his school's founder, and to one of Zeno's two most prominent students. It was not an easy task to carve out his own position between the extremes defined by Aristo and Chrysippus; Seneca's attempt to do so may not have led to major philosophical breakthroughs, but it does deserve to be recognized for what it was.

But does Seneca's position on the issue 'Why physics?' have any philosophical interest for us *at all*? I think it does. Provisionally, I think we can identify two distinct answers to this challenge. Why physics? One answer is, 'in order to know nature, our own and the rest of it, as a means to the end of living well'. For this instrumental purpose it may well be true that there are limits to the level of detail one needs and the depth of scientific exploration which it is appropriate for anyone but a technical wizard to engage in. The other answer is, 'because we just are built for knowing things and the world of nature is the grandest and best

[38] Cooper 2004: esp. p. 333.

possible object of knowledge'. Like Aristotle, Seneca and other Stoics did think that actualizing our nature is a self-justifying thing to do *and* that we do it best when we come to know the best possible object. In short, we do it because any normal human being does just love to do it.

So we need to do physics to learn to navigate in the moral world and we want to do physics because, well, it is in us to do so, to explore and savour the body of knowledge which reflects our best selves and our kinship with the gods. And for this latter purpose there does not seem to be and indeed ought not to be any limitation on how deep and thorough our knowledge should be. And sure enough, when Seneca is focused on the instrumental case for physics he emphasizes the limitations and (as we have seen) when he focuses on the intrinsic worth of the study he does not. And it is when he is emphasizing the intrinsic worth of physics as an object of study for humans that the theological side becomes most pertinent. It is the divinity of the object of our study and the divinity in our own rational nature which help to make 'big picture' physics such a natural (that is, self-justifying) fit for human beings.

Let me close, though, with a reminder. Seneca also notices a connection between these two motivations, something more than mere compatibility and pragmatic convergence. He thought that the study of physics, especially but not only theology and cosmology, revealed something critically important about our place in the universe. As physical entities we are small and transitory. As thinking beings, as rational animals, we have no limits except those of the divine cosmos. Faced with a choice between allegiance to something limited and transitory and allegiance to something enormous and long-lasting, we ought to know what to do. And Seneca, like Plato and like Aristotle, and indeed like many cosmologists and physicists even today, did know what to do. The unremitting and dedicated study of physics calls to us as rational animals in a way that puts our whole life in perspective and provides a rich and inclusive framework for all those little issues (like my life as a whole and my death) that ethics so mundanely and consistently deals with.

REFERENCES

Annas, Julia (2007) 'Ethics in Stoic Philosophy', *Phronesis,* 52: 58–87.

Betegh, Gábor (2003) 'Cosmological Ethics in the *Timaeus* and Early Stoicism', *Oxford Studies in Ancient Philosophy,* 24: 273–302.

Brennan, Tad (2005) *The Stoic Life* (Oxford: OUP).

Cooper, John (2004) 'Moral Theory and Improvement: Seneca', *Knowledge, Nature and the Good: Essays on Ancient Philosophy* (Princeton: Princeton UP), ch. 12.

Inwood, Brad (2005) *Reading Seneca* (Oxford: OUP).

Ioppolo, Anna Maria (1980) *Aristone di Chio e lo stoicismo antico* (Naples: Bibliopolis).

Ludlam, Ivor (2003) 'Two Long-Running Stoic Myths: A Centralized Orthodox Stoic School and Stoic Scholarchs', *Elenchos,* 24: 33–55.

Menn, Stephen (1995) 'Physics as a Virtue', *Proceedings of the Boston Area Colloquium in Ancient Philosophy*, 11: 1–34.

Nussbaum, Martha (1995) 'Commentary on Menn', *Proceedings of the Boston Area Colloquium in Ancient Philosophy*, 11: 35–45.

Rist, John (1969) *Stoic Philosophy* (Cambridge: CUP).

Sedley, David (1989) 'Philosophical Allegiance in the Greco-Roman World', in M. Griffin and J. Barnes (eds.), *Philosophia Togata* (Oxford: OUP), ch. 4.

——(2003) 'The School, from Zeno to Arius Didymus', in B. Inwood (ed.), *The Cambridge Companion to the Stoics* (Cambridge: CUP), ch. 1.

White, Nicholas (1985) 'The Role of Physics in Stoic Ethics', *Southern Journal of Philosophy*, 23 suppl.: 57–74.

9

Stoic Philosophical Theology and Graeco-Roman Religion

Keimpe Algra

1

One of Horace's *Satires* describes the poet's journey from Rome to Brindisi. At the end of the poem we are told that the inhabitants of the village of Gnatia, which is near Brindisi, claimed that at its temple threshold frankincense could melt without fire. Horace's reaction is straightforward and clear: 'Let the Jew Apella believe this; not I. For I have learnt that the gods lead a care-free life, and if nature produces anything marvellous, it is not sent down by surly gods.'[1] Here a local religious or quasi-religious—or perhaps we should merely say folkloristic—belief is regarded through the spectacles of philosophy, in this case Epicurean philosophy. Criticizing or reinterpreting elements of traditional religion was not new with Epicureanism. Attempts to assess religion from the point of view of philosophy, or to explain its existence and origins, go back at least to the period of the Sophists. And in so far as the philosophers claimed to be able to pronounce themselves authoritatively on the nature of god or gods, i.e. in so far as they came up with what we might call a philosophical theology, the question of the relation between this 'natural' theology and the complex edifice of the tradition was bound to arise in various ways.[2]

The contents of this article partly overlap with my *Conceptions and Images: Hellenistic Philosophical Theology and Traditional Religion* (limited edn.: Amsterdam: Dutch Royal Academy of Sciences (KNAW), 2007). I have greatly benefited from the constructive comments and suggestions made by my Utrecht colleagues Jaap Mansfeld, Teun Tieleman, Maarten van Houte, Claartje van Sijl, and Irene Conradie and by the participants of the conference in Mexico City.

[1] Horace, *Sat.* 1. 5. 100–4; the words *deos didici securum agere aevum* are a quotation from Lucretius 5. 82.

[2] On the use of the term 'natural' theology in this connection, see Algra (2004: 173 n. 1). By 'the complex edifice of the tradition' I mean the broad and flexible amalgam of cult practices, beliefs,

In this connection Stoicism presents a particularly interesting case. On the one hand Stoic cosmo-theology was itself much more than just a value-free theory about the world. Chrysippus seems to have warned explicitly against regarding philosophy (including theology) as just an intellectual exercise or pastime.[3] The moral and psychological commitment involved on the part of adherents of Stoicism came close to, and hence was likely to some extent to compete with, the kind of commitment required by religion. At the same time we know that the Stoics were willing to link their philosophical monotheism or pantheism with at least parts of traditional polytheism. They were prepared to call their one cosmic god by many traditional names—Zeus, Hera, etc.—thus calling forth further reflection on the question of the relation between philosophy and traditional religion.

This question could be rephrased in terms of what has become known as *theologia tripertita*, a concept usually connected with the name of the Roman first-century BC antiquarian Varro. In his *De Civitate Dei* Augustine provides a paraphrase of Varro's argument, which claims that there are basically three kinds of theology, with three corresponding types of gods: the theology of the city, i.e. of traditional cult, the theology of myth, i.e. the traditional stories, and the theology of the philosophers: 'They call the theology that is used chiefly by poets "mythical", that used by philosophers "physical", and that used by city-states "civil"' (Augustine, *Civ. Dei* 6. 5).

Varro himself (as paraphrased by Augustine, *Civ. Dei* 6. 6) appears to have argued that the theology 'of the city', i.e. traditional cult, contained elements of both myth and natural theology, being thus a mixture of the serious and the frivolous. Augustine disagrees: the gods of pagan cult are basically identical with, i.e. no better than, the gods of myth.[4]

Although Varro became the eponymous hero of this tripartition, the idea itself was not original with him. A number of sources—some of them going back at least to early Hellenistic times—provide independent evidence for the currency of this distinction, which seems to a large extent to have reflected actual practice, i.e. a real division of territory between myth, cult, and philosophy.[5] In principle there are three ways in which the philosophical part of the *theologia tripertita*—i.e. 'physical' or philosophical

and myths, with a special role being accorded to ancient poets like Homer, Hesiod, and the poets from the Orphic tradition.

[3] See Chrysippus *ap*. Plutarch, *SR* 1033c–d (*SVF* 3. 702).

[4] Augustine, *Civ. Dei* 6. 7: 'Quid enim aliud ostendunt illa simulacra formae aetates sexus habitus deorum? Numquid barbatum Iovem, imberbem Mercurium poetae habent, pontifices non habent?'

[5] On the concept of the *theologia tripertita* in Varro and in the preceding tradition see Lieberg (1982).

theology—could position itself *vis-à-vis* traditional religion (i.e. 'mythical' and 'civil' theology):

(1) by providing a rational, independent, alternative to the conceptions of the gods that can be found in myth or cult, without explicitly criticizing the tradition;

(2) by explicitly criticizing (aspects of) the conceptions of the gods that can be found in myth or cult; or

(3) by appropriating or reinterpreting, and thus to a greater or lesser extent taking over, the conceptions of the gods that can be found in myth or cult.

These approaches differ in virtue of the various kinds of connection they establish between philosophical theology and the tradition: no explicit connection (as in case 1), a negative connection (case 2), or a positive connection (case 3).[6] As a classification of approaches this scheme may serve as a useful analytical framework. Yet it should as such not hide the complexities of historical reality. For one thing the boundaries between philosophy and religion are not always easy to draw, especially where in the earlier period philosophy was not a clearly defined category at all.[7] For another, as the example of Heraclitus (but also, as we shall see, that of Stoicism) shows, one and the same author could in principle combine the appropriation of some elements of traditional religion with the outright rejection of others. Finally, the boundaries between the three approaches outlined above are themselves not always clear. Given the fact that a philosophical theology, or a philosophical religion, will almost inevitably borrow some of its language and concepts from the tradition, the borderline between an *alternative* philosophical religion (i.e. what I labelled approach 1) and a philosophical *reinterpretation* of traditional religion (approach 3) is

[6] Examples of (1): the earliest Ionian philosophers (who seem to have introduced a new, cosmological, conception of the divine without, for all we know, disparaging traditional religion) and Aristotle (in so far as he seems to have believed to be able to combine his philosophical conception of god as a first unmoved mover with a positive commitment to the cult of the traditional gods); examples of (2): Xenophanes' attacks on anthropomorphic conceptions of the gods and Heraclitus' critique of various aspects of traditional cult and worship, such as the habit of praying to cult statues; examples of (3): Heraclitus' reinterpretation of some rituals connected with the cult of Dionysus, which he appears to explain, by means of etymologization, as intimating the identity of Dionysus and Hades, i.e. presumably of life and death (fr. DK 22 B 15), and the Derveni papyrus in which some Orphic ritual practices and an Orphic poem are interpreted in a wider, physical or cosmological, context. On Heraclitus and religion, see Osborne (1997), and Adomenas (1999). For text, translation, commentary, and an extensive study of the religious and philosophical aspects of the Derveni papyrus, see now Betegh (2004: esp. 349–72) on the status of the text as a 'commentary'.

[7] Thus it appears that the author of the Derveni papyrus is best seen, not as a philosopher (whatever that may have meant at the time), but as a kind of priestly figure—explaining the tradition by taking the perspective of, and applying elements taken from, contemporary philosophy. For some of the problems involved in using the concept of 'philosophy' for the Presocratics, and for further references, see Curd (2002).

not always easy to draw. How, for example, should we classify Cleanthes' hymn to Zeus?

In the Hellenistic period the debate on the relation between the three divisions of the *theologia tripertita* appears to have intensified. In part this may have been due to the fact that philosophers now had a larger audience—in Cicero's day, for example, many members of the Roman elite claimed allegiance to one of the philosophical schools—and that philosophy itself was seen more and more as an all-encompassing 'way of life' available to every educated citizen, rather than as an 'arm-chair' pastime for specialists.[8] For the Stoics especially, as already noted above, philosophy was more than just a matter of wearing the right clothes (e.g. the philosopher's cloak) or adopting the right lifestyle (*diagôgê*) of intellectual curiosity. It was a matter of taking on a new view of oneself and of the world. Accordingly, the question of how to combine such a philosophical world view with the concepts, rites, and duties involved in traditional religion will have become more pressing, especially in a Roman context, where the relation between Greek philosophy and the indigenous *mos maiorum* had been a controversial issue from the start.[9] This is consistent with the fact that one of the recurrent themes in the polemics between Stoics and Epicureans in the first century BC, to which both Cicero's *philosophica* and the surviving fragments of Philodemus' works testify, is the extent to which each of these schools was able to make sense of, or to salvage, the tradition—be it the sociopolitical or the religious tradition.[10]

In this chapter I want to examine some aspects of the philosophical attitude of the Stoics, early as well as later, *vis-à-vis* traditional religion. I shall first address some preliminary issues, such as the epistemological basis of Stoic philosophical theology (section 2), and the Stoic view or views of the development of human civilization (sections 3 and 4). In the course of this discussion we shall be mainly dealing with the Stoic attitude towards *myth*. Whereas some aspects of this attitude—in particular the Stoic practice of etymologization and allegorizing readings—have been given considerable attention in recent scholarly literature,[11] this does not apply to the subject of sections 5 and 6, where I shall turn to the

[8] Cf. Sedley 1989; Hadot 1993: 291–304.

[9] On the tension between Greek philosophy and the *mos maiorum* in Rome, the tendency to subordinate the former to the latter, and the notorious expulsions of philosophers in 161 and (probably) 154 BC, see Lévy (1996: 14–15).

[10] See e.g. Philodemus *Piet.* col. 10, line 8–col. 11, line 5 (ed. Henrichs 1974: 20–1), where the Stoics are accused of acknowledging only one god, if any god at all, and 'not leaving us those gods of the form like that in which they are universally worshipped', whereas the Epicureans themselves claim to allow 'as many gods as the Hellenic people affirm, but also many more'. See on this passage Obbink (2002: 209–10), and my comments in Algra (2003*b*). See also the remains of Philodemus' *De Stoicis*, where one of the points being debated is which of the two schools, Stoicism or Epicureanism, manages to stay closer to conventional morality, and where Philodemus replies to attacks on Epicurean hedonism, and its allegedly scandalous and anti-social consequences, by pointing to the equally shocking quasi-cynicism of Zeno's *Politeia*.

[11] See e.g. Hadot 1987; Long 1992; Algra 2001; Brisson 2004; Gourinat 2005.

Stoics' attitude towards various aspects of traditional *cult*.[12] Conclusions will be drawn in section 7.

2

One of the things that makes Stoic philosophical theology philosophical is its epistemological basis. It is a theology based on what are taken to be secure and evident starting points. Whereas ancient sceptics—both the Academic sceptics and the Neopyrrhoneans—argued that no such starting points were available, that rational philosophical theology was an impossibility,[13] and that we should accept the tradition, simply because it is the tradition (or because it is convenient),[14] the Stoics boasted a foundationalist epistemology offering a secure basis for their theology: the natural concept, or preconception (*prolêpsis*) of god, which in principle any human being was capable of forming on the basis (directly or indirectly) of experience. According to the Stoics the natural conception of god is somehow formed in the minds of men on the basis of repeated experience of the world around them and its rational structure, more particularly as a result of a feeling of awe *vis-à-vis* celestial phenomena or a feeling of gratitude for the good things in life.[15] The fact that the Stoics thus did not confine their criteria of truth to what can be observed in the strict sense, but granted a criterial status to preconceptions as well, sets their system off from later forms of empiricism. It allows them to make what they take to be well-founded assertions about the existence of the gods—something which, for example, would be anathema to someone like Hume or to philosophers of the Wiener Kreis. To put it differently,

[12] An older but still useful overview, which restricts itself to 2nd-cent. AD theories, is Clerc (1915).

[13] In the case of the Academics (see Cicero, *ND* 3): because the arguments of their chosen opponents (mainly the Stoics) could be shown to be inconclusive; in the case of Neopyrrhonean scepticism (see Sextus, *M* 9.1–191): because all attempts to arrive at a rational defence of either the existence or the non-existence of god or gods could be shown to be inconclusive, so that a general *epochê* should result.

[14] See Cicero *ND* 3. 6 where the Academic spokesman Cotta is made to say: 'You [i.e. the Stoic Balbus] are a philosopher, and I ought to receive from you a rational account of your religion, whereas I must believe the word of our ancestors even without any rational account (*nulla ratione reddita*).' For the Neopyrrhonean counterpart to this position, see Sextus, *M* 9.49 'For perhaps the Sceptic, as compared with philosophers of other views, will be found in a safer position, since in conformity with his ancestral customs and the laws, he declares that the gods exist, and performs everything which contributes to their worship and veneration, but, so far as regards philosophical investigation (ὅσον ἐπὶ τῇ φιλοσόφῳ ζητήσει) he declines to commit himself rashly.' In these cases, in other words, the tradition was salvaged by showing the epistemological impossibility of a rational alternative. According to the sceptics there could be no connection between the accepted tradition and philosophical *truth* of any kind.

[15] On the aetiology of man's conception of god or gods, see Cleanthes *ap*. Cicero *ND* 2. 13–15 (= *SVF* 1. 528); Aetius 1. 6 (= *SVF* 2. 1009); and Dio Chrysostomus, *Or*. 12, discussed below, pp. 245–7. See also Algra (2003*a*: 158) on Persaeus' reinterpretation of Prodicus' suggestion that the men of old considered as gods those who had discovered what was useful and beneficial.

the Stoic brand of empiricism did not involve the claim that the 'verificationist challenge' to theology could only be met by adducing strictly empirical evidence for god's existence; the evidence of natural conceptions, that were formed on the basis of experience and that were supposed to be reliable in virtue of the structural affinity between human rationality and the rationality of the world, was also given a rightful place.

According to the Stoics our natural conception of god involves an entity who is emphatically not human in shape, but nevertheless rational and providential. It is perhaps not too bold to assume that the contents of the preconception of god according to the Stoics can be summarized by the brief description provided by Diogenes Laertius (7. 147): 'an immortal living being, rational, perfect and thinking in happiness, unreceptive of anything bad and provident with regard to the cosmos and the things therein'.[16] However, such a preconception provides only the basic characteristics of god. It does not tell us where this god is to be found, or how he relates to the rest of the world. Moreover in the course of one's lifetime—and indeed in the course of the history of humankind—the right preconception itself may well become blurred or partly blotted out by the intrusion of mistaken beliefs. This is why we need philosophical theology with its various proofs concerning the existence and nature of the gods—proofs which may be said to confirm, strengthen, and further articulate our original preconception.[17] The Stoic versions of what we nowadays call the 'argument from design', for example, may be seen as attempts to lay out explicitly the kind of implicit argument that is supposed to be at the basis of our preconception.

The foundationalist epistemology behind Stoic theology can be—and has indeed sometimes been—contrasted with the approach of Plato and the Platonic tradition. Plato's Socrates had claimed, in the *Phaedrus* (246c–d), that a divine attribute like immortality is something we basically just make up (*πλάττομεν*) and ascribe to the gods without any sound reasoning (*οὐδ' ἐξ ἑνὸς λόγου λελογισμένου*)—the idea behind this being that no sound and solid knowledge of god is possible for humans.[18] This quasi-scepticism *in theologicis* left its traces with later so-called Middle Platonists, who as a rule claimed that the nature of god was inaccessible.[19] The Stoics, by contrast, were in a position to claim (as indeed did the Epicureans) that not only god's or the gods' existence, but also their main attributes could be established beyond doubt on the basis of our preconception. Yet the contrast should not be overemphasized. The

[16] DL 7. 147: *Θεὸν δ' εἶναι ζῷον ἀθάνατον, λογικόν, τέλειον καὶ νοερὸν ἐν εὐδαιμονίᾳ, κακοῦ πάντος ἀνεπίδεκτον, προνοητικὸν κόσμου τε καὶ τῶν ἐν κόσμῳ.*

[17] On the Stoic 'proofs' in relation to preconception, see Schofield (1980: 305).

[18] Compare Plato's famous claim, *Tim.* 28c, that 'to discover the maker and father of this universe were a task indeed; and having discovered him, to declare him unto all men were a thing impossible'.

[19] See Runia 2002.

confidence on the Stoics' part merely concerns the main characteristics of god or gods.[20] The further job of articulating these basic features into a full-blown philosophical theology was believed to be a hard one: theology was conceived as a kind of initiation in the mysteries, to be put at the end of the philosophical curriculum. Neither the Stoics nor, for that matter, the Epicureans believed complete knowledge of the divine nature could easily be, or had in fact been, achieved.[21] Both Stoic and Epicurean theology thus combine what one might call epistemological optimism on the one hand and epistemological modesty on the other. For this reason the occurrence of expressions of quasi-scepticism or epistemological modesty *in theologicis* in such later Stoics as Seneca should not be regarded automatically as due to the influence of Platonism. They take up a strand that was present in Stoicism all along.

3

As we shall see in a moment, the epistemological basis of Stoic theology was relevant to the school's philosophical attitude towards traditional religion in its various forms. But first a possible objection to the claim that the Stoics addressed the issue philosophically at all has to be faced: doesn't at least some of the evidence indicate that the Stoics adhered to a simple non-reflective form of conservatism in so far as the practices of religion were concerned? In an article devoted to Cynicism and religion, M.-O. Goulet-Cazé characterized the early Stoic attitude, as exemplified by Chrysippus,[22] as 'extremely conservative', thus opposing it to the straightforward rejection of traditional religion by at least some Cynics.[23] It is certainly true that some sources suggest that at least in some contexts some Stoics were prepared to talk about philosophy and traditional religion as two separate realms, with the implication that the former could leave the latter completely intact. As an example we may adduce Epictetus. Having laid out the essence of true, philosophical, piety—consisting in having the right opinions about

[20] Accordingly, the claim in Runia (2002: 283) that 'Stoics and Epicureans argue with confidence about the nature of god (or the gods)' is in need of some qualification.

[21] See Plutarch *SR* 1035a–b (on Chrysippus relegating theology to the final stage of the curriculum); Epihanius, *Adv. Her.* 3. 2. 9 (*SVF* 1. 538, on Cleanthes); Etym. Magn. *s.v.* τελετή (*SVF* 2. 1008). On the Epicurean side Philodemus, *On Piety,* col. 23 Obbink, acknowledges the limitations of Epicurean theology, arguing that no one has as yet been able to offer absolute certainty about the nature of the gods.

[22] Cf. Goulet-Cazé (1996: 67). The texts on which she bases her claim—the common Stoic accounts of Stoic ethics in Stobaeus, probably derived from Arius Didymus (*SVF* 3. 604), and Diogenes Laertius (*SVF* 3. 608)—contain no name-label and hence cannot securely be ascribed specifically to Chrysippus, although their contents may in fact well go back to him.

[23] Perhaps Antisthenes fr. A 181 Giannantoni comes close to formulating what we might call the mainstream Cynic attitude: 'god is not known through an image, is not seen with the eyes, resembles nothing'; and 'according to tradition there are many gods, but according to nature there is only one'.

god and gods and submitting oneself to everything that happens—his *Manual* continues as follows:

But it is always appropriate to make libations, and sacrifices, and to give of the first fruits after the manner of our fathers (κατὰ τὰ πάτρια), and to do all this with purity, and not in a slovenly or careless fashion, nor, indeed, in a niggardly way, nor yet beyond our means. (*Ench.* 31. 5)

As we shall see, Seneca in some contexts also claims that the elements of philosophical truth on the one hand and the tradition on the other (i.e. what he calls *res* and *mos*) should coexist.[24] This is an attitude familiar from other ancient philosophers: overall, the religious tradition was strong and influential, and few people were prepared to question its value openly or explicitly.

Yet the nature and extent of this Stoic conservatism should, I think, be further specified. First of all, the evidence on which Goulet-Cazé bases her claims concerning Chrysippus does not straightforwardly vindicate the latter's acceptance of traditional religion in its traditional appearance. It consists of some fragments which claim that the sage will be a true priest, truly pious, etc. These statements, however, are specimens of a larger group of so-called *paradoxa* about the Stoic sage: *only* the sage is *truly* free, *truly* king, etc. What such statements claim is that, in virtue of his perfect rationality, only the sage is in the strictest sense entitled to these labels or predicates. Accordingly, what the remark about the sage as priest suggests is not that the sage will in the relevant respects perform the actions of ordinary priests, but that being the only one who is capable of *true* piety (piety being defined as 'knowledge of how to worship the gods', ἐπιστήμη θεῶν θεραπείας),[25] he alone will be able to act as a *true* priest, i.e. presumably by performing prayers and rites with the proper rational attitude. In other words, these fragments rather suggest a specific way in which traditional religion may be interpreted and cultivated by the sage, or even by the Stoic philosopher in general. They thus seem to hint at what one might call a two-tier theory (implying a different attitude on the part of ordinary people on the one hand, and of the philosophical elite on the other) of the kind we find in some later philosophers like Porphyry, rather than at an unqualified conservatism.[26]

Secondly, although Epictetus' suggestion of a peaceful *Nebeneinander* of philosophical theology and conventional religion may have worked for all practical purposes in most circumstances, we should not think that the Stoics altogether refrained from linking the two domains at a theoretical level. In fact

[24] Compare the words of the Stoic spokesman Balbus in Cicero *ND* 2. 71: 'It is our duty to revere and worship these gods under the names which custom has bestowed upon them.'

[25] For the definition see e.g. Sextus, *M* 9. 123 (*SVF* 2. 1017); see also the texts printed as *SVF* 3. 264, 273, and 604.

[26] One may compare the two different views of the legitimacy of animal sacrifice taken in Porphyry's *De Abstinentia*: it is allowed for the common man, whereas the true philosopher should steer clear of performing animal sacrifices altogether; on which see Bruit Zaidman (2001: 201–10).

we find examples of both philosophical critique and philosophical adaptation of elements of the traditions of myth, popular belief, and cult. It is here that the Stoic theory of preconceptions appears to provide the appropriate background, and that a connection with the issues discussed in the previous section can be made. For the very same epistemological basis which supported their philosophical theology allowed the Stoics to connect it with some aspects of the tradition. After all, preconceptions were thought to be natural, acquired on the basis of ordinary experience, and thus in principle available to anyone, including those who had been responsible for constructing the religious tradition. Pre-philosophical rationality appears to have been conceived of as having such preconceptions as its material, and this applies both to the human individual and to humankind as a whole. Although one would certainly like to have more explicit information on how exactly the theory of preconception connects with the Stoic practice of interpretation of the tradition, there is no reason to doubt that there was a connection, and that this connection was at least part of the *rationale* behind the Stoic preoccupation with (wholly or partially correct) views of primitive man as embedded in myth and early poetry.[27] The truth inherent in these views could be brought out by means of the allegorical interpretation of myths and by the etymological interpretation of divine names.[28] This explains the many early Stoic attempts—continued on a larger and more systematic scale in Cornutus, who presents himself as explaining what he calls a *palaia theologia*, a pristine

[27] Explicit, i.e. non-circumstantial, evidence for the connection between preconceptions and (aspects) of traditional belief: Cicero, *ND* 2. 5: 'quod nisi cognitum conprehensumque animis haberemus, non tam stabilis opinio permaneret nec confirmaretur diuturnitate temporis nec una cum saeclis aetatibusque hominum invereterari posset . . . opinionis enim commenta delet dies, naturae iudicia confirmat'; see also Dio Chrysostomus, *Or.* 12. 39–48 on the *emphutos ennoia* on the one hand, and myth, religious customs and laws and visual images on the other as its three (intrinsically vaguer) 'interpreters', on which see further below, pp. 245–7; and Strabo's claim (10. 3. 23) that *πᾶς δὲ ὁ περὶ θεῶν λόγος ἀρχαίας ἐξετάζει δόξας καὶ μύθους, αἰνιττομένων τῶν παλαιῶν ἃς εἶχον ἐννοίας φυσικὰς* (as so much in Strabo a Stoicizing passage, but with no explicit reference to the Stoics). See also Frede 1989: 2088–9).

[28] As Long (1992) has argued, this does not imply that the Stoics thought of early poetry as being self-consciously allegorical. They rather thought of myth, the subject of much early poetry, as being by its own nature a primitive expression of truths that could be phrased in a more articulate way by philosophy. Allegorical interpretation thus turns out to be not so much a matter of laying bare the hidden meaning intended by the poet, but a translation of the inevitably rude, plastic, and inarticulate language of early humankind into the more sophisticated language of philosophy. I here largely follow this view, because I believe it is accurate for the early Stoics at least. Yet the overall picture appears to be less neat: Seneca's critique of Chrysippus' etymologizations (*Ben.* 1. 3. 2–10) seems to presuppose that it was the poet (Hesiod, in this case) who was thought to have imposed the relevant names (note, by the way that Seneca is the only Stoic who appears to be critical of such etymologizations or allegorical interpretations). And Chaeremon (test. 12 in van der Horst (1984)), on whom see below, pp. 247–8, claims that allegories and myths were devised by the early Egyptian priests to reveal their wisdom to the uninitiated. In both cases the implication is that the use of etymologically significant names or allegory is a device that is deliberately or consciously applied. For some further theorizing about what may and may not count as allegory, see now Goulet (2005). On etymologization and allegorization in early Stoicism, see Gourinat (2005).

theology—to salvage aspects of cosmological or theological myth, as found in Homer and Hesiod.[29]

On the other hand, the mere fact that preconceptions were thought to be natural did not in itself imply that they were also thought to be fully and universally shared. Like the Epicureans, the Stoics seem to have believed that the natural conception of god could be, and had indeed been, contaminated with strange and wrongheaded extras that could be found even in the very same early poets who also transmitted some correct mythical conceptions.[30] This means that there were also parts of the tradition which had to be repudiated, because they reflected not so much a pristine unadulterated rationality, but the influence of external irrational and corrupting factors.[31] Cicero's Stoic spokesman in *ND* 2 vividly describes this process:

> Do you see . . . how from a true and valuable philosophy of nature has been evolved this imaginary and fanciful pantheon? The perversion has been a fruitful source of false beliefs, crazy errors and superstitions hardly above the level of old wives' tales. We know what the gods look like and how old they are, their dress and their equipment, and also their genealogies, marriages and relationships and all about them is distorted into the likeness of human frailty. (*ND* 2. 70)

Apparently what made such mistaken beliefs especially dangerous was that, once they had been formed, they were likely to keep misleading people, in so far as they became 'canonized' in the traditions of myth and cult. According to the Stoic account in Cicero, *ND* 2. 45, one of the reasons why people distort their original preconception is their daily experience of the culture that surrounds them: it is the *consuetudo oculorum*

[29] Zeno's interpretations of Hesiod were probably provided in the context of his physical or cosmological treatise *On the Whole*, on which see Algra (2001); Philodemus and Cicero provide evidence of a similar procedure being adopted in Chrysippus' theological *On Gods* (see below, n. 36). But of course in Stoicism the line between physics *stricto sensu* and theology (the latter being part of physics *sensu lato*) is hard to draw.

[30] On such early forms of corruption in Hesiod, see Cornutus, *Epidr*. 31. 12–17: *ἀλλὰ τῆς μὲν Ἡσιόδου τελειοτέρα ποτ' ἂν ἐξήγησις σοι γένοιτο, τὰ μέν τινα, ὡς οἴμαι, παρὰ τῶν ἀρχαιοτέρων αὐτοῦ παρειληφότος, τὰ δὲ μυθικώτερον ἀφ' αὑτοῦ προσθέντος, ᾧ τρόπῳ καὶ πλεῖστα τῆς παλαιᾶς θεολογίας διεφθάρη*. Ps.-Plutarch, *Placita* 1. 6 (*SVF* 2. 1009) speaks of *τὸ ὑπὸ τῶν ποιητῶν πεπλασμένον*. Of course corruption of preconceptions, as conceived by the Stoics, need not be wholesale. Thus the 2nd-cent. BC Stoic Antipater of Tarsus claimed (*ap*. Plutarch, *SR* 1052b) that philosophers who divest the gods of beneficence are in *partial* (*ἀπὸ μέρους*) conflict with our preconception of them.

[31] At first sight we might think that this idea of corruption does not sit comfortably with the Stoic commitment to general providence. However, we should bear in mind, first, that according to the Stoics providence concerns the gift of reason as such, not the use we make of it; secondly, that reason, when not yet stabilized in the form of the knowledge (*epistêmê*) of the sage, is fickle and easily led astray. In the pre-rational stages of our life we are all the more prone to forming wrongheaded value judgements based on pleasure and pain (or fear) which may disturb the natural process of *oikeiôsis*. For a discussion of such sources of corruption in the treatment of young children, see Vegetti (1983). For a further source of corruption, the *consuetudo oculorum*, see n. 32.

which leads them to endorse the idea that the gods are anthropomorphic in shape.[32]

It appears, then, that mainstream Stoicism was committed to an interesting combination of primitivism (the 'natural' world view of the people of old inevitably got corrupted), and progressivism (the subsequent development of philosophy can remedy this, and show us what can and cannot be salvaged). Accordingly, the Stoic attitude towards the tradition was decidedly mixed. There was no desire either to defend or to reject it as such or *in toto*.[33] The Stoic spokesman Balbus in Cicero's *ND* in this connection explicitly advocates differentiating between superstition and religion (*ND* 2. 69–72), and in general the desire, on the Stoics' part, to merge traditional polytheism and philosophical monotheism seems to have had its limits.[34] The evidence suggests that the Stoics were selective in what they adopted. Thus, as I have tried to show elsewhere, Zeno's interpretations of Hesiod focused primarily if not exclusively on the cosmogonical myths in the first part of the *Theogony*.[35]

Finally, it is worth noting that the term used by our sources to indicate the interpretative procedure applied by the Stoics is 'appropriation',[36] which suggests that they adopted elements of myth to support their philosophy—especially their cosmo-theology—rather than the other way round.

4

All this is not to say that the philosophical context of this Stoic willingness to appropriate elements of myth, or more generally of the religious tradition, is altogether clear. As so often, we are hampered by the scarcity of the evidence, while the evidence that happens to be available for the most part concentrates on individual examples of Stoic interpretations rather than on the underlying assumptions or the methodological background. In what follows I shall try to shed some further light on the matter by addressing a few points which are particularly controversial or otherwise in need of further clarification.

First of all, one might well ask why the Stoics thought they needed myth at all to underpin views which they could defend by philosophical arguments. I think the

[32] As Pease notes in his commentary (ad loc.), 'the *consuetudo oculorum* is, naturally, the daily sight of anthropomorphic images of the gods'.

[33] That their primary aim was just to defend the tradition has been suggested by Zeller, whose views are discussed and rejected by Most (1989: 2020).

[34] *Contra* Brisson (2004: 54), who speaks of the 'acceptance of the existence of all traditional divinities and the allegorical interpretation of their nature'.

[35] See Algra (2001).

[36] Philodemus, *Piet.* PHerc. 1428, col. vi, 16–26 (text in Henrichs 1974: 17) uses the term συνοικειοῦν to refer to Chrysippus' practice in the second book of his *On the Gods*; Cicero *ND* 1. 41 (probably going back to the same source) uses *accommodare* in the same connection: *in secundo [scil. libro] autem volt Orphei Musaei Hesiodi Homerique fabellas accommodare ad ea quae ipse primo libro de deis immortalibus dixerat.*

outlines of an answer to this question have been provided by Teun Tieleman in his study of Chrysippus' use of poets, proverbs, conventions of ordinary language, and common conceptions in the argument of his *On the Soul.*[37] A passage of Chrysippus' *On Lives* quoted by Plutarch (*SR* 1036 e = SVF 2. 270) suggests that Chrysippus allowed the use of (merely) persuasive arguments (*πιθανὰ*) on both sides in dialectical discussions, although he also warned against the dangers of adducing persuasive arguments for the case of the opponents. The catalogue of Chrysippus' writings, moreover, contains a title *πιθανὰ λήμματα εἰς τὰ δόγματα* (DL 7. 199) which seems to refer to a set of (merely) persuasive arguments that could be used to support Stoic tenets. According to Chrysippus, apparently, the true cognitions (*katalêpseis)* of the non-sage (unlike true knowledge or *epistêmê*) were still capable of being overturned, and should accordingly be further strengthened not just by philosophical argument, but also by adducing what is merely 'persuasive' (*πιθανόν*), though not clinching, evidence, including the evidence of some comparatively vague and inarticulate traditional notions that could be unearthed, for example, through etymology.

Secondly, on this interpretation of the procedure of adducing evidence that is only *πιθανόν*, we can explain why the practice of appropriation in the sense sketched above does not—at least not in principle: the actual Stoic practice may well have been less neat—involve circularity, contrary to what has been argued by Glenn Most.[38] For on the one hand, as noted, the practice of allegorical interpretation itself should not be thought of as providing 'proof' in the sense of clinching evidence, for the Stoic theory. On the other hand, the reliability of the allegorizations involved was not thought to be secured (again: at least not in theory) by the truth of the very Stoic theology which it was supposed to underpin, but by what the Stoics could claim were independent epistemological reasons legitimating their reading of certain myths as conveying physical or proto-physical concepts or arguments. In other words, we are not dealing with a pair of claims justifying each other in any strong sense.

Thirdly, it has recently been suggested that there is an important continuity between the way the Stoics dealt with early myths and poetry and the way in which some later Neoplatonists and early Christian thinkers dealt with their foundational (sometimes sacred) texts.[39] Although *de facto* historical influence of Stoic allegorical interpretations on these later schools cannot and should not be excluded, we should also note that the philosophical or methodological context was, by and large, different. Later philosophical interpretations of the contents of the Bible, but also Neoplatonic interpretations of the texts of Plato or the Chaldaean oracles, often assumed that such truth claims as could be made for their sacred texts were ultimately based on some form of divine revelation. The Stoics, however, ascribed (as indeed did the Epicureans) what they saw

[37] Tieleman (1996: 264–87), discussing the role of the *πιθανὸν* in Stoic philosophical method.

[38] Most (1989: 2020).

[39] This is a central claim of Boys-Stones (2001).

as the correct elements in the religious tradition to the same epistemological basis, natural conceptions rooted in experience, which supported their own philosophies.[40] In their view the people of old had had the right preconception of the one cosmic god, and on the basis of that preconception they had devised significant names for a number of subordinate gods, which represented parts or aspects of that single cosmic deity.

Fourthly, it is important to recall that the Stoics did not believe that preconceptions as such offered *complete* knowledge: as noted, they thought that in the end preconceptions had to be further articulated and inserted into the larger and coherent framework of (Stoic) philosophy.[41] Accordingly, and given the empirical and non-revelatory basis of preconceptions, there is no reason to assume that they ascribed a *complete* wisdom, or a *complete* philosophy-cum-theology to the people of old and the myths and cult-practices they invented.[42]

At this point it is unfortunate that the evidence we have on the Stoic view, or views, of cultural development is scarce and controversial. One thing that is certain, however, is that some kind of original moral corruption was assumed to have occurred in the course of the development of civilization—a corruption subsequently to be remedied by philosophy. But a corruption of what? Sextus *M* 9. 27–8 speaks of 'some of the later Stoics' who declared that the first men (the γηγενεῖς) greatly surpassed the men of today in comprehension (συνέσει) and that the keenness of their intelligence (ὀξύτης τῆς διανοίας) served as some kind of sixth sense by means of which they easily perceived the divine.

[40] Hadot (1987: 23–4) ascribes to the Stoics the belief in a παλαιὸς λόγος revealed by the gods; see also his n. 42 'pour les stoïciens, le langage lui-même est en quelque sorte une révélation naturelle que l'on peut interpréter par l'etymologie (notamment celle qui porte sur les noms des dieux)'. There is no evidence for the idea of such a special divine revelation, it makes no sense within the context of Stoic epistemology, and the Stoics did not need it.

[41] Stoicism speaks of the philosophical articulation of original concepts (διάρθρωσις τῶν ἐννοιῶν). The catalogue of Chrysippus' ethical writings contains a special section devoted to the articulation of ethical concepts (DL 7. 199–200), including not only books on definition, logical division, genera and species, etc., but also on etymology, proverbs, and the interpretation of poetry. On the underlying view of the relation between poetry, myth, etymology, and proverbs on the one hand and philosophical method on the other, see Tieleman (1996: 229–33 and 264–73). Note that the Epicureans as well, even if their view of the relation between preconceptions and reason differed, seem to have believed that philosophy could add to the knowledge encompassed in preconceptions. Thus the Epicurean spokesman in Cicero, *ND* 1. 45 claims that the preconceptions themselves are sufficient to ground a pious religion; yet 'the mind strives to strengthen this belief by trying to discover the form of god, the mode of his activity, and the operation of his intelligence'. Philodemus tries to excuse the weaknesses, or the incompleteness, of such Epicurean attempts to offer additional proofs of the nature of the gods by claiming that in this area no one has been able to offer absolute certainty, on which see above, n. 21.

[42] *Contra* Gourinat (2005: 27), who speaks of 'une vérité physique qu'ils connaissaient parfaitement' and of 'la connaissance scientifique de la nature, commune aux philosophes stoïcienns et aux anciens mythographes'. Boys-Stones (2001) does not offer an unambiguous answer to the question whether or not the Stoics claimed (as I think they did not) that the people of old had a complete philosophy at their disposal.

'Comprehension' and 'keenness of intellect' are not the same thing as complete knowledge or wisdom, and their existence among the earliest men was most likely supposed to be the result of, or concomitant with, the absence of external corrupting factors. In that case the assumption may have been that their keen and uncorrupted experience of nature allowed these people easy access to a proper conception of god, a conception not immediately blurred by rival conceptions furnished by the images of culture and the fanciful stories of other people. In this sense this report about 'later Stoics' is perhaps to be connected with another piece of evidence: a summary of Posidonius' view in Seneca *Ep.* 90, 6, which speaks of a second phase, after an initial Golden Age, when vice crept in (*subrepentibus vitiis*). Unlike our passage in Sextus, this testimony seems to focus on *moral* corruption, but we should of course realize that in Stoicism intellectual corruption and moral corruption are two sides of the same coin. And here again, we have no reason to infer that it was assumed that the earliest phase was one in which the *logos* was present in its purest form and completely.[43] At any rate, Posidonius seems to have reserved the term 'wise' exclusively for the first inventors and lawgivers, not for early people in general.[44] Seneca objects even to this: the inventions of the mechanical arts should not be ascribed to philosophy, and in general the period when men were 'fresh from the gods' (*recentes a diis, Ep.* 90. 44) was not a period of sages. Instead he claims that the first humans displayed what were merely approximations of virtue, and hence of philosophy in the true sense (*omnibus his virtutibus habebat similia quaedam rudis vita, Ep.* 90. 46). It appears, then, that neither Posidonius, nor the anonymous report on 'later Stoics' in Sextus, nor yet Seneca, who may here represent mainstream Stoicism, implies the existence of complete knowledge or philosophical wisdom among early humankind.

Finally, in his version of the *Kulturentstehungslehre* Seneca speaks of the first people as being *recentes a diis*. This appears to be a Stoic adaptation of a traditional motif—at any rate, the motif occurs in Plato as well.[45] It might be supposed that the underlying idea in Seneca is that being close to a conflagration, either at the beginning or at the end of a cosmic period, would involve an *increased* rationality on the part of humans—an interpretation which might get some further support from Sextus' claim about the 'younger Stoics' quoted above. I doubt, however, whether this is likely to have been the view of mainstream Stoicism. After all, their claim that at the event of a conflagration god remains on his own, without matter or evil, does not involve the claim that god is more rational during a conflagration than during a particular world order.[46] I would suggest that what

[43] *Contra* Most (1989: 2020).

[44] Frede (1989) suggests that Posidonius introduced a more loose, non-technical, sense of 'philosophy' to describe the kind of wisdom that was to be found in the early period of human history, and that it is thanks to this changed concept of philosophy that the Stoic Chaeremon could argue in the 1st cent. AD that early Egyptian priests had been philosophers.

[45] To the references listed in Boys-Stones (2001: 19 n. 24) one should add *Tim.* 40d.

[46] For a different view, however, see Most (1989: 2021–2).

the Stoics *may* have claimed instead—and what may thus also have been the point of Seneca's qualification *recentes a diis*—is that the rationality of early humankind was still relatively uncorrupted.

5

In the previous sections I have mainly dealt with the way in which Stoic theology relates—via Stoic epistemology and the Stoic *Kukturentstehungslehre*—to one particular aspect of the religious tradition: (cosmological) myth. In what follows I want to cover some less familiar ground and turn to the Stoic practice of critique and appropriation of cult and cult practices, including the representation of the gods in the form of statues and other images.

The cult-centred nature of ancient religion did not involve a particularly strong or coherent conceptual or dogmatic core. Yet of course every cult has *some* conceptual content,[47] and it is this explicit or implicit conceptual content which the Stoics in some cases drew out and confronted with their own philosophical theology. The result is a similar combination of critique and adaptation as we encountered in the case of myths. Let us first turn to the critique, and start with the founder of the school, Zeno of Citium. His *Politeia* describes a utopian 'city of sages'. In this city, we are told by various sources, there should be no temples or statues of the gods. One of our sources, Clement of Alexandria, renders Zeno's own reasons:

> There will be no need at all to build sanctuaries. For a sanctuary that is not worth much at all should not be regarded as sacred. But the work of craftsmen and mechanics is not worth much and not sacred. (Clement, *Strom.* 5. 12. 76; SVF 1. 264)

Plutarch interprets Zeno's words as meaning that no temples should exist at all, and characteristically constructs a contradiction between this part of the Stoic teaching and Stoic practice:

> The Stoics, while applauding this [i.e. Zeno's words] as correct, attend the mysteries in temples, go up to the Acropolis, do reverence to statues, and place wreaths upon the shrines, though these are works of builders and mechanics. Yet they think that the Epicureans are confuted by the fact that they sacrifice to the gods, whereas they are themselves worse confuted by sacrificing at altars and temples which they hold should not exist and should not be built. (Plutarch, *SR* 1034B–C; tr. Cherniss, slightly modified)

[47] For a rather extreme claim in this respect, compare Plutarch who claimed that 'nothing does more to reveal what was in the mind of the ancients than the rites of initiation and the ritual acts that are performed in religious services with symbolic intent' (fr. 157. 23–5 Sandbach). The only Stoic who explicitly makes a comparable claim on the conceptual content of rites and rituals is Chaeremon, on whom see below, n. 78.

We should note that, strictly speaking, Zeno's words imply no such thing. There is no need to take them as a straightforward and general prohibition: Zeno is merely claiming that building sanctuaries is superfluous in a city of sages—presumably given the other means available to these people to honour the gods. We may perhaps connect this text with the doxographical testimony in Epiphanius (*Adv.Her*.3.2.9 = SVF 1. 146) which tells us that Zeno said we shouldn't build sanctuaries, but that we should instead have the divine solely in our own mind (ἐν μόνῳ τῷ νῷ), and with the repeated claim of such later Stoics as Seneca and Epictetus that the only proper way to honour the gods is by our own spiritual attitude, i.e. by imitating them through becoming virtuous.[48] When taken in this way, Zeno's words are perfectly compatible with a view that condones traditional cult in everyday life.

Another instance of a critical attitude towards certain elements of cult can be found in two Stoic fragments preserved in Philodemus' *On Piety*. In the first one Philodemus quotes Chrysippus' *On Gods*. In the course of an argument aiming to show how various gods from the Greek pantheon are in reality aspects of the one cosmic Zeus, Chrysippus makes two significant remarks. First, he points out that the gods, just like cities or virtues, are neither male nor female, but that they have only been given male or female *names*.[49] He then goes on to claim that it is childish (παιδαριωδῶς) to speak of the gods, or to paint or sculpt them, as if they were human in form, just as it is childish to do so in the case of cities, rivers, or places.[50] Similar remarks are made by Chrysippus' pupil Diogenes of Babylon, in a passage also quoted by Philodemus in the same context.[51]

Nevertheless it looks as if Chrysippus did not offer this critique in an unqualified form, as something that should lead to an absolute prohibition. Several sources inform us of his mentioning a picture—variously located at Argos or in Samos—in which Hera and Zeus were depicted as involved in a particular sexual act; and one of these sources, Origenes *Cels*. 4. 48 (*SVF* 2. 1074) claims that according to Chrysippus this picture should be interpreted as a representation of how the divine *spermatikoi logoi* fertilize matter, i.e. as an interpretation of what was traditionally known as the *hieros gamos* between Zeus

[48] Cf. Epictetus, *Ench*. 31, 1: 'in piety towards the gods . . . the chief element is this, to have the right opinions about them . . . and to have set yourself to obey them and to submit to everything that happens'. Seneca claims that we do not honour the gods by bloody offerings, but by our right and virtuous intention (*Benef*. 1. 6. 3). God seeks no servants, and the proper cult of god is to know him, and to imitate him (*Ep*. 95. 47 and 95. 50).

[49] Philodemus, *Piet*. col. V, 8–14 Henrichs.

[50] Ibid., col. V, 28–35 Henrichs (*SVF* 2. 1076): κα[ὶ παι]δαριωδῶς λέγεσ[θαι] [κ]α[ὶ] γράφε[σ]θαι κα[ὶ] πλάτ[τε]σθαι [θε]οὺς ἀνθρω[ποειδεῖς], ὃν τρόπον καὶ π[ό]λεις καὶ ποταμοὺς καὶ τόπους κτλ.

[51] Philodemus *Piet*. col. VII, 24–8 Henrichs (*SVF* 3, Diog. Bab. 33): [καὶ] παιδα[ρι]ῶδες εἶν[αι] θεοὺς ἀ[ν]θρωποε[ι]δεῖς λ[έγει]ν καὶ ἀδύνατον.

and Hera.[52] Apparently in some contexts even anthropomorphic depictions of the gods, however inadequate and 'childish' they may otherwise have been, could be seen as conveying some relevant conceptual information on the gods and their nature.[53]

Seneca appears to have been more elaborate and explicit in his dealings with traditional cult than any other Stoic we know. In the sixth book of the *De Civitate Dei* Augustine quotes from his lost treatise *On Superstition*.[54] In this work Seneca—apparently partly influenced by Varro (on whom more below)—attacked representations of the gods ('images of the cheapest inert material'),[55] which in many cases would be regarded as monsters, should they be suddenly brought to life. In addition, he chastized the ritual practices connected with the veneration of Jupiter at the Roman Capitol:

> One servant informs Jupiter of the names of his worshippers, another announces the hours; one is his bather, another his anointer, that is, he gestures with empty hands to imitate the act of anointing. There are women who are hairdressers for Juno and Minerva; while standing far away from the temple as well as from the image they move the fingers as if they were dressing the hair, and there are others who hold a mirror. There are men who summon the gods to give bonds for them, and some who offer them lawyers' briefs and explain their case. An expert leading actor in the mimes, now a decrepit old man, used to act a mime each day in the Capitol—as if the gods would enjoy the performance of a player when men had ceased to do so. Every kind of artisan is there to devote his time to the immortal gods. (Seneca *ap.* Augustine, *Civ. Dei* 6. 10)

If Seneca is here critical of an overly anthropomorphic conception of god and of the childish rituals to which it gives rise, he does not believe that a Stoic should straightforwardly advocate their abolition. Although the sage should not regard such rites as part of his personal worship, he should 'go through the motions of feigned conformity' (*in animi religione non habeat, sed in actibus fingat*): 'For he says: "the wise man will observe all these rites as being enjoined by the laws, not as being pleasing to the gods"' (ibid.).[56]

[52] The evidence is printed as *SVF* 2. 1071–4. It is not unlikely, by the way, that the original picture of Zeus and Hera was based on the story as told in *Iliad* 14. 153–353. On the influence of Homer's depiction of the *hieros gamos*, and on artistic representations of the scene in the Heraion of Samos and elsewhere, see Burkert (1985: 132).

[53] There is some additional evidence that, for this purpose, gods could be described as male or female, according to their supposed active or passive role: Servius, *In Aen.* 4. 638 (*SVF* 2. 1070) claims this as the Stoic position and adds as an example a Virgilian (and meteorological: what now descends is rain) version of the *hieros gamos* in *Georg.* 2. 326 (*coniugis in gremium laetae descendit*); see also Cicero, *ND* 2. 66 (*SVF* 2. 1075): *effeminarunt autem eum [scil. aera] Iunonique tribuerunt, quod nihil est eo mollius.*

[54] Augustine, *Civ. Dei* 6. 10: *nam in eo libro quem contra superstitiones condidit multo copiosius atque vehementius reprehendit ipse civilam istam et urbanam theologian quam Varro theatricam atque fabulosam.*

[55] Augustine, *Civ. Dei* 6. 10: *sacros immortales in materia vilissima atque immobili dedicant.*

[56] Augustine, *Civ. Dei* 6. 10: *quae omnia sapiens servabit tamquam legibus iussa, non tamquam deis grata.*

With regard to what Augustine, possibly quoting Seneca himself, calls 'this obscure throng of gods, assembled through long years by ancient superstition', Seneca claims that 'we shall invoke them, but with the reservation in mind that their worship belongs rather to custom than to truth' (*adorabimus ut meminerimus cultum eius magis ad morem quam ad rem pertinere*). The position taken here by Seneca—with its distinction between *religio animi* and *actus*, and between *mos* and *res*, implying a rigid separation between the various elements of the *theologia tripertita*—calls for some further comment. The notion of a *religio animi* reminds us of what he elsewhere has to say about true, i.e. philosophical, religion:

It is foolish to pray for this (i.e. a *bona mens*), if you can acquire it from yourself. We do not need to lift our hands up towards heaven, or beg the keeper of a temple to let us approach the ear of the idol . . . God is near you, he is with you, he is within you. (*Ep.* 41. 1)

In this respect also his letter 95 offers some quotable jewels: 'god is worshipped by those who know him' (*deum colit qui novit, Ep.* 95. 47); 'god seeks no servants' (*non quaerit ministros deus*, ibid.); 'whoever imitates the gods, worships them sufficiently' (*satis illos coluit, quisquis imitatus est, Ep.* 95. 50). In the passages quoted above, Seneca seems to entertain the idea that this properly Stoic religion can be combined with a general adherence to the religious customs of the state. These passages, in other words, represent a blend of philosophical elitism (in stressing the importance of the *religio animi*), criticism (in putting *res* above *mos*), and general conservatism (in nevertheless leaving room for *mos*). It is to be noted, however, that elsewhere, most notably in letter 95 from which I just quoted, we find Seneca taking a decidedly more negative position towards the tradition:

Precepts are commonly given as to how the gods should be worshipped. But let us forbid lamps to be lighted on the Sabbath, since the gods do not need light, neither do men take pleasure in soot. Let us forbid men to offer morning salutation and to throng the doors of temples; mortal ambitions are attracted by such ceremonies, but God is worshipped by those who truly know Him. Let us forbid bringing towels and flesh-scrapers to Jupiter, and proffering mirrors to Juno; for God seeks no servants. Of course not; he himself does service to mankind, everywhere and to all he is at hand to help. (*Ep.* 95. 47)

Seneca's position, we may conclude, does not appear to be fully consistent. As is more often the case in his writings, different contexts may have given rise to different emphases.[57] Letter 95, which focuses on the difference between *praecepta* and *decreta*, offers religious precepts as an example of *praecepta* that may lead us astray and that are incompatible with the true *decreta* of Stoic philosophy. Its perspective is predominantly philosophical and moral and hence it takes the stricter view. On the other hand, the treatise *On Superstition* apparently had

[57] On different views, on Seneca's part, taken in different contexts on the question of the afterlife and immortality of the soul, see Hoven (1971: 108–26); on different views on the feasibility of prayer, see Richards (1964).

to discuss the question of how in practice one should deal with the ingrained superstitions of traditional cult, and in such a context a more conciliatory position may have been called for. Anyway, Seneca does appear to remain within the boundaries set by the earlier Stoic philosophical tradition: like Zeno and Chrysippus he is predominantly critical of traditional cult, including the use of cult statues, but as in their case critique does not necessarily imply radical rejection.

In his *On Superstition* Seneca was indebted to Varro's *Antiquitates Rerum Divinarum*, published shortly before 45 BC.[58] Our source Augustine tells us, moreover, that Seneca was more critical of the *urbana theologia* than Varro had been,[59] which is understandable, given the fact that Varro wrote as an antiquarian, professedly aiming to strengthen traditional religion, whereas Seneca wrote as a philosopher for whom protecting or reviving the religious tradition was not a goal in itself. Varro was, of course, not a Stoic in any proper sense. He appears to have been a follower of Antiochus of Ascalon. Yet the fragments show that his theology was at least indebted to Stoicism, no doubt via Antiochus. This, as well as his influence on Seneca, warrants his inclusion in this brief overview.

Varro claimed, we are told, that if he were to found the city anew, he would 'consecrate the gods and give them names according to the principles of nature, rather than as they are now'.[60] In other words, and in the terms of his own tripartite theology, he would replace the religion of the city by the religion of the philosophers, for 'the only ones who have discovered what god really is, are those who have adopted the view that he is the soul which governs the world by a movement that accords with reason'—surely a Stoic-sounding claim.[61] Yet, since he is living in an old country, 'he says that he must keep the traditional account of the names and surnames, and that the object of his writing and research is to persuade the people to worship rather than disregard these'. He thus adds the conservative slant we are familiar with by now while in passing revealing his own agenda, i.e. his ambition to strengthen and revive traditional religion.[62]

When it comes to the question of the legitimacy of having (anthropomorphic) images of the gods (*Civ. Dei* 4. 31), Varro praises the early Romans for having

[58] That Varro's work was available in 45 is clear from Cicero (*Acad. Post.* 8 ff.). The work was dedicated to Caesar as *pontifex maximus*; on its probable date of publication, see further Cardauns (1978: 86). The fragments of the *Antiquitates Rerum Divinarum* have been collected in Cardauns (1976).

[59] Augustine, *Civ. Dei* 6. 10: *Libertas sane quae huic* (sc. *Varroni*) *defuit, ne istam urbanam theologiam theatricae simillimam aperte sicut illam reprehendere auderet, Annaeo Senecae . . . non quidem ex toto verum ex aliqua parte non defuit.*

[60] Varro *ap.* Augustine, *Civ. Dei* 4. 3 (= fr. 12 Cardauns): *si eam civitatem novam constitueret, ex naturae potius formula deos nominaque eorum se fuisse dedicaturum.*

[61] Varro *ap.* Augustine, *Civ. Dei* 4. 31 (= fr. 13 Cardauns): *hi soli ei videantur animadvertisse quid esset deus qui crediderint eum esse animam motu ac ratione mundum gubernantem.*

[62] Varro *ap.* Augustine, *Civ. Dei* 4. 31 (= fr. 13 Cardauns); *acceptam ab antiquis nominum et cognominum historiam tenere, ut tradita est, debere se dicit, et ad eum finem illa scribere ac perscrutari ut potius eos magis colere quam despicere vulgus velit.*

worshipped the gods without any image for more than 170 years. If this usage had continued to his own days, he adds, our worship of the gods would be more devout (*castius dii observarentur*). In this sense those who first set up images did religion no favour: they 'diminished reverence and added error' (*et metum dempsisse et errorem addidisse*), since gods in the shape of senseless images might easily inspire contempt (*existimans deos facile posse in simulacrorum stoliditate contemni*). Yet also in this case Varro's critique is not unqualified. He appears to have been capable of a more positive attitude elsewhere. In the seventh book of the *City of God* Augustine refers to Varro again, this time as defending the physical, i.e. philosophical, interpretations (*interpretationes physicas*) of the representations of gods in the form of images. According to Varro the ancients designed the images, attributes, and ornaments of the gods 'so that men who had approached the mysteries of the doctrine, when they considered these visible things, might gain mental insight into the world and its parts, that is, the true gods'.[63] In addition, Varro also has something to say about the anthropomorphic shape of the cult statues as such. The basically inadequate human shapes of these images, he claims, were called forth by the consideration that 'the mortal mind that is in the human body is very much like the immortal mind'.[64] The latter consideration—an anthropomorphic image of a god, though as such misleading, may indirectly represent an important and true aspect of god, namely his rationality—was shared by two later authors, a Stoic and a half-Stoic: Epictetus and Dio Chrysostomus.[65] It seems to constitute the background to their remarkably positive assessments of one particular and very famous cult statue: the statue of Zeus in Olympia by Pheidias.

6

The critical attitude with regard to the fashioning and use of anthropomorphic images of the gods which we find—though, as we saw, not in an unqualified form—in Chrysippus, Diogenes of Babylon, Seneca, and Varro is no doubt consistent with what we might call the pantheistic aspect of Stoicism, which sees

[63] Varro *ap.* Augustine, *Civ. Dei* 7. 5 (= fr. 225 Cardauns): *antiquos simulacra deorum et insignia ornatusque finxisse, quae cum oculis animadvertissent hi qui adissent doctrinae mysteria possent animam mundi et partes eius, id est deos veros, animo videre*. Such claims concerning the attributes of cult statues remind one of Porphyry's interpretation of the iconography of these statues in his *On Images*, the fragments of which have been edited by Bidez (1913). One may also compare the Stoic Chaeremon, as quoted below, n. 79, with whose work, incidentally, Porphyry was acquainted.

[64] Varro *ap.* Augustine, *Civ. Dei* 7. 5 (= fr. 225 Cardauns): *qui simulacra specie hominis fecerunt, hoc videri secutos quod mortalium animus, qui est in corpore humano, simillimus est immortalis animi*.

[65] For a recent assessment of Dio's work, see the essays in Swain (2000). Apart from Stoic elements, Dio also shows traces of Cynicism and he carries his philosophical convictions lightly. In the present case (*Or.* 12. 8), however, it is clearly Stoicism, not Cynicism, which provides the philosophical background.

god as the formative force within the cosmos, a force not confined to any shape (apart perhaps from the spherical shape of the cosmos as a whole), let alone to a human shape.[66] Yet as we have seen, even from the pantheistic perspective, aspects of god could in some contexts be described in an anthropomorphic way: despite his own general injunctions against anthropomorphism Chrysippus allowed the representation of a male Zeus and female Hera in a cosmological context (referring to their active and passive roles); in another context, moreover, he appears to have been prepared to compare god (this time conceived as identical with the cosmos) to a human being in so far as he has both body and soul (god's providence, to which god 'retires' at the event of a conflagration, being the soul of the cosmos).[67]

In addition, however, Stoic theology contains a more clearly theistic strand as well. The cosmic god is not *eo ipso* an impersonal god. On the contrary, seen as an immanent formative principle, god could be labelled, as Zeno put it, not just 'craftsmanlike', but actually 'a craftsman',[68] or even a 'father'.[69] Indeed precisely because god's rationality—or, for that matter, the rationality of the cosmos—was thought not to differ in kind from human rationality, god qua rational and providential governing principle could be viewed as a 'person' with purposes and intentions: providence could be identified as 'the will of Zeus'.[70] So although the Stoic god should not be conceived anthropomorphically in any *physical* sense related to his shape, he does resemble humans in so far as his providential rationality and its various qualities are concerned. He is virtuous and happy and as such an example to be followed.[71] That this may involve a degree of anthropomorphism in our descriptions may be illustrated by a quotation from Chrysippus' *On Nature*:

> as it befits Zeus to glory in himself and in his way of life and to be haughty and, if it should be said this way, to carry his head high and plume himself, and to boast since he lives in a way worth boasting about, so does this befit all good men, since they are in no way surpassed by Zeus. (Chrysippus *ap*. Plutarch, *SR* 1038c)

The structural resemblance between human and divine rationality not only allowed a certain form of personalistic theism in thinking and speaking about

[66] See DL 7. 147; Aetius 1. 6; with my remarks in Algra (2003*a*: 166). See also Lactantius, *De Ira* 18 (*SVF* 2. 1057).

[67] Cf. Plutarch, *Comm. Not.* 1077d (*SVF* 2. 1064); Diogenes of Babylon *ap*. Philodemus (*SVF* 3, Diog. Bab. 33).

[68] Cicero, *ND* 2. 58: *ipsius vero mundi, qui omnia complexu suo coercet et continet, natura non artificiosa solum, sed plane artifex ab eodem Zenone dicitur*. Cf. DL 7. 137: δημιουργὸς τῆς διακοσμήσεως, and 7. 147, quoted in n. 69.

[69] Cf. DL 7. 147: εἶναι δὲ τὸν μὴν δημιουργὸν τῶν ὅλων καὶ ὥσπερ πατέρα πάντων

[70] See DL 7. 147. On providence as the will of Zeus, cf. Calcidius *In Tim.* 144 (= *SVF* 2. 933). For a comparison between the *anima mundi* and ourselves in this connection, see Cicero, *ND* 2. 58.

[71] On god's wisdom being comparable to that of the sage, see Plut. *Comm. Not.* 1076a (= *SVF* 3. 246): ὠφελεῖσθαι τε γὰρ ὁμοίως ὑπ' ἀλλήλων τὸν Δία καὶ τὸν Δίωνα, σοφοὺς ὄντας, ὅταν ἕτερος θατέρου τυγχάνῃ κινουμένου. On his happiness, see Stob. *Ecl.* 2. 98. 17 ff., Wachsmuth (= part of *SVF* 3. 54).

god, it also provided the basis for the claim that the sage leads a life in which what the Stoics call 'the god inside', our own *daimôn*, is in agreement with the 'will of the orderer of the universe' and has thus become 'like god'.[72] It is against this background that we should explain Epictetus' at first sight perhaps somewhat surprising reference to Pheidias' statue of Zeus in Olympia. In *Discourse* 2. 8 he is considering the case of a student who still lacks confidence in what he has learnt and agreed to, i.e. who has not completely internalized the Stoic teachings and is still working on his own self. This process is compared to the polishing and finishing of a statue:

> When the statue is finished and polished I will show it to you. What do you think of it? A lofty air, say you? Certainly not. For the Zeus in Olympia does not show a proud look, does he? No, but his gaze is steady, as befits one who is about to say 'No word of mine can be revoked or proved untrue [Homer, *Iliad* 1. 526]'. Of such character will I show myself to you—faithful, reverent, noble, unperturbed. (Epictetus, *Diss*. 2. 8. 25–7; tr. based on Oldfather)

The metaphor of the self as a statue to be polished is of course what induces the comparison with the fashioning of Pheidias' Zeus—we are dealing, so to speak, with a plastic way of describing the process of 'becoming like god'.[73] Nevertheless it provides us with yet another instance of a Stoic assessment of an anthropomorphic cult statue as being more than just 'childish': the way in which Pheidias has managed to convey the moral, exemplary aspect of Zeus apparently compensates for the fact that the Stoic Zeus strictly speaking does not look like a man.

Epictetus' contemporary Dio of Prusa, later also known as Dio Chrysostomus, devotes a whole speech to this same statue in Olympia, which at the time enjoyed great fame.[74] The central question of Dio's Olympian speech, apparently triggered by the presence of the statue itself, is what it is that is capable of moulding and forming man's conception of the deity (*Or*. 12. 26). The answer is rather complex (*Or*. 12. 39–47), but basically isolates four factors that may be at work:

(1) the natural preconception of god (the terms used are *ἔμφυτος ἐπίνοια* and *ἔννοια*, *Or*. 12. 27, 39–40);[75]
(2) myth and custom (*Or*. 12. 40);

[72] DL 7. 88: *εἶναι δ' αὐτὸ τοῦτο τὴν τοῦ εὐδαίμονος ἀρετὴν καὶ εὔροιαν βίου, ὅταν πάντα πράττηται κατὰ τὴν συμφωνίαν τοῦ παρ' ἑκαστῳ δαίμονος πρὸς τὴν τοῦ⟨τῶν⟩ὅλων δοικηιοῦ βούλησιν.*

[73] On the idea of godlikeness, see Sedley (1999). The comparison of the inner self with a statue to be polished may be traditional; it recurs e.g. in Plotinus 1. 6. 9.

[74] Dio himself calls it 'the most beautiful and the most dear to the gods' (*Or*. 12. 25), and Epictetus claims that 'you travel to Olympia to behold the work of Pheidias, and each of you regards it as a misfortune to die without seeing such sights' (*Diss*. 1. 6. 23–4). See also the descriptions in Pausanias 5. 11. 1–11 and Strabo 8. 3. 30.

[75] The word *ἔμφυτος* here means 'natural', or 'growing naturally', not 'innate', as is made clear in the various passages where the *ἐπίνοια* is claimed to originate from experience; see e.g. 12. 39: *ἔμφυτον . . . ἐπίνοιαν, ἐξ αὐτῶν γιγνομένην τῶν ἔργων καὶ τἀληθοῦς*.

(3) the laws (*Or.* 12. 40); and
(4) painted and sculpted images of the gods (*Or.* 12. 44).[76]

In the end, Dio claims, it falls to philosophy to interpret the data provided by these various sources (*Or.* 12. 47). He thus ranges himself squarely in the Stoic tradition which assumes, as we saw, that philosophy can not only strengthen and systematize preconceptions (factor 1), but also (with the help of these same preconceptions as a criterion), criticize or adapt aspects of myth (factor 2) and cult (factor 3), including images (factor 4). Dio also pronounces himself on the way these various sources are related. Factors (2) and (3)—roughly covering what according to the *theologia tripertita* are the gods of the poets and the gods of the city—are said to be ultimately based on factor (1), but at the same time to be lacking in clarity. Apparently the purity of the original preconception is almost inevitably compromised in the transmission through myth and cult (*Or.* 12. 41–3). In its turn factor (4) is presented as rooted in factor (2): the poets' image making is said to have been the earlier (*Or.* 12. 45–6), and later on Pheidias himself is made to claim that he drew the conception of his statue from Homer's poetry (*Or.* 12. 62, 73–4), selecting from the many beautiful images of the gods to be found in his epics one that is not dreadful, but 'peaceful and gentle'.

Dio's story—which is by and large built up from Stoic elements—thus has it that the original preconceptions are blurred in the theology of the city and in the theology of the poets, and a fortiori in the derived theology of the sculptor—cult images are, if I may use a Platonic formula, at various removes from the truth. Even so, they as well, according to Dio, can contain an adumbration of theological truth. In order to show how, he introduces the *persona* of Pheidias himself who, after having been praised for his beautiful statue, is asked to account for the method he adopted in fashioning it. Pheidias' 'defence' is a rather complex set-piece which I cannot fully paraphrase or analyse here.[77] For our present purpose we may just single out two reasons which he adduces for the anthropomorphic shape of his Zeus. One of them concerns our inability to portray the god's rationality otherwise than in a human body—a point, we may recall, that was already made by Varro:

> For mind and intelligence in and of themselves no statuary or painter will ever be able to represent; for all men are utterly incapable of observing such attributes with their eyes or of learning them by inquiry. But as for that in which this intelligence manifests itself, men, having no mere inkling thereof but actual knowledge, fly to it for refuge, attributing to god a human body as a vessel to contain intelligence and rationality in their lack of a

[76] Here we may compare the remark of Cicero's Stoic spokesman Balbus in *ND* 2 on the *consuetudo oculorum,* on which see above, pp. 233–4.

[77] Note e.g. that the Zeus which is here being spoken of appears to be a kind of hybrid between the Zeus of the traditional Greek pantheon and the cosmic Zeus of the Stoics. But the overall setting of the speech makes it quite clear that, for Dio, it is the philosophical question whether and to what extent the statue can be said to represent the Stoic Zeus that is at the focus of attention.

better illustration, and in their perplexity seeking to indicate that which is invisible and unportrayable by means of something portrayable and visible, using the function of a symbol and doing so better than certain barbarians, who are said to represent the divine by animals . . . (Dio Chrysostomus, *Or.* 12. 59)

Secondly,

. . . certainly no one would maintain than it had been better that no statue or picture of gods should have been exhibited among men, on the ground that we should look only at the heavens. For although the intelligent man does indeed reverence all those objects, believing them to be blessed gods that he sees from a great distance, yet on account of our belief in the divine all men have a strong yearning to honour and worship the deity from close at hand, approaching and laying hold of him with persuasion by offering sacrifice and crowning him with garlands. (Dio Chrysostomus, *Or.* 12. 60)

Dio compares the latter attitude to that of children wanting to be close to their parents, thus unwittingly giving a positive twist to Chrysippus' claim that the use of statues is 'childish'. However that may be, each of his two defences of the anthropomorphic character of Pheidias' statue links up with the theistic side of Stoic theology: god's rationality is like human rationality, and we want to conceive of him as an object of worship near to us (rather than as a remote astral or cosmic god) and as a father. According to Dio, this is why we are allowed to depict him in human form.

7

Now that we have reviewed the main evidence, from Zeno to Dio Chrysostomus, we may try to draw some threads together. First of all we may note that the Stoic attitude towards the religious tradition cannot be captured in a simple formula: it was neither one of radical rejection, nor one of unqualified conservatism. On the one hand the Stoics felt free to criticize aspects of the tradition that were at odds with their philosophical theology. On the other, they were prepared to argue, presumably on the basis of their theory of preconceptions, that the narrative elements of myth, but also the visual representations of cult, could be thought to contain adumbrations of theological truth. Chaeremon, who was a contemporary of Seneca and himself an Egyptian priest (*hierogrammateus*, or sacred scribe), though no doubt Hellenized and possibly even of Greek origin, was even prepared to claim that also the *rites* and *rituals* performed in connection with cult statues were indicative of some allegorical truth.[78] Between rejection and acceptance a whole gamut of possible reactions was available, and reactions

[78] Porphyry, *De Abstinentia* 4. 6 (= Chaeremon fr. 10 in the edn. of van der Horst 1984): ὧν ἕκαστον οὐ τῦφος ἦν, ἀλλά τινος ἔνδειξις φυσικοῦ λόγου. I have not encountered this view among other Stoics.

by individual Stoics to individual religious phenomena may have varied according to circumstances or personal leanings.

Dio Chrysostomus explicitly points out—and other Stoics must have believed this as well—that both myth and visual representations remain inferior to our natural preconception in clarity and truthfulness; in its turn, as we have seen, the preconception itself should be at the basis of a further articulation by means of philosophical arguments. Yet even the inferior, and sometimes partly mistaken, representations of myth and cult could be put to use to convey at least some aspects of the cosmic god: the pedagogical purpose of myths and allegories for the uninitiated was stressed by Chaeremon,[79] and the pedagogical usefulness of cult statues was emphasized by Dio Chrysostomus. The same may go for the way in which Epictetus exploits the anthropomorphic traits of Pheidias' Zeus and for Chrysippus's use of the Samian picture of Zeus and Hera. Like the mythical examples which were inserted into the cosmological writings of early Stoics, such references to elements of the tradition may have been accorded the status of persuasive (*pithanon*) evidence, helping the uninitiated to form a proper (philosophical) conception of god.

However, even in cases where no such justification could be given, the Stoics seem to have displayed a certain tolerance towards the tradition. Seneca's 'let us forbid' (*Ep.* 95. 47, quoted above) remains an exception, even within the context of his own work. One would like to have more explicit information about the reasons behind this apparent *laissez-faire* approach. Perhaps a properly philosophical justification was given, for example by labelling participation in cult as essentially morally indifferent: what really counts is our rational and philosophically pious attitude. Or perhaps we are merely dealing with a concession to convention, an instance of the widespread conservatism we encounter in other philosophers as well.

We do know, at any rate, that the Stoics criticized their Epicurean opponents precisely for defending a theology that demolished the basic presuppositions of the tradition, arguing that the Epicureans made traditional cult a pointless exercise in so far as they claimed that the gods have no dealings with our world, nor, in fact, with the universe at large.[80] Probably in response to this Stoic critique, the Epicureans devised a whole range of fairly radical reinterpretations which gave new meanings to the various aspects (prayer, sacrifice, language) of traditional cult.[81] On the Stoic side, we do not find such detailed reinterpretations, at least

[79] Chaeremon T. 12 in the edn. of van der Horst (1984).

[80] Thus those Epicureans who did pray, sacrifice, etc. were said to refute their own philosophy; see Plutarch, *SR* 1034c. Posidonius even claimed that Epicurus was, if not professedly, at least in actual fact an atheist (Cicero, *ND* 1. 123). Philodemus' *On Piety* is basically a defensive work, trying to answer such allegations on the part of the Stoics.

[81] Philodemus, *Piet.* (cf. Obbink 1996) mentions Epicurean reinterpretations of the veneration of statues (col. 32 Obbink and elsewhere), but also of oaths, traditional expressions such as 'if the

not on a similar scale.[82] Maybe the Stoics thought they did not need them, because their philosophical theology allowed them to make sense of at least part of traditional cult in its actual form.

REFERENCES

Adomenas, M. (1999) 'Heraclitus on Religion', *Phronesis,* 45(2): 87–113.

Algra, K. A. (2001) 'Comments or Commentary? Zeno of Citium and Hesiod's *Theogonia*', *Mnemosyne,* 54: 562–81.

——(2003*a*) 'Stoic Theology', in B. Inwood (ed.), *The Cambridge Companion to the Stoics* (Cambridge: CUP), 153–79.

——(2003*b*) 'Hellenistic Philosophy and Some Science', *Phronesis,* 48(1): 71–88.

——(2004) 'Eternity and the Concept of God in Early Stoicism', in G. van Riel and C. Macé (eds.), *Platonic Ideas and Concept Formation in Ancient and Medieval Thought* (Leuven: Leuven UP), 173–91.

Betegh, G. (2004) *The Derveni Papyrus: Cosmology, Theology and Interpretation* (Cambridge: CUP).

Bidez, J. (1913) *Vie de Porphyre le philosophe néo-platonicien, avec les fragments des traités PERI AGALMATON et DE REGRESSU ANIMAE* (Ghent; repr. Hildesheim: Weidmann, 1964).

Boys-Stones, G. R. (2001) *Post-Hellenistic Philosophy: A Study of its Development from the Stoics to Origen* (Oxford: OUP).

Brisson, L. (2004) *How Philosophers Saved Myths: Allegorical Interpretations in Classical Mythology* (Chicago: University of Chicago Press).

Bruit Zaidman, L. (2001) *Le Commerce des dieux: Eusebeia, Essai sur la piété en Grèce ancienne* (Paris: Découverte).

Buckert, W. (1985) *Greek Religion* (Cambridge, MA: Harvard University Press) = English translations of Buckert, W. (1977) *Griechische Religion der archaischen klassischen Epoche* (Stuttgart: Kohlhammer Verlag).

Cardauns, B. (1976) *M. Terentius Varro Antiquitates Rerum Divinarum* (2 vols. Wiesbaden: Akademie der Wissenschafteb und der Literatur).

——(1978) 'Varro und die römische Religion: Zur Theologie, Wirkungsgeschichte und Leistung der *Antiquitates Rerum Divinarum*', *ANRW* II.16.1 (Berlin), 80–103.

Clerc, C. (1915) *Les Théories relatives au culte des images chez les auteurs grecs du IIme siècle après J.-C.* (Paris: Fontemoing).

Curd, P. (2002) 'The Presocratics as Philosophers', in A. Laks and C. Louguet (eds.), *Qu'est-ce que la philosophie présocratique? What is Presocratic Philosophy* (Villeneuve d'Ascq: Presses Universitaires du Septentrion), 115–40.

Frede, M. (1989) 'Chaeremon der Stoiker', *ANRW* II.36.3 (Berlin), 2067–103.

gods are friendly', and various terms traditionally referring to the relation between the gods and us ('affinity', 'alienation', cols. 37–9; 'benefits' and 'harm', cols. 40–1 and 81).

[82] On what appears to be a Stoic reinterpretation of what is at stake in petitionary prayer, see Algra (2003*a*: 174–7).

Goulet, R (2005) 'La Méthode allégorique chez les stoïciens', in J.-B. Gourinat (ed.), *Les Stoïciens* (Paris: Librairie Philosophique J. Vrin), 93–119.

Goulet-Cazé, M.-O. (1996) 'Religion and the Early Cynics', in R. Bracht Branham and M.-O. Goulet-Cazé (eds.), *The Cynics: The Cynic Movement in Antiquity and its Legacy* (Berkeley-Los Angeles and London: University of California Press), 47–80.

Gourinat, J.-B. (2005) '*Explicatio fabularum*: La Place de l'allégorèse dans l'interprétation stoïcienne de la mythologie', in G. Dahan and R. Goulet (eds.), *Allégorie des poètes, allégorie des philosophes: Études sur la poétique et l'herméneutique de l'allégorie de l' Antiquité a la Réforme* (Paris: Librairie Philosophique J. Vrin), 9–34.

Hadot, P. (1987) 'Théologie, exégèse, révélation, écriture, dans la philosophie grecque', in M. Tardieu (ed.) *Les Règles d'interprétation* (Paris: Cerf), 13–34.

—— (1993) *Exercices spirituels et philosophie antique*, rev. edn. (Paris: Études Augustiniennes).

Henrichs A. (1974) 'Die Kritik der stoischen Theologie im PHerc. 1428', *Cronache Ercolanesi,* 4: 5–32.

Horst, P. W. van der (1984) *Chaeremon: Egyptian Priest and Stoic Philosopher. The Fragments Collected and Translated with Explanatory Notes* (Leiden: Brill).

Hoven, R. (1971) *Stoïcisme et stoïciens façe au problème de l'au-delà* (Paris: Belles Lettres).

Lévy, C (1996) 'Philosopher à Rome', in C. Lévy (ed.), *Le Concept de nature à Rome: La Physique* (Paris: Presses de l'École normale supérieure).

Lieberg, G. (1982) 'Die Theologia tripertita als Formprinzip antiken Denkens', *Rheinisches Museum,* 125: 25–53.

Long, A. A. (1992) 'Stoic Readings of Homer', in R. Lamberton and J. J. Keaney (eds.), *Homer's Ancient Readers: The Hermeneutics of Greek Epic's Earliest Exegetes* (Princeton: Princeton UP), 41–66; repr. in A. A. Long, *Stoic Studies* (Cambridge: CUP, 1996), 58–85.

Most, G. W. (1989) 'Cornutus and Stoic Allegoresis: A Preliminary Report', *ANRW* II.36.3 (Berlin), 2014–65.

Obbink, D. (1996) *Philodemus On Piety, Part I. Critical Text with Commentary* (Oxford: OUP).

—— (2002) ' "All Gods are True" in Epicurus', in D. Frede and A. Laks (eds.), *Traditions of Theology: Studies in Hellenistic Theology, its Background and its Aftermath* (Leiden: Brill), 183–222.

Osborne, C. (1997) 'Heraclitus', in C. C. W. Taylor (ed.), *Routledge History of Philosophy* (London and New York: Routledge), i. 88–128.

Richards, W. J. (1964) *Gebed by Seneca, die Stoïsyn* (Groningen: VRB Publishers; written in Afrikaans, with a Summary in English).

Runia, D. T. (2002) 'The Beginnings of the End: Philo of Alexandria and Hellenistic Theology', in D. Frede and A. Laks (eds.), *Traditions of Theology: Studies in Hellenistic Theology, its Background and its Aftermath* (Leiden: Brill), 281–316.

Schofield, M. (1980) 'Preconception, Argument and God', in M. Schofield, M. Burnyeat, and J. Barnes (eds.), *Doubt and Dogmatism: Studies in Hellenistic Epistemology* (Oxford: OUP), 283–308.

Sedley, D. N. (1989) 'Philosophical Allegiance in the Greco-Roman World', in M. Griffin and J. Barnes (eds.), *Philosophia Togata: Essays on Philosophy and Roman Society* (Oxford: OUP), 97–119.

—— (1999) 'The Ideal of Godlikeness', in G. Fine (ed.), *Oxford Readings in Philosophy: Plato 2: Ethics, Politics, Religion and the Soul* (Oxford: OUP), 309–28.

Swain, S. (ed.) (2000) *Dio Chrysostom: Politics, Letters and Philosophy* (Oxford: OUP).

Tieleman, T. L. (1996) *Galen and Chrysippus on the Soul: Argument and Refutation in the De Placitis, Books II–III* (Leiden: Brill).

Vegetti, M. (1983) 'Passioni e bagni caldi: Il problema del bambino cattivo nell' antropologia stoica', in M. Vegetti, *Tra Edipo e Euclide: Forme del sapere antico* (Milan: Il Saggiatore), 71–90.

Index Nominum

Index Locorum

Made in the USA
Las Vegas, NV
27 September 2026

47ec5aee-9c7c-4bb2-baf0-f00ac35ac721R01